2D PAGE: Sample Preparation and Fractionation

METHODS IN MOLECULAR BIOLOGY™

John M. Walker, SERIES EDITOR

460. **Essential Concepts in Toxicogenomics,** edited by *Donna L. Mendrick and William B. Mattes,* 2008
459. **Prion Protein Protocols,** edited by *Andrew F. Hill,* 2008
458. **Artificial Neural Networks:** *Methods and Applications,* edited by *David S. Livingstone,* 2008
457. **Membrane Trafficking,** edited by *Ales Vancura,* 2008
456. **Adipose Tissue Protocols,** *Second Edition,* edited by *Kaiping Yang,* 2008
455. **Osteoporosis,** edited by *Jennifer J. Westendorf,* 2008
454. **SARS- and Other Coronaviruses:** *Laboratory Protocols,* edited by *Dave Cavanagh,* 2008
453. **Bioinformatics, Volume 2:** *Structure, Function, and Applications,* edited by *Jonathan M. Keith,* 2008
452. **Bioinformatics, Volume 1:** *Data, Sequence Analysis, and Evolution,* edited by *Jonathan M. Keith,* 2008
451. **Plant Virology Protocols:** *From Viral Sequence to Protein Function,* edited by *Gary Foster, Elisabeth Johansen, Yiguo Hong, and Peter Nagy,* 2008
450. **Germline Stem Cells,** edited by *Steven X. Hou and Shree Ram Singh,* 2008
449. **Mesenchymal Stem Cells:** *Methods and Protocols,* edited by *Darwin J. Prockop, Douglas G. Phinney, and Bruce A. Brunnell,* 2008
448. **Pharmacogenomics in Drug Discovery and Development,** edited by *Qing Yan,* 2008
447. **Alcohol:** *Methods and Protocols,* edited by *Laura E. Nagy,* 2008
446. **Post-translational Modification of Proteins:** *Tools for Functional Proteomics, Second Edition,* edited by *Christoph Kannicht,* 2008
445. **Autophagosome and Phagosome,** edited by *Vojo Deretic,* 2008
444. **Prenatal Diagnosis,** edited by *Sinhue Hahn and Laird G. Jackson,* 2008
443. **Molecular Modeling of Proteins,** edited by *Andreas Kukol,* 2008
442. **RNAi:** *Design and Application,* edited by *Sailen Barik,* 2008
441. **Tissue Proteomics:** *Pathways, Biomarkers, and Drug Discovery,* edited by *Brian Liu,* 2008
440. **Exocytosis and Endocytosis,** edited by *Andrei I. Ivanov,* 2008
439. **Genomics Protocols,** *Second Edition,* edited by *Mike Starkey and Ramnanth Elaswarapu,* 2008
438. **Neural Stem Cells:** *Methods and Protocols, Second Edition,* edited by *Leslie P. Weiner,* 2008
437. **Drug Delivery Systems,** edited by *Kewal K. Jain,* 2008
436. **Avian Influenza Virus,** edited by *Erica Spackman,* 2008
435. **Chromosomal Mutagenesis,** edited by *Greg Davis and Kevin J. Kayser,* 2008

434. **Gene Therapy Protocols:** *Volume 2, Design and Characterization of Gene Transfer Vectors,* edited by *Joseph M. LeDoux,* 2008
433. **Gene Therapy Protocols:** *Volume 1, Production and In Vivo Applications of Gene Transfer Vectors,* edited by *Joseph M. LeDoux,* 2008
432. **Organelle Proteomics,** edited by *Delphine Pflieger and Jean Rossier,* 2008
431. **Bacterial Pathogenesis:** *Methods and Protocols,* edited by *Frank DeLeo and Michael Otto,* 2008
430. **Hematopoietic Stem Cell Protocols,** edited by *Kevin D. Bunting,* 2008
429. **Molecular Beacons:** *Signalling Nucleic Acid Probes, Methods and Protocols,* edited by *Andreas Marx and Oliver Seitz,* 2008
428. **Clinical Proteomics:** *Methods and Protocols,* edited by *Antonio Vlahou,* 2008
427. **Plant Embryogenesis,** edited by *Maria Fernanda Suarez and Peter Bozhkov,* 2008
426. **Structural Proteomics:** *High-Throughput Methods,* edited by *Bostjan Kobe, Mitchell Guss, and Huber Thomas,* 2008
425. **2D PAGE:** *Sample Preparation and Fractionation, Volume 2,* edited by *Anton Posch,* 2008
424. **2D PAGE:** *Sample Preparation and Fractionation, Volume 1,* edited by *Anton Posch,* 2008
423. **Electroporation Protocols,** edited by *Shulin Li,* 2008
422. **Phylogenomics,** edited by *William J. Murphy,* 2008
421. **Affinity Chromatography,** *Methods and Protocols, Second Edition,* edited by *Michael Zachariou,* 2007
420. **Drosophila,** *Methods and Protocols,* edited by *Christian Dahmann,* 2008
419. **Post-Transcriptional Gene Regulation,** edited by *Jeffrey Wilusz,* 2008
418. **Avidin-Biotin Interactions,** *Methods and Applications,* edited by *Robert J. McMahon,* 2008
417. **Tissue Engineering, Second Edition,** edited by *Hannsjörg Hauser and Martin Fussenegger,* 2007
416. **Gene Essentiality:** *Protocols and Bioinformatics,* edited by *Andrei L. Osterman,* 2008
415. **Innate Immunity,** edited by *Jonathan Ewbank and Eric Vivier,* 2007
414. **Apoptosis and Cancer:** *Methods and Protocols,* edited by *Gil Mor and Ayesha B. Alvero,* 2008
413. **Protein Structure Prediction, Second Edition,** edited by *Mohammed Zaki and Chris Bystroff,* 2008
412. **Neutrophil Methods and Protocols,** edited by *Mark T. Quinn, Frank R. DeLeo, and Gary M. Bokoch,* 2007
411. **Reporter Genes for Mammalian Systems,** edited by *Don Anson,* 2007
410. **Environmental Genomics,** edited by *Cristofre C. Martin,* 2007
409. **Immunoinformatics:** *Predicting Immunogenicity In Silico,* edited by *Darren R. Flower,* 2007

METHODS IN MOLECULAR BIOLOGY™

2D PAGE: Sample Preparation and Fractionation

Volume 1

Edited by

Anton Posch

Bio-Rad Laboratories GmbH, Munich, Germany

Editor
Anton Posch
Bio-Rad Laboratories GmbH
Munich, Germany

Series Editor
John M. Walker
School of Life Sciences
University of Hertfordshire
Hatfield, Herts., UK

ISBN: 978-1-58829-722-8   e-ISBN: 978-1-60327-064-9
ISSN: 1064-3745

Library of Congress Control Number: 2007942891

©2008 Humana Press, a part of Springer Science+Business Media, LLC
All rights reserved. This work may not be translated or copied in whole or in part without the written permission of the publisher (Humana Press, 999 Riverview Drive, Suite 208, Totowa, NJ 07512 USA), except for brief excerpts in connection with reviews or scholarly analysis. Use in connection with any form of information storage and retrieval, electronic adaptation, computer software, or by similar or dissimilar methodology now known or hereafter developed is forbidden.
The use in this publication of trade names, trademarks, service marks, and similar terms, even if they are not identified as such, is not to be taken as an expression of opinion as to whether or not they are subject to proprietary rights.
While the advice and information in this book are believed to be true and accurate at the date of going to press, neither the authors nor the editors nor the publisher can accept any legal responsibility for any errors or omissions that may be made. The publisher makes no warranty, express or implied, with respect to the material contained herein.

Cover illustration: Figure 3, Chapter 14, "The Terminator: A Device for High-Throughput Extraction of Plant Material," by B. M. van den Berg, Volume 2.

Printed on acid-free paper

9 8 7 6 5 4 3 2 1

springer.com

Preface

This book, split into two volumes, presents a broad coverage of the principles and recent developments of sample preparation and fractionation tools in Expression Proteomics in general and for two-dimensional electrophoresis (2-DE) in particular. 2-DE, with its unique capacity to resolve thousands of proteins in a single run, is still a fundamental research tool for nearly all protein-related scientific projects. The methods described here in detail are not limited to 2-DE and can also be applied to other protein separation techniques.

Because each biological sample is unique, a suited sample preparation strategy has to consider the type of sample as well as the type of biological question being addressed. The complex nature of proteins often requires a multitude of sample preparation options. In addition, sample preparation is not only a prerequisite for a successful and reproducible Proteomics experiment, but also the key factor to meaningful data evaluation. Interestingly, not much attention was paid to this area during Proteomics methodology development and therefore this book is intended to explain in depth how proteins from various sources can be properly isolated and prepared for reproducible Proteome analysis.

The application of fractionation and enrichment strategies has become a major part of sample preparation. The number of possible different proteins in a cell or tissue sample is believed to be in the several hundreds of thousands, spanning concentration ranges from the level of a single molecule to micromolar amounts, and no single analytical method developed today is capable of resolving and detecting such a diverse sample. Sample fractionation reduces the overall complexity of the sample, and enriches low abundance proteins relative to the original sample. Proteins that may originally have been undetectable are thus rendered amenable to analysis by 2-DE and a broad variety of gel-free mass spectrometry-based technologies.

This book is for students of Biochemistry, Biomedicine, Biology, and Genomics and will be an invaluable source for the experienced, practicing scientist, too.

Anton Posch

Contents

Preface v
Contributors xi

PART I: SAMPLE PREPARATION BASICS

1. Mechanical/Physical Methods of Cell Disruption and Tissue Homogenization 3
   Stanley Goldberg

2. Bacteria and Yeast Cell Disruption Using Lytic Enzymes 23
   Oriana Salazar

3. Sample Solublization Buffers for Two-Dimensional Electrophoresis 35
   Walter Weiss and Angelika Görg

4. Quantitation of Protein in Samples Prepared for 2-D Electrophoresis 43
   Tom Berkelman

5. Removal of Interfering Substances in Samples Prepared for Two-Dimensional (2-D) Electrophoresis 51
   Tom Berkelman

6. Protein Concentration by Hydrophilic Interaction Chromatography Combined with Solid Phase Extraction 63
   Ulrich Schneider

PART II: PROTEIN LABELING TECHNIQUES

7. Difference Gel Electrophoresis Based on Lys/Cys Tagging 73
   Reiner Westermeier and Burghardt Scheibe

8. Isotope-Coded Two-Dimensional Maps: Tagging with Deuterated Acrylamide and 2-Vinylpyridine 87
   Pier Giorgio Righetti, Roberto Sebastiano, and Attilio Citterio

9. Stable Isotope Labeling by Amino Acids in Cell Culture (SILAC) 101
   Albrecht Gruhler and Irina Kratchmarova

10. ICPL—Isotope-Coded Protein Label 113
    Josef Kellermann

11. Radiolabeling for Two-Dimensional Gel Analysis 125
    Hélian Boucherie, Aurélie Massoni, and Christelle Monribot-Espagne

Part III: Fractionation of Proteins by Chemical Reagents and Chromatography

12. Sequential Extraction of Proteins by Chemical Reagents 139
    Stuart J. Cordwell

13. Reducing Sample Complexity by RP-HPLC: Beyond the Tip of the Protein Expression Iceberg 147
    Gert Van den Bergh and Lutgarde Arckens

14. Enriching Basic and Acidic Rat Brain Proteins with Ion Exchange Mini Spin Columns Before Two-Dimensional Gel Electrophoresis 157
    Ning Liu and Aran Paulus

15. Reducing the Complexity of the *Escherichia coli* Proteome by Chromatography on Reactive Dye Columns 167
    Phillip Cash and Ian R. Booth

16. Reducing Sample Complexity in Proteomics by Chromatofocusing with Simple Buffer Mixtures 187
    Hong Shen, Xiang Li, Charles J. Bieberich, and Douglas D. Frey

17. Fractionation of Proteins by Immobilized Metal Affinity Chromatography 205
    Xuesong Sun, Jen-Fu Chiu, and Qing-Yu He

18. Fractionation of Proteins by Heparin Chromatography 213
    Sheng Xiong, Ling Zhang, and Qing-Yu He

Part IV: Fractionation of Proteins by Electrophoresis Methods

19. Fractionation of Complex Protein Mixtures by Liquid-Phase Isoelectric Focusing 225
    Julie Hey, Anton Posch, Andrew Cohen, Ning Liu, and Adrianna Harbers

20. Microscale Isoelectric Focusing in Solution 241
    Mee-Jung Han and David W. Speicher

21. Prefractionation, Enrichment, Desalting and Depleting of Low Volume and Low Abundance Proteins and Peptides Using the MF10 257
    Valerie Wasinger, Linda Ly, Anna Fitzgerald, and Brad Walsh

22. Sample Prefractionation in Granulated Sephadex IEF Gels 277
    Angelika Görg, Carsten Lück, and Walter Weiss

23. Free-Flow Electrophoresis System for Plasma Proteomic Applications 287
    Robert Wildgruber, Jizu Yi, Mikkel Nissum, Christoph Eckerskorn, and Gerhard Weber

24. Protein Fractionation by Preparative Electrophoresis 301
    Michael Fountoulakis and Ploumisti Dimitraki

PART V: ENRICHMENT STRATEGIES FOR ORGANELLES, MULTIPROTEIN COMPLEXES, AND SPECIFIC PROTEIN CLASSES

25. Isolation of Endocitic Organelles by Density Gradient
    Centrifugation.. 317
    Mariana Eça Guimarães de Araújo, Lukas Alfons Huber, and Taras Stasyk

26. Isolation of Highly Pure Rat Liver Mitochondria with the Aid of
    Zone-Electrophoresis in a Free Flow Device (ZE-FFE)............. 333
    Hans Zischka, Josef Lichtmannegger, Nora Jaegemann, Luise Jennen, Daniela Hamöller, Evamaria Huber, Axel Walch, Karl H. Summer, and Martin Göttlicher

27. Isolation of Proteins and Protein Complexes
    by Immunoprecipitation... 349
    Barbara Kaboord and Maria Perr

28. Isolation of Phosphoproteins 365
    Lawrence G. Puente and Lynn A. Megeney

29. Glycoprotein Enrichment Through Lectin Affinity Techniques 373
    Yehia Mechref, Milan Madera, and Milos V. Novotny

30. Isolation of Bacterial Cell Membranes Proteins Using Carbonate
    Extraction.. 397
    Mark P. Molloy

31. Enrichment of Membrane Proteins by Partitioning
    in Detergent/Polymer Aqueous Two-Phase Systems 403
    Henrik Everberg, Niklas Gustavsson, and Folke Tjerneld

32. The Isolation of Detergent-Resistant Lipid Rafts
    for Two-Dimensional Electrophoresis 413
    Ki-Bum Kim, Jae-Seon Lee, and Young-Gyu Ko

33. Isolation of Membrane Protein Complexes by Blue Native
    Electrophoresis.. 423
    Veronika Reisinger and Lutz A. Eichacker

34. Tissue Microdissection... 433
    Heidi S. Erickson, John W. Gillespie, and Michael R. Emmert-Buck

Index.. 449

Contributors

Lutgarde Arckens • *Katholieke Universiteit Leuven, Leuven, Belgium*
Juan A. Asenjo • *University of Chile, Santiago, Chile*
Tom Berkelman • *Bio-Rad Laboratories, Inc., Hercules, California*
Charles J. Bieberich • *University of Maryland Baltimore County, Baltimore, Maryland*
Ian R. Booth • *University of Aberdeen, Aberdeen, Great Britain*
Hélian Boucherie • *Centre National de la Recherche Scientifique, Bordeaux, France*
Phillip Cash • *University of Aberdeen, Aberdeen, Great Britain*
Jen-Fu Chiu • *The University of Hong Kong, Hong Kong, China*
Attilio Citterio • *Polytechnic of Milano, Milan, Italy*
Andrew Cohen • *Bio-Rad Laboratories, Inc., Hercules, California*
Stuart J. Cordwell • *The University of Sydney, Sydney, Australia*
Mariana Eça Guimarães De Araújo • *Innsbruck Medical University, Innsbruck, Austria*
Ploumisti Dimitraki • *Academy of Athens, Athens, Greece*
Christoph Eckerskorn • *BD GmbH, Martinsried, Germany*
Lutz A. Eichacker • *Ludwig-Maximalians-University Munich, Munich, Germany*
Michael R. Emmert-Buck • *National Institutes of Health, Bethesda, Maryland*
Heidi S. Erickson • *National Institutes of Health, Bethesda, Maryland*
Henrik Everberg • *Lund University, Lund, Sweden*
Anna Fitzgerald • *Minomic, Chatswood West, Australia*
Michael Fountoulakis • *F. Hoffmann-La Roche Ltd, Basel, Switzerland*
Douglas D. Frey • *University of Maryland Baltimore County, Baltimore, Maryland*
John W. Gillespie • *National Cancer Institute at Frederick, Frederick, Maryland*
Stanley Goldberg • *Glen Mills, Inc., Clifton, New Jersey*
Angelika Görg • *Technical University of Munich, Munich, Germany*
Martin Göttlicher • *GSF-National Research Center for Environment and Health, Neuherberg, Germany*
Albrecht Gruhler • *Novo Nordisk A/S, Måløv, Denmark*

NIKLAS GUSTAVSSON • *Lund University, Lund, Sweden*
DANIELA HAMÖLLER • *GSF-National Research Center for Environment and Health, Neuherberg, Germany*
MEE-JUNG HAN • *The Wistar Institute, Philadelphia, Pennsylvania*
ADRIANA HARBERS • *Bio-Rad Laboratories, Inc., Hercules, California*
QING-YU HE • *The University of Hong Kong, Hong Kong, China*
JULIE HEY • *Bio-Rad Laboratories, Inc., Hercules, California*
EVAMARIA HUBER • *GSF-National Research Center for Environment and Health, Neuherberg, Germany*
LUKAS ALFONS HUBER • *Innsbruck Medical University, Innsbruck, Austria*
NORA JAEGEMANN • *GSF Research Centre for Environment and Health, Institute of Toxicology, Munich-Neuherberg, Germany*
LUISE JENNEN • *GSF-National Research Center for Environment and Health, Neuherberg, Germany*
BARBARA KABOORD • *Pierce Biotechnology, Inc., Rockford, Illinois*
JOSEF KELLERMANN • *Max-Planck Institute of Biochemistry, Martinsried, Germany*
KI-BUM KIM • *Korea University, Seoul, Korea*
YOUNG-GYU KO • *Korea University, Seoul, Korea*
IRINA KRATCHMAROVA • *University of Southern Denmark, Odense, Denmark*
JAE-SEON LEE • *Korea University, Seoul, Korea*
XIANG LI • *University of Maryland Baltimore County, Baltimore, Maryland*
JOSEF LICHTMANNEGGER • *GSF-National Research Center for Environment and Health, Neuherberg, Germany*
NING LIU • *Bio-Rad Laboratories, Inc., Hercules, California*
CARSTEN LÜCK • *Technische Universitat Munchen, Freising, Germany*
LINDA LY • *University of NSW, Kensington, Australia*
MILAN MADERA • *Indiana University, Bloomington, Indiana*
AURÉLIE MASSONI • *Centre National de la Recherche Scientifique, Bordeaux, France*
YEHIA MECHREF • *Indiana University, Bloomington, Indiana*
LYNN A. MEGENEY • *The Ottawa Hospital, Ottawa, Canada*
MARK P. MOLLOY • *Australian Proteome Analysis Facility Ltd, Sydney, Australia*
CHRISTELLE MONRIBOT-ESPAGNE • *Centre National de la Recherche Scientifique, Gif-sur-Yvette, France*
MIKKEL NISSUM • *BD GmbH, Martinsried, Germany*
MILOS V. NOVOTNY • *Indiana University, Bloomington, Indiana*
ARAN PAULUS • *Bio-Rad Laboratories, Inc., Hercules, California*
MARIA PERR • *Pierce Biotechnology, Inc., Rockford, Illinois*
ANTON POSCH • *Bio-Rad Laboratories GmbH, Munich, Germany*

Contributors

LAWRENCE G. PUENTE • *The Ottawa Hospital, Ottawa, Canada*
VERONIKA REISINGER • *Ludwig-Maximalians-University Munich, Munich, Germany*
PIER GIORGIO RIGHETTI • *Polytechnic of Milano, Milan, Italy*
ORIANA SALAZAR • *University of Chile, Santiago, Chile*
BURGHARDT SCHEIBE • *GE Healthcare Europe GmbH, Munich, Germany*
ULRICH SCHNEIDER • *Siegfried-Biologics GmbH, Berlin, Germany*
ROBERTO SEBASTIANO • *Polytechnic of Milano, Milan, Italy*
HONG SHEN • *University of Maryland Baltimore County, Baltimore, Maryland*
DAVID W. SPEICHER • *The Wistar Institute, Philadelphia, Pennsylvania*
TARAS STASYK • *Innsbruck Medical University, Innsbruck, Austria*
KARL H. SUMMER • *GSF-National Research Center for Environment and Health, Neuherberg, Germany*
XUESONG SUN • *The University of Hong Kong, Hong Kong, China*
FOLKE TJERNELD • *Lund University, Lund, Sweden*
GERT VAN DEN BERGH • *Katholieke Universiteit Leuven, Leuven, Belgium*
AXEL WALCH • *GSF-National Research Center for Environment and Health, Neuherberg, Germany*
BRAD WALSH • *Minomic, Chatswood West, Australia*
VALERIE WASINGER • *University of NSW, Kensington, Australia*
GERHARD WEBER • *BD GmbH, Martinsried, Germany*
WALTER WEISS • *Technical University of Munich, Munich, Germany*
REINER WESTERMEIER • *GE Healthcare Europe GmbH, Munich, Germany*
ROBERT WILDGRUBER • *BD GmbH, Martinsried, Germany*
SHENG XIONG • *Jinan University, Guangzhou, China*
JIZU YI • *BD Diagnostics, Franklin Lakes, New Jersey*
LING ZHANG • *Jinan University, Guangzhou, China*
HANS ZISCHKA • *GSF-National Research Center for Environment and Health, Neuherberg, Germany*

I

Sample Preparation Basics

1

Mechanical/Physical Methods of Cell Disruption and Tissue Homogenization

Stanley Goldberg

Summary

This chapter covers the various methods of mechanical cell disruption and tissue homogenization that are currently commercially available for processing minute samples (<1 ml) to larger production quantities. These mechanical methods of lysing do not introduce chemicals or enzymes to the system. However, the energies needed when using these "harsh" methods can be high and destroy the very proteins being sought.

The destruction of cell membranes and walls is effected by subjecting the cells (1) to shearing by liquid flow, (2) to exploding by pressure differences between inside and outside of cell, (3) to collision forces by impact of beads or paddles, or (4) a combination of these forces. Practical suggestions to optimize each method, where to acquire such equipment, and links to reference sources are included.

Key Words: Cell disruption; bead mills; BioNeb cell disruption; cell disruption bomb; douce tissue grinder; Dyno-mill; French press; Gaulin high pressure homogenizer; high pressure homogenizers; Megatron; Microfluidics; mixer-mill; mortar; pestle; nitrogen Parr bomb; opposed jet homogenization; Parr nitrogen bomb; Polytron; Potter-Elvehjem tissue grinders; pressure bomb; Sonicator; Sonitube; tissue grinders; tissue homogenization; ultrasonic processor.

1. Introduction

The need to release cell components without introducing encumbering chemicals or enzymes suggests the use of mechanical methods of lysing. The destruction of cell membranes and walls by these "harsh" methods is effected by subjecting the cells (1) to shearing by liquid flow, (2) to exploding by pressure differences between inside and outside of cell, (3) to collision forces by impact of beads or paddles, or (4) a combination of these forces.

Generally speaking, any of the techniques described here can, to some degree, disrupt any cells or tissues. For more difficult materials, just the increase of motivation force or time of exposure will improve breakage. However, use of excessive force is limited because of the generation of detrimental heat and/or shear that can ruin the desired proteins. In addition, excess force will accelerate wear and ultimately damage the equipment.

By judicious use of the equipment one can select from a gentle nicking of the cell to release intact organelle up to a vigorous action to release membrane bound proteins. Some methods are suitable to handle tissues only, others for free cells only, and some are suitable for both. Some techniques are capable of processing only small quantities of material whereas others are limited to handling larger amounts.

Tissues that are difficult to break down include heart muscle, lung, intestine, and skin. On the other hand, some fragile mammalian cells can be broken by just a moderate shaking of the suspended cells. Free cells that are difficult to process include those that are extremely small size (below 0.25 micron) bacteria, and the tough yeasts and spores. Plant materials and seeds will need higher energy inputs for proper maceration. Table 1 provides an overview of the mechanical methods with at least one example of commercial equipment and related websites whereas Table 2 describes suitable subjects and capacity for each method.

2. Bead Impact Methods—Shaking Vessel
2.1. Theory

All bead devices open the cells or homogenize tissues by throwing the beads (also called "grinding media") against the cells/tissue. Also the accelerated beads generate strong shear in the liquid buffer surrounding the cells/tissues, which also pulls them apart. Two methods to accelerate the grinding media (beads) are (1) by shaking the entire container or (2) by a spinning agitator within a container (see next section). The shaking container method is usable for tissues as well as free cells. For extremely small samples of 0.2 ml to somewhat larger quantities of 50 ml the shaking of the vessel is the method of choice. The motion can be of differing geometries depending upon what equipment is selected. Shaking can only be done in batch operation thus limiting the amount of materials than can be processed. The equipment is very low cost, durable, and simple to operate requiring minimal training.

Materials needed to operate any of these shaking include the cells or tissue, grinding media (beads), liquid phase such as buffer, the container, and the

Table 1
Methods Overview: Techniques and Trade Names and Websites

Technique	Trade name(s)	Websites
Bead Impact – shaking vessel	MIXER MILL (1)	www.RETSCH.de
	BEAD BEATER™ (2)	www.BIOSPEC.com
Bead Impact – agitator shaft	DYNO®-MILL (3)	www.WAB.ch
		www.GLENMILLS.com
Rotor / Stator – shear by spinning shaft	POLYTRON® (4)	www.KINEMATICA.ch
Mortar / Pestle – shear by mechanical pressure	Potter-Elvehjem Tissue Grinders (5)	www.WHEATON.com
High pressure batch – liquid expansion	FRENCH® PRESS (6)	www.THERMO.com
High pressure batch – gas expansion	PARR® BOMB (7)	www.PARRINST.com
High pressure flow – high velocity liquid shear	APV® GAULIN (8)	www.APV.com
High pressure – opposed liquid streams	MICROFLUIDIZER® (9)	www.MICROFLUIDICSCORP.com
Droplet – low pressure flow droplet nebulizing	BioNeb® (10)	www.GLAS-COL.com
Ultrasonic – shear by collapsing bubbles	SONCIATOR® (11) SONITUBE®	www.MISONIX.com

Table 2
Suitable subjects and capacity for each method

Type of Biomaterial to be Lysed →	Bacteria	Yeast, Algae, Fungus, Spores	Seeds	Plants	Tissues	Capacity
Technique ↓						
Bead Impact—shaking vessel	Y	Y	Y	Y	Y	S/M
Bead Impact—agitator shaft	Y	Y	?	Y	Y	M/L
Rotor/Stator—shear by spinning shaft	N	N	Y	Y	Y	S/M/L
Mortar/Pestle—shear by mechanical pressure	Y	Y	Y	Y	Y	S/M
High-pressure batch—liquid expansion	Y	Y	?	?	?	S/M
High-pressure batch—gas expansion	Y	N	N	Y	Y	S/M
High-pressure flow—high velocity shear	Y	Y	N	N	N	S/M/L
High-pressure flow—opposed liquid streams	Y	Y	N	N	N	S/M/L
Droplet—low-pressure droplet nebulizing	Y	Y	N	N	N	S/M
Ultrasonic—shear collapsing bubbles	Y	Y	N	Y	Y	S/M/L

Quantities:
S = Small 0.1mL to 25mL
M = Medium 10mL to 500mL
L = 250mL to many liters

Suitability
Y - general good practice
N - not recommended
? - not known or marginal success

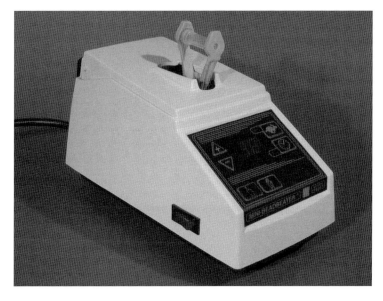

Fig. 1. Bead impact—shaken vessel—Mini Bead Beater-1 equipment.

equipment. Variables include the bead selection (density, diameter, and quantity), speed of agitation, cell concentration, and duration of run. (Fig. 1 and 2)

2.2. Practical Aspects

- Denaturing of proteins because of high temperature or excessive shear is to be considered in all mechanical disruption/homogenization equipment. Also, the wear of the grinding media and container into the samples need to be evaluated.

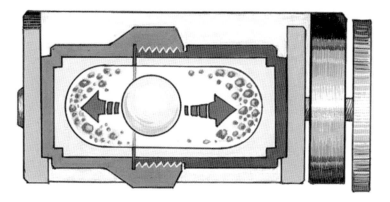

Fig. 2. Bead impact—shaken vessel—(inside view) beads moving in jar.

- Hints for successful temperature control include (1) Prechilling of samples and containers; (2) Runs of short duration with rest time to allow for rechilling samples on ice; (3) Use of fewer beads and/or extra buffer to act a heat sink; (4) Reduced degree of shaking vigor.
- Detrimental contamination of the batch because of wear of either the grinding media or container walls is usually rare because of the insignificant amounts involved. A hint to mitigate any contamination problem is to use inert materials of construction such as using beads and containers of zirconium oxide stabilized with yttria (95%/5%; specific gravity 6.0).
- Beads size: For small diameter cells (e.g., bacteria) use beads of 0.10–0.5 mm diameter. For larger cells (e.g., yeast, algae, hyphae) use beads of 0.5–1.25 mm. Glass (specific gravity 2.5) is a good starting material because of low cost. For homogenizing plant or animal tissues that have been previously chopped with a razor, beads of 1.0–5.0mm diameter are used.
- Bead density: If additional energy is needed to improve breakage/homogenization of tough cells, then higher density materials are used. These material types include ceramic (zirconium oxide family) with specific gravities from 3.8–6.0, stainless steel of specific gravity 7.0+, and tungsten carbide of specific gravity 14.2+. Also, use of larger diameter beads from 2 to 20 mm can improve breakage. Recently, success has been reported with SiC grit with its sharp edges, and with stainless steel ballcones that have a wedged edge at the equator.
- Time savings can be achieved by ganging several samples into 96 well titer plates rather then running one sample at a time. The larger models can accommodate these plates.
- A simple way to evaluate the suitability of bead shaking can be done as follows. In a test tube place some glass beads and buffered cells/tissues. Hold the tube against a vortex shaker for 1–3 min. If breakage/homogenization is realized, then bead shaking has promise. Switching to suitable equipment as describe above will reduce repetitive strain to the technician holding the test tubes.

3. Bead Impact Methods—Stirred Agitated Beads

3.1. Theory

For modest sample quantities of 50ml and scaling up to industrial amounts of several thousand liters, the agitation of the beads by a turning agitator within the vessel is the method of choice. In this class of agitated bead mills, the beads and the cell suspension are loaded into a chamber. Into the mix is placed one or more spinning discs that accelerate the beads. The beads striking the cells combine with the shearing by the moving liquid phase to disrupt the cells.

These units are normally used for disruption of free microorganism cells (bacteria, yeasts, hyphea, and mycelia) and not for tissue samples. For lesser quantities, see previous section on shaking container method. Materials needed to operate the agitated bead mills include the cells/tissue, grinding media

Fig. 3. Bead impact—agitated beads: Dyno-Mill with 600 ml chamber equipment.

(beads), liquid phase such as buffer, cooling ice, or jacket fluids, and the mill equipment. Variables include the bead selection (density, diameter, and quantity), speed of agitation, cell concentration, and duration of run. (Fig. 3 and Fig. 4)

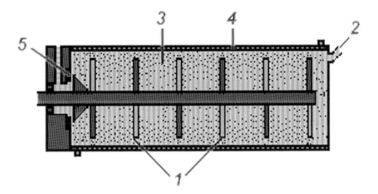

Fig. 4. Bead Impact—agitated stirred beads—inside view: (1) agitators (2) input (3) beads (4) cooling jacket (5) outlet to retain beads.

3.2. Practical Aspects

- Denaturing of proteins because of high temperature or excessive shear is to be considered in all mechanical disruption/homogenization equipment. Also, the wear of the grinding media and container into the samples need to be evaluated.
- Hints for successful temperature control include (1) Prechilling of samples to below 5 °C. (2) Reduce residence time by increased feed rates and then do a second pass with interstage cooling. (3) Slower tip speed of agitator discs to 6 m/sec. (4) Lower the concentration of cells if viscosity increases contribute to excessive heating.
- Detrimental contamination of the batch because of wear of either the grinding media or container walls is usually rare because of the insignificant amounts involved. A hint to mitigate any contamination problem is to use inert materials of construction. For example when iron would be harmful, do not use steel beads, switch to glass or ceramics. The least wear is reportedly seen when using ceramic beads and containers that are fabricated from zirconium oxide stabilized with yttrium (95%/5%; specific gravity 6.0).
- For small diameter cells (e.g., *Escherichia coli* of 0.25–1.0 μm) use beads of 0.10–0.5 mm diameter. For larger cells (e.g., yeast, algae, hyphae) use beads of 0.5–1.25 mm. Glass (specific gravity 2.5) is a good starting material because of low cost. If additional energy is needed to improve breakage/homogenization, then use ceramic (zirconium oxide family) with specific gravities from 3.8 to 6.0.
- A quick preliminary test can be run with standard lab equipment. Load a beaker with the buffered cells/tissue along with some beads. Place on a magnetic stirrer (or use an overhead stirrer) and spin the beads. If some breakage is seen then the bead mill is a good candidate for processing the cells in question.

4. Rotor—Stator Homogenizer

4.1. Theory

Rotor/stator homogenizers consist of a rapidly spinning paddle contained within an open-ended tube with slots near the working end. The turning paddle pushes liquid out the slots, creating a low-pressure region that draws fresh suspension up from the open end. As the tissue is pulled up, there is a stretching action. Then, as the material is forced between the narrow gaps, there is a cutting action. The working part of the equipment that contacts the samples is called the generator.

The rotor/stator design consists of two or more coaxial interlocking rows of teeth. The internal rotor(s) is driven with a motor running at speeds between 3,000 rpm for the larger, industrial units and 27,000 rpm for the smaller lab units. The size of the motor is chosen based on the size of the equipment and application. Rotor tip speeds of up to 50 m/s can be achieved. Usually the gap between rotor and stator is around 300–500 μm. (Fig. 5 and Fig. 6)

Mechanical/Physical Methods of Cell Disruption

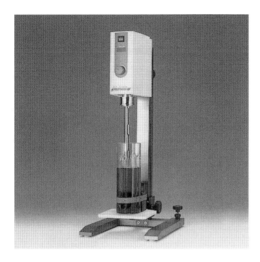

Fig. 5. Rotor-stator—shear by spinning shaft—Polytron equipment.

4.2. Practical Aspects

- Choice of the rotor/stator geometry depends upon the application. Some examples of specialized generators include extended knives to cut into a larger tissue sample, low foaming generators, and easy-cleaning units that may be autoclaved.
- Next choose the correct size generator for the sample volume to be homogenized. Factory tables provide the volume ranges suitable for each generator thus selecting

Fig. 6. Rotor-stator—shear by spinning shaft—inside view.

the correct parts is quite easy. For example, a sample of size from 0.1 to 25 ml would be best processed with a generator of 5 mm diameter. As the quantity of material to be processed at one time increases, larger sizes of generators and motors are needed.
- Interchangeable generators are designed to fit onto only certain motors. Therefore, one must decide which motor is the most suitable for the particular range of applications and then be sure to select only those generators designed to fit onto that motor.
- The homogenizers currently available can process sample volumes anywhere from less than 0.5ml to 150,000 liters. Batch equipment (e.g., POLYTRON®) is used for small sample quantities of less than 0.5 ml though larger units can handle several hundred liters. For larger industrial in-line systems for either continuous or recirculation flow there are single-, double-, or even triple-stage rotor/stator configurations (MEGATRON®).

5. Mortar and Pestle Tissue Grinders—Shear By Mechanical Pressure

This class of simple devices disrupts cells and homogenizes tissues by the pressure and friction generated when a moving pestle pinches the samples against the wall of the mortar. Though usually manual, there are electrically driven units available. The previous equipment class of rotor/stator equipment is an alternative design of this technique. Selection of proper mortar depends upon samples to be processed. Materials of construction are of glass, stainless steel, Teflon, plastics. Some units available include: Wheaton Potter-Elvehjem Tissue Grinders 2-ml samples and larger.

One practical application of using these devices for bacteria is freezing the sample with liquid nitrogen.

6. High Pressure Batch—Expanding Fluids

6.1. Theory

There are two widely used cell disruption methods that employ rapidly expanding fluids from within the cell to explode the cell membranes. The FRENCH® Press uses liquid under pressure, and the Parr Cell Disruption Bomb uses compressed gases. Because these are bath operations they are only suitable for small quantities of less than about a liter.

6.1.1. The French Press

The French Press design consists of a stainless steel cylinder (called pressure cell) fitted with an exit valve at one end and a piston/plunger at the other end. Up to 35 ml of suspended cells in buffer are loading into the pressure

Mechanical/Physical Methods of Cell Disruption

cell. With the exit valve closed the piston is pressed against the liquid by a hydraulic press, called the French Press. Once a suitable pressure (up to 40 kpsi) is achieved throughout the liquid and within the cell body, then the outlet valve is opened to allow the cell suspension to drip out at a slow rate of about one ml/min (9–20 drops/min). This exposure to atmospheric pressure, being much lower than the pressure that was forced within the microorganism's body causes the liquid to rush out, thereby rupturing the cell membrane. Normally bacteria (at 40 kpsi) and yeast (at 20kpsi) are handled in the French Press, not tissues, plants, or seeds. (Fig. 7)

6.1.2. The Parr Cell Disruption Bomb

The Parr cell disruption bomb is a pressure vessel into which the sample to be disrupted is placed along with a dip tube fitted with an exit valve. Suitable gas such as nitrogen at 2 kpsi is forced into the bomb and dissolved into the cells. When the exit valve is opened the gas pressure is suddenly released causing the nitrogen to come out of the solution within the cells as expanding bubbles. This action stretches the membranes of each cell until they rupture and

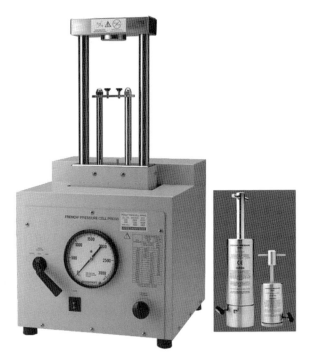

Fig. 7. High pressure batch—liquid expansion French Press, Pressure Cells 35ml and 3.7 ml equipment.

release the contents of the cell. Although sometimes referred to as "explosive decompression," nitrogen decompression is actually a gentle method.

This method is suited for treating mammalian and other membrane-bound cells, for treating plant cells, for releasing virus from fertilized eggs, and for treating fragile bacteria. It is not recommended for untreated bacterial cells, unless using various pretreatment procedures to weaken the cell wall. Yeast, fungus, spores, and other materials with tough walls do not respond well to this method. (Fig. 8 and Fig. 9)

6.2. Practical Aspects

6.2.1. French Press

- Keeping the samples cold is achieved by prechilling of the pressure cell. After the cells exit the valve, immediately cool them by keeping the collection beaker on ice. There are no provisions to chill them during the disruption process.

Fig. 8. High pressure batch—gas expansion—Parr bomb equipment.

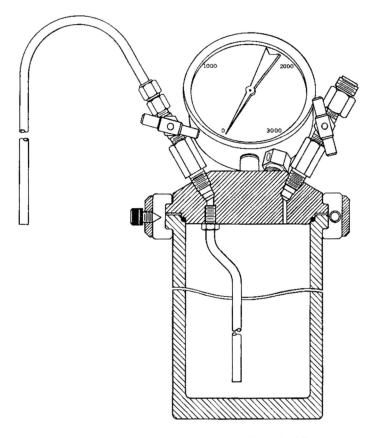

Fig. 9. High pressure batch—gas expansion—inside view.

- Removal of air before closing the exit valve will minimize the amount of oxygen degradation of the released proteins.
- During the paced release of materials, the pressure in the pressure cell will drop. This needs to be balanced by periodically repressurizing the system by running the hydraulic press

6.2.2. Parr Cell Disruption Bomb

- Individual cells such as lymphocytes, leukocytes, tissue culture cells, or very fragile bacterial cells will not require pretreatment. Tissues must usually be preminced to ensure that they not plug the exit dip tube and discharge valve.
- The intended use of a homogenate generally determines the composition of the suspending medium. Isotonic solutions are commonly used. Solutions with higher concentrations will tend to stabilize the nucleus and organelles. Conversely, very

dilute solutions will prestretch the cells by osmotic pressure and will render them more susceptible to disruption by the bomb method.
- Very small quantities of calcium chloride, magnesium acetate, or magnesium chloride added to the suspending medium will stabilize the nuclei when differential rupture is desired. Ratios of approximately 10 ml of suspending medium to 1 g of wet cells are commonly used to prepare the cell suspension.
- Small sample quantities can be held inside of a smaller test tube or beaker placed in the bomb. The inner container should be approximately twice the volume of the suspension to be treated, and the dip tube adjusted to reach the bottom of the container.
- To cool a small inner vessel, it can be floated on ice water within the bomb. The dip tube will hold it in place. Alternatively, the entire external unit may be chilled.
- Degree of disruption can be controlled by the amount of gas pressure introduced. The greater the pressure, the more homogenization. Use of moderate pressures will reduce the disruptive forces and thus leave nuclei, active mitochondria, and other organelles intact.

7. High Pressure Flow – Shear through a Valve or Tube
7.1. Theory

In high-pressure homogenizers the liquid stream of suspended cells is forced at high pressure down a narrow channel or across the small gap of a valve. This accelerates its speed, thereby stretching and shearing cells. In some designs the moving stream is subsequently and abruptly impacted against an obstacle to further damage the cells' membrane. Two versions of impact are (1) where the stream is directed to slam against an impingement wall (trade name Gaulin®), and (2) where the stream is split into two legs of a "Y," and these lines are then directed at one another in an interaction chamber where they collide, further disrupting the cells (trade name Microfluidics®). These devices are used for free-suspended cells, not for tissue samples.

7.1.1. High Pressure Valve with Impingement Wall—Gaulin

The construction of the high-pressure homogenizer consists of a positive displacement pump, a homogenizing valve, and sometime an impingement wall or ring. Though pumps may consist of one-, two-, three-, or five-plungers, most smaller laboratory homogenizers and those for cell breakage have only one plunger.

Attached to the pump is a homogenizing valve assembly that may consist of one or two stages. For cell disruption a single-stage valve is needed, typically consisting of three parts: a seat (bottom part), a valve (top part), and an impact (wear) wall or ring.

By adjusting the gap or clearance between the valve and seat, the flow area in the homogenizing valve is controlled. When the flow area is reduced, pressure within the pump discharge manifold increases. When the flow area is increased, the pressure is reduced. The high-pressure generated by the pump is converted to fluid velocity and heat as the fluid is discharged from the restricted area in the homogenizing valve. For cell disruption the microorganisms are disrupted because of various mechanisms associated with the fluid velocity. At 100 MPa the fluid velocity can be as high as 450 m/sec (Fig. 10 and Fig. 11)

7.1.2. High Pressure Flow Narrow Tubes or Opposed Jets—Microfluidics

These processors disrupt cells by applying a combination of shear and impact forces onto the cells. The media that contains the cells is forced through the narrow channels of the proprietary interaction chamber of the processor. Inside these channels, with typical dimensions of 75–300 microns, the fluid achieves velocities up to 400 m/s. High pressures (up to 275 MPa) are required to generate the high velocities inside the interaction chamber. The resulting shear rates can be up to 10,000,000 s^{-1} and are the highest commercially available. An alternative configuration splits the stream of pressurized cell suspension fluid into two legs. These are then directed at one another in an interaction chamber where they collide, disrupting the cells.

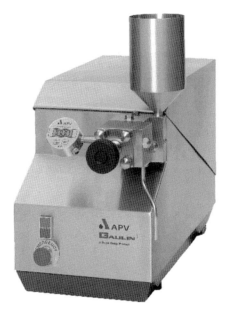

Fig. 10. High pressure flow—liquid shear—APV Gaulin equipment.

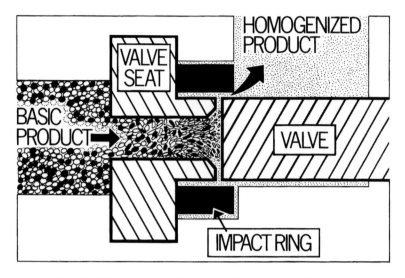

Fig. 11. High pressure flow—liquid shear—inside view.

This equipment is suitable for a variety of free cells (bacteria, yeasts, and mammalian cells), but not seeds, tissue samples, or plant materials. The design of identical fluid channels in both laboratory scale and large-scale units allows for direct scale up from the smallest laboratory unit (14-ml batch) directly to large production units (tens of liters per minute) (Fig. 12).

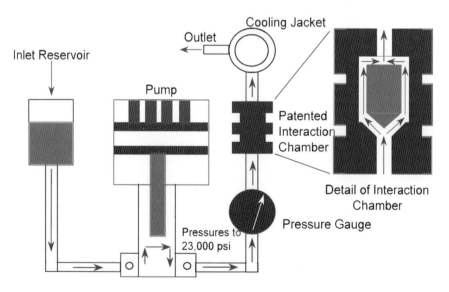

Fig. 12. High pressure flow—opposed liquid streams—Microfluidizer inside view.

7.1.3. Practical Aspects Using High Pressure Valve with Impingement Wall—Gaulin

- To process a small sample size, first a liquid compatible with the continuous phase of the slurry is added to the feed hopper of the homogenizer. The machine is started and the pressure is set. When the liquid level reaches the very bottom of the feed hopper, the cell slurry can be added quickly. The cell slurry will push the liquid ahead of it. When the slurry is observed in the discharge, a sample can be taken. In this way limited sample is not lost while waiting for pressure to rise to operational levels.
- If the product is very sensitive to heat, then it may be necessary to cool the equipment before introducing cell suspension and to cool the suspension immediately after it discharges from the homogenizer. A cooling coil is connected to the discharge tube of the homogenizer and is immersed in an ice water bath. Because the homogenizer is a positive displacement pump there must be no valves or restrictions in this discharge line that could potentially shut off flow. Rapid chilling will minimize losses because of the temperature rise of 2.5 °C per 10 MPa of pressure generated during the nearly adiabatic heating during pressurization.
- To improve breakage, higher pressures are used up to the pump's limits. High pressure can shorten valve life.

7.1.4. Practical Aspects—High Pressure Narrow Tubes/Opposed Jets

- The material should be fully thawed before processing; ice may plug the chambers.
- Multiple passes decrease the particle artifacts' size. Purification methods to be used after cell disruption may determine the desired particle size.
- Optimization with respect to process pressure and number of passes may be needed to increase the yield and facilitate subsequent purification steps.
- Most models are autoclavable and air driven, though some are electric-hydraulic systems. In many circumstances, especially when samples are processed multiple times, the Microfluidizer processors require sample cooling before and/or after processing.

8. Low Pressure—Shear By Droplet's Impingement

Invented at Indiana University the BioNeb disruption system disrupts cells by a low-pressure droplet method. In the process of droplet formation, large molecules or cells suspended in the liquid being nebulized are forcefully distributed from the liquid into the forming secondary droplet. This creates a transient laminar flow in the microcapillary "nebulization" channel formed between the surface of the liquid and the forming secondary droplet. The laminar flow in the capillary channel exerts sufficient shearing forces to break cells. The shearing force created depends on the gas pressure applied (10–250 psi), the type of gas (nitrogen, argon, etc.) and the viscosity of the liquid. By varying these parameters it is possible to precisely regulate the magnitude of the force applied during nebulization. (Fig. 13)

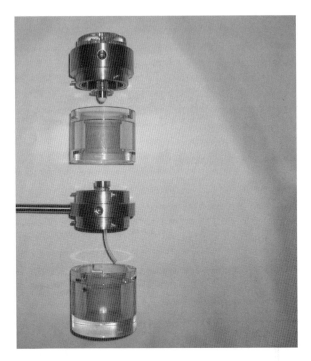

Fig. 13. Droplet low pressure nebulizer—BioNeb A equipment.

9. Ultrasonic Processors—Shear By Collapsing Bubbles

9.1. Theory

The use of sound waves in fluids can disrupt cells. The operation starts with normal electrical current (50Hz or 60Hz) being transformed to 20,000 Hz. This electrical signal is fed to a piezo-electric crystal causing it to oscillation at this high frequency. The vibrations move a titanium metal HORN about 5–15 microns. The shape of the horn amplifies this motion to 100–150 microns per cycle. By placing the horn's end—the tip—into fluid, the tip moves the liquid forward (away) and then retracts (back) quicker than the liquid can return. During the return stroke, the pressure in the system drops below the vapor pressure of the liquid so boiling occurs ("cavitation"). As the liquid flows back, the bubbles collapse. This bubble collapsing imparts the energy needed to disrupt the cells. (Fig. 14)

9.2. Practical Aspects

- Denaturing of proteins because of high temperature or excessive shear is often noted. The equipment has ON-OFF-ON cycle adjustments available. During the OFF periods, the samples can cool. Lowered amplitude settings may reduce protein damage.

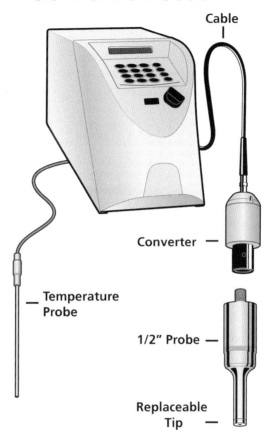

Fig. 14. Ultrasonic—shear by collapsing bubbles—Sonicator equipment.

- Hint to improve bubble formation is to keep the system as cold as possible.
- Time savings can be achieved by ganging several samples into 96-well titer plates rather then running one sample at a time. Multi models can accommodate these plates. Also, some models have several tips fitted on a single horn that can handle several samples simultaneously.
- Larger quantities up to 100 L/h flow rates can be processed in the SONITUBE. Here the entire pipe of 15 inches (35 cm) oscillates and ultrasonically processes all fluids pumped through.

Acknowledgments

Superb assistance was rendered by the following people who are most conversant in their noted equipment areas. Bead milling by Tim Hopkins (BIOSPEC PRODUCTS) and Norbert Roskosch (W. A. BACHOFEN AG);

Rotor/Stators by Roger Munsinger (KINEMATICA USA/AG); High-pressure flowing liquid by William Pandolfe (INVESYS) and Thomai Panagiotou (MICROFLUIDICS); High-pressure batch gas by Deanna Shepard (PARR INSTRUMENTS); Low-pressure flowing gas by Lee Clark (GLAS-COL); and ultrasonic processors by Marc Lusting and Andrea Coppola (MISONIX).

My book editor Dr. Anthon Posch has been exceptionally supportive in both technical as well as motivational matters. My wife Robin and two sons Evan and Adam are the sparkle of my life.

References

1. Reinkemeier, M., Rocken, W., and Leitzmann, C. (1996) A rapid mechanical lysing procedure for routine analysis of plasmids from lactobacilli, isolated from sourdoughs. *Int J Food Microbiol* **29,** 93–104.
2. Rezwan, M., Lanéelle, M. A., Sander, P., and Daffé, M. (2007) Breaking down the wall: Fractionation of mycobacteria. *J Microbiol Methods,* **68**(1), 32–9.
3. Jung, J., Xing, X., and Matsumoto, K. (2001) Kinetic analysis of disruption of excess activated sludge by Dyno Mill and characteristics of protein release for recovery of useful materials. *Biochem. Eng. J.* **8,** 1–7.
4. Lizotte, E., Tremblay, A., Allen, B. G., and Fiset, C. (2005) Isolation and characterization of subcellular protein fractions from mouse heart. *Anal Biochem* **345,** 47–54.
5. Lindeskog, P., Haaparanta, T., Norgard, M., Glaumann, H., Hansson, T., and Gustafsson, J. A. (1986) Isolation of rat intestinal microsomes: partial characterization of mucosal cytochrome P-450. *Arch Biochem Biophys* **244,** 492–501.
6. Nally, J. E., Whitelegge, J. P., Aguilera, R., Pereira, M. M., Blanco, D. R., and Lovett, M. A. (2005) Purification and proteomic analysis of outer membrane vesicles from a clinical isolate of Leptospira interrogans serovar Copenhageni. *Proteomics* **5,** 144–152.
7. Autuori, F., Brunk, U., Peterson, E., and Dallner, G. (1982) Fractionation of isolated liver cells after disruption with a nitrogen bomb and sonication. *J Cell Sci* **57,** 1–13.
8. Kelly, W. J., and Muske, K. R. (2004) Optimal operation of high-pressure homogenization for intracellular product recovery. *Bioprocess Biosyst Eng* **27,** 25–37.
9. Mehlhorn, I., Groth, D., Stockel, J., Moffat, B., Reilly, D., Yansura, D., Willett, W. S., Baldwin, M., Fletterick, R., Cohen, F. E., Vandlen, R., Henner, D., and Prusiner, S. B. (1996) High-level expression and characterization of a purified 142-residue polypeptide of the prion protein. *Biochemistry* **35,** 5528–5537.
10. Vinatier, J., Herzog, E., Plamont, M. A., Wojcik, S. M., Schmidt, A., Brose, N., Daviet, L., El Mestikawy, S., and Giros, B. (2006) Interaction between the vesicular glutamate transporter type 1 and endophilin A1, a protein essential for endocytosis. *J Neurochem* **97,** 1111–25.
11. Jaki, B. U., Franzblau, S. G., Cho, S. H., and Pauli, G. F. (2006) Development of an extraction method for mycobacterial metabolome analysis. J Pharm Biomed Anal 41, 196–200.

2

Bacteria and Yeast Cell Disruption Using Lytic Enzymes

Oriana Salazar

Summary

Enzymatic methods provide a convenient alternative for overcoming technical disadvantages of mechanical disruption. Protocols for protein extraction from bacteria and *Saccharomyces cerevisiae* using lytic enzymes are presented in this chapter. Adaptation of the yeast protocol to a microtiter plate format makes this protocol amenable for proteomic applications and high-throughput screening of libraries expressing genetic variants in yeast. This methodology can also be applied to bacteria.

Key Words: Cell lysis; yeast-lytic endoglucanase; bacteriolytic enzyme; lysozyme; microtiter-plate format.

1. Introduction

The breakage of microbial cells using nonmechanical methods is attractive as it offers the prospect of releasing enzymes under conditions that are gentle, do not subject the proteins to heat or shear, and are not energy-intensive (quiet to the user). Several methods are available including osmotic shock, freezing followed by thawing, cold shock, desiccation, and chemical lysis. Among these, lytic enzymes provide selective, controlled, and gentle procedures for effective cell disruption. The wall is enzymatically degraded to release wall-associated and intracellular products. This is followed by centrifugation for sedimentation of the cell debris and proteins released from the intracellular space are now in the supernatant, where they can be assayed by an appropriate method, i.e., electrophoresis, enzyme activity assays, immunological detection, or other methods for quantitative or qualitative analyses.

1.1. Methods for Bacterial Lysis

Based on the cell wall structure, bacteria are divided in Gram-positive and Gram-negative bacteria. In both types of cells the peptidoglycan layer, a polymer of *N*-Acetyl-D-glucosamine units β-(1,4) linked to *N*-acetylmuramic acid, is responsible for the strength of the wall (reviewed in *(1)*). The cell wall of Gram-positive bacteria is composed of multiple layers of peptidoglycan associated by a small group of amino acids and amino acid derivatives. They form the glycan-tetrapeptide, which is repeated many times through the wall. Penta-glycine bridges associate tetrapeptides of adjacent polymers. To the outer side, peptidoglycan is connected to teichoic acids and polysaccharides. Gram-negative bacterial cells have a two-layer wall structure with a periplasmic space between them: an outer membrane composed of proteins, phospholipids, lipoproteins, and lipopolysaccharides, covers the inner, rigid peptidoglycan layer. Chemical composition and structure of the peptidoglycan in Gram-negative and Gram-positive bacteria is similar, except the peptidoglycan is much thinner in the former.

Enzymes that digest peptidoglycan of bacteria are collectively called murein hydrolases *(2)*. Based on their bond specificity they are classified as: (1) glycosidases, which split polysaccharide chains (lysozymes or muramidases and glucosaminidases), (2) endopeptidases, which split polypeptide chains, and (3) amidases, which cleave the junction between polysaccharides and peptides. Among them lysozyme, the best-described muramidase, *(3)* has high activity. Extensive hydrolysis of peptidoglycan by lysozyme results in cell lysis and death in a hypoosmotic environment. Some lysozymes can kill bacteria by stimulating autolysin activity upon interaction with the cell surface *(4)*. In addition, a nonlytic bactericidal mechanism involving membrane damage without hydrolysis of peptidoglycan, has been reported for c-type lysozymes, including human lysozyme *(5)* and hen egg white lysozyme (HEWL) *(6,7)*.

The presence of the outer membrane in Gram-negative bacteria makes more difficult the access of lytic enzymes. A single enzyme, for instance lysozyme, can lyse Gram-positive bacteria, but pretreatment with a detergent (e.g., Triton X-100) or cation chelating (as EDTA) is usually necessary to remove the outer membrane of Gram-negative cells.

The most studied lysozyme is HEWL, which is widely available and inexpensive. Although Gram-positive bacteria are generally sensitive to HEWL because their peptidoglycan is directly exposed, some are intrinsically resistant because of a modified peptidoglycan structure *(8)*. However, this peptidoglycan is sensitive to ch-type lysozymes, including mutanolysin (a muramidase from *Streptomyces globisporus (9)*), cellosyl from *Streptomyces coelicolor (10)* and the lysozyme from the fungus *Chalaropsis punctulata (11)*. Mutanolysin is a 23 kDa *N*-Acetyl muramidase, like lysozyme, that cleaves the N-acetylmuramyl-β(1-4)-N-acetylglucosamine linkage of the bacterial cell wall polymer peptidoglycan-polysaccharide. Its carboxy terminal moieties are involved in the recognition and binding of unique cell wall polymers. Mutanolysin and lysozyme are occasionally used together for lysis of *Listeria* and other Gram-positive bacteria such as *Lactobacillus* and *Lactococcus*. Lysostaphin has a narrow antibacterial spectrum but a high activity against *Staphylococcus aureus* *(12)*, which have shown to be lysozyme-resistant. Lysostaphin is a zinc endopeptidase with a molecular weight of approximately 25 kDa. Because lysostaphin cleaves the polyglycine cross-links in the peptidoglycan layer of the cell wall of *Staphylococci* species is very useful for cell lysis *(13)*. When Streptococcal and Staphylococcal cells are pretreated with additional enzymes in conjunction with lysozyme, protein yields can be significantly improved.

1.2. Methods for Yeast Cell Lysis

The yeast cell wall is a highly dynamic structure, which is responsible for protecting the cell from rapid changes in external osmotic potential. In recent years several comprehensive reviews on this subject have been published *(14,15)*. The walls are composed mostly of mannoprotein and fibrous β-1,3- glucans with some branches of β-1,6- glucans. The glucans are essential components, responsible for the mechanical strength, shape, and elasticity of the cell wall. The β-1,3- glucan-chitin complex is the major constituent of the inner wall and forms the fibrous scaffold of the wall. On the outer surface of the wall are mannoproteins, which are densely packed and limit wall permeability to solutes. β-1,6-Glucan links the components of the inner and outer walls. All these components are covalently linked to form macromolecular complexes, which are assembled in unit modules, each built around a molecule of β-1,3- glucan. A minority of the modules have chitin (a β-1,4-linked polymer of N-acetylglucosamine) chains attached to the β-1,3- or β-1,6-glucan. The modules are associated by noncovalent interactions in the glucan-chitin layer and by covalent cross-links in the mannoprotein layer, including disulfide bonds between mannoproteins.

Enzyme systems for yeast cell lysis are usually a mixture of several different hydrolases *(16)*, including one or more β-1,3- glucanase (lytic and nonlytic), protease, β-1,6- glucanase, mannanase, or chitinase (which acts on the bud scars). They act synergistically in lysis of the cell wall *(17)*. Enzymatic cell lysis of yeast begins with binding of the lytic protease to the outer mannoprotein layer of the wall. The protease opens up the protein structure, releasing wall proteins and mannans and exposing the glucan surface below. The glucanase then attacks the inner wall and solubilizes the glucan. In vitro, this enzyme cannot lyse yeast in absence of reducing agents, such as dithiothreitol or β-mercaptoethanol, probably because the breakage of disulphide bridges between mannose residues and wall proteins is necessary for appropriate exposition of the inner glucan layer. When the combined action of the protease and glucanase has opened a sufficiently large hole in the cell wall, the plasma membrane and its contents are extruded as a protoplast. In osmotic support buffers containing 0.55 to 1.2 M sucrose, mannitol, or sorbitol, the protoplast remains intact, but in dilute buffers it lyses immediately, releasing cytoplasmic proteins and organelles, which may lyse. Meanwhile, proteins released from the wall and the cytoplasm could be subject to attack by product-degrading protease contaminants in the lytic system or in the yeast cells themselves.

From a commercial point of view, the most important yeast lytic endoglucanases are produced by different strains of *Cellulomicrobium cellulans* (former *Oerskovia xanthineolytica,* also known as *Arthrobacter luteus)*, which produces high concentrations of the extracellular lytic enzyme system *(18–20)*. The purified endoglucanase component (BGLII) supported the release of virus-like particles from viable *Saccharomyces cerevisiae* cells *(21)*. The enzyme from *C. cellulans* was cloned and over-expressed as a recombinant protein in *E. coli (22)*. The 40-kDa polypeptide has a catalytic domain and a putative mannose-binding domain in the carboxy end; a linker of 30 amino acids connects both domains. BGLII degrades soluble (laminarin) and insoluble glucans (curdlan, Zymosan A). Despite this, the enzyme can hydrolyze yeast cell wall and its activity seems to be lower than that of β-1,3-glucanase from *C. cellulans* strain 21606, also known as *Arthrobacter luteus* 73/14 *(23,24)*. BGLII performance in laminarin and yeast has been improved by random mutagenesis and screening of the catalytic/lytic activity *(25)*.

There are several yeast lytic enzymes offered in the market. However, many of them contain significant amounts of protease; in fact, proteases are required for efficient yeast cell lysis. They are frequently used for molecular biology procedures, such as nucleic acids extraction or protoplast production for yeast transformation. However, these systems are rarely applicable for protein extraction from yeast, because contaminant proteases might degrade proteins released. In the protocol presented here a recombinant pure lytic endoglucanase is used for protein extraction from *S. cerevisiae*. The main advantages of a purified endoglucanase are that it does not produce negative effects on the integrity of the polypeptides that are being extracted, and it avoids contamination that would complicate the downstream steps. Release of intracellular recombinant β-galactosidase from viable *S. cerevisiae* cells by purified BGLII is shown in Fig. 1, demonstrating that substantial disruption of the yeast cell wall can be achieved in absence of proteases. Adaptation of the protocol to a microtiter plate format permits the application to areas where a massive analysis of samples is required, as in proteomics and high-throughput screening of gene libraries of variants constructed in yeast.

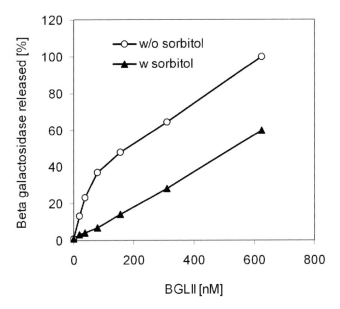

Fig. 1. Release of intracellular (β-galactosidase from viable *S. cerevisiae* cells. Lytic activity was measured using purified (β-1,3-glucanase in a suspension of fresh *S. cerevisiae* cells (2×10^8 cells/ml), by 75 min at 30 °C, with or without 1.4 M sorbitol. After the lysis, (β-galactosidase was quantified in the supernatant fraction by measuring *o*-nitrophenol released from the hydrolysis *o*-nitrophenyl- β(-D-galactopyranoside. (β-galactosidase released was calculated as the percentage produced by the highest enzyme concentration used without sorbitol.

2. Materials

2.1. Bacterial Lysis

1. Hen egg white Lysozyme (HEWL) (Sigma L7651).
2. Mutanolysin (Sigma M-9901).
3. Protease inhibitor cocktail (Sigma P8465).
4. Wash buffer A: 10 mM Tris-HCl, pH 8.0.
5. Wash buffer B: 50 mM Tris-HCl, pH 8.0, 10 mM EDTA, 5 % (w/v) Triton X-100, 0.1 M NaCl, protease inhibitor cocktail.
6. Wash buffer C: 20 mM Tris-HCl, pH 8.0, 2 mM EDTA, 1.2 % (v/v) Triton X-100.

2.2. Yeast Lysis

1. Purified yeast lytic β-1,3-endoglucanase. A recombinant endoglucanase is recommended (BGLII from *C. cellulans* strain DSM10297 is used in our laboratory, but QUANTAZYME ylg™ (Krackeler Scientific, Inc.) is also available.

2. *S. cerevisiae* strain rhSOD 2060 411 SGA122 (MATα, *leu2*; Chiron Corporation) (other strains used with similar results are: INVSc'1 (*his3Δ1*, *leu2 trp1-289 ura3-52*; Invitrogen); DSY-5 (MATa *leu2 trp1 ura3-52 his3 pep4 prb1*; Dualsystems Biotech).
3. YPD: 10 g/L yeast extract, 20 g/L tryptone, 2 % (w/v) glucose.
4. Wash buffer D: 100 mM sodium phosphate pH 7.0.
5. Resuspension buffer: 100 mM sodium phosphate pH 7.0, 50 mM β-mercaptoethanol.

3. Methods

3.1. Bacterial Cell Disruption

Protocols described below have been useful for breaking down: (1) lysozyme–sensitive Gram-positive bacteria, such as *Micrococcus lysodeikticus*, *lactobacilli*, and *Listeria monocytogenes*; (2) *E. coli* and other Gram-negative organisms (i.e., *Salmonella typhimurium*, *Pseudomonas aeruginosa*, and *Yersinia enterocolitica*) and (3) less sensitive or lysozyme-resistant Gram-positive organisms, such as streptococci, staphylococci, and others.

3.1.1. Gram-Positive Bacteria

1. Obtain a bacterial pellet by centrifugation of a culture at 6,000 g for 1–5 min (*see* Note 1).
2. Wash cells twice in wash buffer A.
3. Resuspend cells in the same buffer plus 20 mg lysozyme/ml and protease inhibitor cocktail (*see* Note 2).
4. Incubate at 20–37 °C for 10–15 min (*see* Notes 3 and 4), or until a clear solution is observed.
5. Centrifuge at 12,000 g for 5 min to sediment the cell debris and proceed with further analytical protocols.

3.1.2. E. coli cells

1. Obtain a bacterial pellet by centrifugation of a culture at 6,000 g for 1–5 min.
2. Wash the cells in wash buffer B (*see* Notes 5, 6 and 7).
3. Suspend washed *E. coli* cells in 3 mL buffer per gram of cells and bring to 20–37 °C.
4. Prepare lysozyme at 10 mg/ml (*see* Note 3).
5. Lysozyme is added to a final concentration of 1–2 mg/ml (*see* Note 8).
6. The mixture is incubated at 20–37 °C for 30–60 min (or overnight) shaking gently (*see* Notes 9 and 10).
7. Centrifuge at 10,000 g for 5 min to sediment the cell debris and proceed with further analytical protocols.

3.1.3. Streptococcal Cells

1. Obtain a pellet from 1.5–2.0 ml of culture by centrifugation at 6,000 g for 1–5 min at 4–25 °C (*see* Notes 1 and 11).
2. Wash the cells by suspension in wash buffer C (*see* Note 7).
3. Suspend bacterial pellet in 200 µl of the enzyme solution (*see* Note 12), containing 20 mg/ml lysozyme and 100 µg/ml mutanolysin in wash buffer C (*see* Note 13).
4. Incubate for at least 30 min at 37 °C or until the solution becomes clear.
5. Centrifuge for a 5–10 min at 12,000 g.
6. Save supernatant and proceed with further analytical protocols.

3.2. Yeast Cell Lysis

Protocols described here consider the use of purified glucanase for *S. cerevisiae* cell lysis. Yeast cells are incubated with the lytic endoglucanase in a suitable buffer (usually sodium phosphate) containing β-mercaptoethanol. After 2–14 h at 30–37 °C, soluble and insoluble fractions are separated by centrifugation and proteins are analyzed in the supernatant. More than 80% of cells are lysed under these conditions, as confirmed by reduction of the optical density of the yeast suspension; most of the intracellular protein is released under these conditions (*see* **Fig. 1**). Although the protocol is quantitative and consistent, however, the actual yields depend on susceptibility of the specific yeast strain to degradation by lytic endoglucanases.

3.2.1. Culture and Harvest of S. cerevisiae Cells

1. *S. cerevisiae* cells are grown until early exponential phase of the growth curve (*see* Notes 14, and 15).
2. Collect the cells by centrifugation by 10–15 min at 12,000 g at 4 °C.
3. Wash the cells by resuspension in wash buffer D and centrifuge again.
4. Store the cell pellet at 4 °C (*see* Notes 16 and 17).

3.2.2. Enzymatic Lysis of S. cerevisiae Cells

1. Resuspend the yeast cells at optical density at 620 nm (OD_{620}) between 1 and 10 (10^7–10^8 cells/ml; *see* Note 18) in resuspension buffer. Incubate at room temperature (RT) by 30 min, mixing by inversion every 5–10 min.
2. Dispense aliquots of 100µl of cells in a transparent flat bottom 96-well microtiter plate (*see* Note 19). Preincubate for 5 min at 30 °C in a dry bath equipped with a thermo-block for microtiter plates. Start the lysis reaction by addition of 100 µl β of 0.2–1 µM lytic enzyme kept at RT (*see* Note 20). Incubate at 30 °C by 2–14 h (*see* Note 21), with permanent shaking of the samples (*see* Note 22).
3. Centrifuge plates at 1,600 g for 15 min, at 4 °C and transfer aliquots of supernatant to a 96-well microtiter plate. Proceed with further analytical protocols.

4. Notes

1. In general, grow bacterial cells until mid-exponential phase for higher susceptibility to lytic enzymes.
2. Protease inhibitor is optional, however is highly recommended.
3. Use fresh lysozyme solution. Dissolve lysozyme in buffer (same buffer as the one used for pellet reconstitution; water can also do for lysozyme) to make a 10 mg/ml solution. Add the enzyme powder to the buffer and allow to dissolve slowly – do not shake or mix. Keep on ice.
4. Less susceptible bacteria usually require more rigorous digestion (increased incubation time, increased incubation temperature, etc.). For better results Gram-positive bacteria are treated with a mix of lysozyme and mutanolysin or lysostaphin, as described in protocol 2.3.1.3 Highly susceptible strains can be totally lysed in 5 min at 4 °C.
5. The activity of lysozyme is a function of both pH and ionic strength. The enzyme is active over a broad pH range (6.0–9.0). At pH 6.2, maximal activity is observed over a wider range of ionic strengths (0.02–0.1 M) than at pH 9.2 (0.01–0.06 M). Ionic strength higher than 0.1 M is inhibitory for HEWL.
6. EDTA is fundamental in the lysis solution of Gram-negative bacteria, because it chelates Ca^{2+} and destabilizes the outer membrane, which is a requisite for lysozyme hydrolytic action.
7. Triton X-100 is a gentle detergent, compatible with most of the downstream steps that allows for the cell membrane disruption. Alternatively, 2% (w/v) sodium dodecylsulphate (SDS) can be used; however, SDS has to be added after the lytic reaction, because inhibits the lysozyme activity. 3[(3-Cholamidopropyl) dimethylammonio]-propanesulfonic acid (CHAPS) is used in many protocols for sample preparation for two dimensional (2D)-electrophoresis; however the effect on lytic enzymes has not been reported.
8. More rapid cell lysis may be obtained by raising the lysozyme concentration (to as much as 10 mg/ml). With higher lysozyme concentrations, satisfactory lysis may be obtained in as little as 5 min even at temperatures as low as 4 °C.
9. Sample preparation and solubilization are crucial factors for the overall performance of the 2D PAGE technique. Protein complexes and aggregates should be completely disrupted in order to avoid appearance of new spots because of a partial protein solubilization. Persistent cloudiness of the lysate may indicate incomplete lysis. Most methods available carry out lysozyme digestion on ice, however, if your protein(s) is fairly thermostable, it is recommended to perform the digestion at 25–30 °C as this is the optimal temperature for lysozyme. Keeping the temperature optimal is an important consideration if lysozyme will be digesting *E. coli* at a pH more acidic than its optimum (pH 9.2). To improve cell lysis protein recovery, extend the incubation time an additional 15–30 minutes and elevate the incubation temperature to 55 °C. Overnight lysis is recommended only when working with strains having less sensitivity to lysozyme. Extended incubation times may result detrimental because of intracellular proteolytic enzymes released during the lysis.

10. The resultant solution will be highly viscous because of the presence of genomic DNA. The viscosity can be reduced by passage through a 21-gauge needle (3–5 passages) or one could add $MgCl_2$ and DNAse I.
11. Grow cells in glucose-supplemented medium, because glucose as the carbon source produces *Streptococci* with higher susceptibility to HEWL, comparing to the same bacteria growth in other carbon sources. The same was observed for bacteria cultured in media with 10 mM L-threonine *(26)*.
12. The amount of HEWL added can be calculated to be 1 mg/1 mg (dry weight) of cells. In the absence of a standard curve relating dry weight to tubidity, it has been found that use of 1 mg of lysozyme per 1.0 optical density unit at 600 nm per ml of culture gives essentially reliable results.
13. For lysis of *Staphylococci*, Lysostaphin (Sigma L-0761) together with HEWL are an effective treatment *(13)*. Labiase from *Streptomyces fulvissimus* is an enzyme preparation useful for the lysis of many Gram-positive bacteria such as *Lactobacillus*, *Aerococcus* and *Streptococcus* . Labiase contains β-N-acetyl-D-glucosaminidase and lysozyme activity *(27)*.
14. It is highly recommended not to use cells grown beyond of the mid-exponential phase, because the susceptibility of the cell wall to be attacked by the lytic enzyme is significantly reduced at the stationary phase of the growth curve.
15. Specific conditions for every strain culture need to be established; however YPD is probably satisfactory for most of the strains.
16. Cells stored for up 2 or 3 days a 4 °C give consistent results and can be lysed with high yield. However, it is recommended to use cells as fresh as possible. Cells stored for longer periods are not recommended, because they become resistant to the lytic action of glucanases.
17. For avoiding proliferation of bacteria and other microorganisms, 0.02% sodium azide can be added to the washing buffer (CAUTION: Sodium azide is toxic and can be absorbed through the skin. After prolonged and successive exposure it could cause death).
18. Minimal yeast cell concentration in the lytic assay strongly depends on the susceptibility of the yeast cell wall to lytic enzymes and the specific activity of the lytic enzyme. Optimal experimental conditions for a particular lytic enzyme-yeast strain system can be evaluated by a succession of experiments in which variables such as cell density; incubation temperature, enzyme concentration and time of incubation are studied.
19. For analysis of a large number of samples, use of a dispenser, multi-channel pipet is highly recommended.
20. In presence of sorbitol the release protein increase linearly with the lytic enzyme concentration (*see* Fig. 1), but without the osmotic support linearity is only until 100-nM lytic enzyme; after that the apparent specific lytic activity decreases, possibly because massive lysis is also releasing proteases that degrade the recombinant enzyme. Practically, this indicates that if the integrity of the recombinant protein is an issue, a high lytic enzyme concentration should be avoided. Alternatively, lysis should be carried out in the same buffer containing 1.4-M sorbitol.

21. In our experience longer periods of incubation with the lytic enzyme do not improve significantly the yield in the protein extraction; on the contrary, detrimental effects on the recovery have been observed. Extended incubations are recommended only in cases of natural resistance to the action of lytic enzymes or to resistance because of aging of the yeast cell.
22. Vigorous shaking of the samples during the incubation with the lytic enzyme is important for maintaining the yeast cells in suspension. Cell sedimentation increases the variability in the assay. Sorbitol in the permeabilization medium notably helps to reduce this effect.

Acknowledgment

The authors would like to thank the CONICYT (Project 1030797) and the Millennium Scientific Initiative (Millennium Institutes) (ICM –P99-031) for financial support.

References

1. Koch, A.L. (1998) Orientation of the peptidoglycan chains in the sacculus of *Escherichia coli*. *Res Microbiol* **149**, 689–701.
2. Shockman, G.D., Daneo-Moore, L., Kariyama, R. and Massidda, O. (1996) Bacterial walls, peptidoglycan hydrolases, autolysins, and autolysis *Microb Drug Resist*. **2**, 95–98.
3. Ibrahim, H. R., Aoki, T. and Pellegrini, A. (2002) Strategies for new antimicrobial proteins and peptides: Lysozyme and aprotinin as model molecules. Curr. Pharm. Design **8**, 671–693.
4. Iacono, V.J., Zove, S.M., Grossbard, B.L., Pollock, J.J., Fine, D.H., Greene, L.S. (1985) Lysozyme-mediated aggregation and lysis of the periodontal microorganism *Capnocytophaga gingivalis* 2010. *Infect Immun*. **47**, 457–464.
5. Laible, N.J. and Germaine, G.R. (1985) Bactericidal activity of human lysozyme, muramidase-inactive lysozyme, and cationic polypeptides against *Streptococcus sanguis* and *Streptococcus faecalis*: inhibition by chitin oligosaccharides. *Infect Immun*.**48**, 720–728.
6. Ibrahim, H.R., Matsuzaki, T. and Aoki, T. (2001) Genetic evidence that antibacterial activity of lysozyme is independent of its catalytic function. *FEBS Lett*. **506**, 27–32.
7. Masschalck, B., Deckers, D. and Michiels, C.W. (2002) Lytic and nonlytic mechanism of inactivation of gram-positive bacteria by lysozyme under atmospheric and high hydrostatic pressure. *J Food Prot*. **65**, 1916–1923.
8. Bera, A., Herbert, S., Jakob, A., Vollmer, W., Gotz, F. (2005) Why are pathogenic staphylococci so lysozyme resistantβ The peptidoglycan O-acetyltransferase OatA is the major determinant for lysozyme resistance of *Staphylococcus aureus*. *Mol. Microbiol*. **55**, 778–787.

9. Shiba, T., Harada, S., Sugawara, H., Naitow, H., Kai Y., Satow, Y. (2000) Crystallization and preliminary X-ray analysis of a bacterial lysozyme produced by Streptomyces globisporus. *Acta Crystallogr D Biol Crystallogr.* **56**, 1462–1463.
10. Rau, A., Hogg, T., Marquardt, R., Hilgenfeld, R. (2001) A new lysozyme fold. Crystal structure of the muramidase from *Streptomyces coelicolor* at 1.65 A resolution. *J Biol Chem.* **276**, 31994–31999.
11. Hash, J.H., Rothlauf, M.V. (1967) The N,O-diacetylmuramidase of *Chalaropsis* species. I. Purification and crystallization. *J Biol Chem.* **242**, 5586–5590.
12. Schindler, C.A. and Schuhardt, V.T. (1964) Lysostaphin: a new bacteriolytic agent for the staphylococcus. *Proc. Natl Acad. Sci.* USA. **51**, 414–421.
13. Malatesta, M.L., Heath, H.E., LeBlanc, P.A., Sloan, G.L. (1992) EGTA inhibition of DNase activity in commercial lysostaphin preparations. *Biotechniques.* **12**, 70–72.
14. Kollar, R., Petrakova, E., Ashwell, G., Robbins, P.W., Cabib, E. (1995) Architecture of the yeast cell wall. The linkage between chitin and β (1–>3)-glucan. *J. Biol. Chem .* **270**, 1170–1178.
15. Lipke, P.N. and Ovalle, R. (1998) Cell wall architecture in yeast: new structure and new challenges. *J. Bacteriol.* **180**, 3735–37340.
16. Bielecki, S. and Galas, E. (1991) Microbial beta-glucanases different from cellulases. *Crit. Rev. Biotechnol.* **10**, 275–304.
17. Obata, T., Fujioka, K., Hara, S. and Namba, Y. (1977) The synergistic effects among β-1, 3-glucanases from *Oerskovia sp.* CK on lysis of viable yeast cells. *Agric. Biol. Chem.* **41**, 671–677.
18. Ventom, A.M. and Asenjo, J.A. (1990) Purification of the major glucanase of *Oerskovia xanthineolytica* LL-G109. *Biotechnol Tech* **4**, 165–170.
19. Ventom, A.M. and Asenjo, J.A. (1991) Characterization of yeast lytic enzymes from Oerskovia xanthineolytica LL-G109. *Enzyme Microbiol. Technol.* **13**, 71–75
20. Ferrer, P. (2006) Revisiting the Cellulosimicrobium cellulans yeast-lytic -1,3-glucanases toolbox: A review. *Microbial Cell Factories* **5**, 10–18.
21. Asenjo, J.A., Ventom, A.M., Huang, R.-B. and Andrews, B.A. (1993) Selective release of recombinant protein particles (VLPs) from yeast using a pure lytic glucanase enzyme. *Bio/technol.* **11**, 214–217.
22. Salazar, O., Molitor, J., Lienqueo, M.E. and Asenjo, J.A. (2001) Overproduction, purification and characterization of β-1,3-glucanase type II in *Escherichia coli. Prot Expres. Purif.* **23**, 219–225.
23. Shen, S.-H., Chrétien, P., Bastien, L. and Slilaty, S.N. (1991) Primary sequence of the glucanase gene from Oerskovia xanthineolytica. *J. Biol. Chem.* **266**, 1058–1063.
24. Scott, J.H. and Scheckman, R. (1980) Lyticase: endoglucanase and protease activities that act together in yeast cell lysis *J. Bacteriol.* **142**, 414–423.
25. Salazar, O., Basso, C., Barba, P., Orellana, C. and Asenjo, J.A. (2006) Improvement of the lytic properties of a β-1,3-glucanase by directed evolution. *Mol. Biotechnol.* **33**, 211–220.

26. Chassy, B.M. and Giuffrida, A. 1980. Method for the lysis of Gram-positive, asporogenous bacteria with lysozyme. *Appl. Environ. Microbiol.* **39**, 153–158.
27. Niwa T., Kawamura, Y., Katagiri, Y., Ezaki, T. (2005) Lytic enzyme, labiase for a broad range of Gram-positive bacteria and its application to analyze functional DNA/RNA. *J Microbiol Methods.* **61**,251–260.

3

Sample Solublization Buffers for Two-Dimensional Electrophoresis

Walter Weiss and Angelika Görg

Summary

Before two-dimensional electrophoresis (2-DE), proteins of the sample must be denatured, reduced, disaggregated, and solubilized. Sample solubilization is usually carried out in a buffer containing chaotropes (typically 9.5 M urea, or 5–8 M urea and 2 M thiourea), 2–4% nonionic and/or zwitterionic detergent(s), reducing agent(s), carrier ampholytes and, depending on the type of sample, protease inhibitors. In this chapter, the major constituents of sample solubilization/lysis buffers will be briefly reviewed, some general sample preparation guidelines will be given, and the most common protein solubilization cocktails will be described.

Key Words: Chaotrope; detergent; buffer; proteome; reductant; sample preparation; two-dimensional electrophoresis.

1. Introduction

To take advantage of the high resolution of two-dimensional electrophoresis (2-DE), proteins of the sample have to be denatured, disaggregated, reduced, and solubilized to achieve complete disruption of molecular interactions, and to ensure that each spot represents an individual polypeptide. Unfortunately, there is no single method of sample preparation that can be universally applied to all kinds of samples analyzed by 2-DE. A variety of "standard" protocols and sample solubilization buffers has been reported, but usually these protocols have to be adapted and further optimized for different types of sample. The fundamental steps in sample preparation are (**1**) cell disruption, (**2**) inactivation or removal of interfering compounds, and (**3**) solubilization of the proteins *(1–3)*. Briefly, cell disruption can be achieved (individually or in combination) by various procedures based on different physico-chemical principles, such as osmotic lysis, freeze-thaw cycling, detergent lysis, enzymatic lysis of the cell wall, sonication, grinding with (or without) liquid nitrogen, high pressure, homogenization with glass beads and a bead beater, rotating blade homogenizers etc. For details see other chapters in this book. During or after cell lysis, interfering compounds such as proteolytic enzymes, salts, lipids, nucleic acids, polysaccharides, plant phenols, and highly abundant proteins have to be removed or inactivated. Following cell disruption and removal of interfering compounds, the individual polypeptides must be denatured and reduced to disrupt intra- and intermolecular interactions, and solubilized while maintaining the inherent charge properties. In the following sections, the major constituents of sample solubilization/lysis buffers will be briefly reviewed and discussed, some general sample preparation guidelines will be given, and the most commonly used sample solubilization cocktails will be described in more detail.

Sample solubilization is usually carried out in in a buffer containing *chaotropes* (e.g., urea or urea/thiourea mixtures), nonionic and zwitterionic detergents (e.g., NP-40, Triton X-100, or CHAPS), reducing agent(s), carrier ampholytes and, depending on the type of sample, inhibitors of proteases, phosphatases, and oxidoreductases. The most popular sample solubilization buffer is based on O'Farrell's lysis buffer and modifications thereof (9.5 M urea, 2–4% CHAPS, 1% dithiothreitol, and 2% [v/v] carrier ampholytes) *(4)*. Unfortunately, this buffer is not ideal for the solubilization of all protein classes, particularly not for membrane or other highly hydrophobic proteins. Solubilization of hydrophobic proteins has been improved with the use of urea/thiourea mixtures as chaotrope *(5)* and new zwitterionic detergents such as sulfobetaines *(6)*.

Urea (Fig. 1) is the chaotrope of choice and is usually included at rather high concentrations (9.5 M) in sample solubilization buffers. Urea is quite efficient in disrupting hydrogen bonds, resulting in protein unfolding and denaturation. In contrast to urea, substituted ureas are superior for breaking hydrophobic interactions and for improved solubilization of hydrophobic proteins *(2,5)*. Unfortunately, their usefulness is somewhat limited because of their poor solubility in water. Currently, the best solution for solubilization of hydrophobic proteins for 2-DE is a combination of 5–8 M urea and 2 M thiourea (Fig. 1), in combination with appropriate detergents (*see* Note 1).

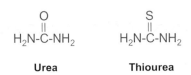

Fig. 1. Structural formula of urea and thiourea chaotropes.

Detergents are amphipatic molecules, containing both a polar (hydrophilic) "head" and a nonpolar (hydrophobic) "tail." The ionic character of the hydrophilic head forms the basis for classification of detergents: nonionic (uncharged), anionic or cationic (charged), and zwitterionic (having both positively and negatively charged groups with a net-charge of zero) (Fig. 2). Nonionic and zwitterionic detergents, or mixtures thereof, are always included in the sample solubilization buffer in concentrations up to 4% to prevent hydrophobic interactions between the hydrophobic protein domains and to avoid loss of proteins because of aggregation and precipitation (*see* Note 2). The most popular nonionic detergents are NP-40 and Triton X-100. Regrettably, these surfactants are not very effective in solubilizing hydrophobic proteins. Zwitterionic detergents such as CHAPS, and sulfobetaines (e.g., SB 3–10 or ASB 14) perform better, and have been shown to solubilize—in combination with urea and thiourea chaotropes—at least several integral membrane proteins *(7)*.

Reduction and prevention of reoxidation of disulfide bonds is also a critical step of the sample preparation procedure. Reducing agents are necessary for cleavage of intra- and intermolecular disulfide bonds to achieve complete protein unfolding and to maintain all proteins in their fully reduced state. The most commonly used reductants are dithiothreitol (DTT) (Fig. 3) and dithioerythritol (DTE) which are applied in excess, i.e., in concentrations up to 100 mM. However, DTT and DTE are not ideally suited for the reduction and solubilization of proteins that contain a high cysteine content, e.g., wool proteins. Hence, Herbert et al. *(8)* have introduced the nonthiol reductant

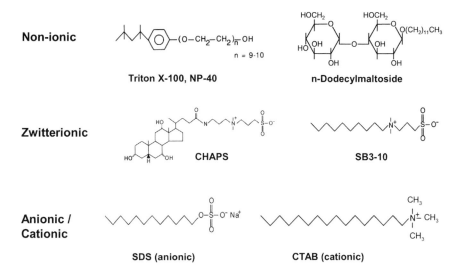

Fig. 2. Commonly used detergents for sample solubilization.

Fig. 3. Cleavage of disulfide bonds by the reducing agent dithiothreitol (DTT).

tributylphosphine (TBP) as an alternative to DTT. TPB is applied in quite low concentrations (2 mM) because of its stoichiometric reaction (*see* Note 3).

Immediately after cell lysis, different enzymes such as proteases, phosphatases, or oxidoreductases may be liberated or activated. Because the cell lysate usually remains in the sample solubilization buffer for up to 1 hour to fully denature, disaggregate, and solubilize the proteins, it is important to inactivate the aforementioned enzymes (in particular proteases) as rapidly and effectively as possible to prevent protein degradation or modification that otherwise may result in artifactual spots and loss of high molecular mass proteins. Because enzymes are less active at lower temperatures, sample preparation at as low a temperature as possible is recommended, preferably at −196 °C (liquid nitrogen). In addition, protease and phosphatase inhibitors may be added (*see* Note 4). Protease inhibitors are usually low-molecular compounds that function by reversibly or irreversibly binding to the (active center of the) protease. Typical examples are phenylmethylsulfonylfluoride (PMSF) and the—less toxic—Pefabloc® (4-(2-Aminoethyl)-benzenesulfonylfluoride-hydrochloride) serin protease inhibitor, or EDTA, which inactivates metalloproteases. Nonetheless, it should be kept in mind that it may be rather difficult to completely inhibit all proteolytic activity, even if a combination of different protease inhibitors is applied (*see* Note 5).

Carrier ampholytes not only aid in improving protein solubility, but also prevent carbamylation of protein amino groups (*see* Note 1). Recommended carrier ampholyte concentration is 2% (v/v).

2. Materials

3-[(3-cholamidopropyl)dimethylammonio]-1-propanesulfonate (*CHAPS*; Roche Diagnostics, Mannheim, Germany); dithiothreitol (*DTT*; Sigma-Aldrich, St. Louis, MO); Pefabloc® proteinase inhibitor (Merck, Darmstadt, Germany); Pharmalyte™ 3–10 carrier ampholytes (GE Healthcare Life Sciences, Freiburg, Germany); Serdolite MB-1 mixed ion exchange resin (Serva, Heidelberg, Germany); sodium dodecyl sulfate (*SDS*; GE Healthcare Life Sciences, Freiburg, Germany); thiourea (Fluka/Sigma-Aldrich, Buchs, Switzerland); urea (GE Healthcare Life Sciences, Freiburg, Germany) (*see* Note 6).

3. Methods

No single method of sample preparation can be universally applied because of the great diversity of sample types and origins. For more detailed guidance, see other chapters in this book. Here, only general recommendations for sample preparation and protein solubilization will be provided:

(1) Sample preparation should be as simple as possible to increase reproducibility; i.e., multi-step procedures should be avoided since they are open to variability.
(2) The sample should be disrupted in such a way as to minimize proteolysis and other modes of protein degradation or modification, because they might result in artifactual spots on 2-DE gels; in particular, proteolytic enzymes in the sample must be inactivated *(9)*.
(3) The efficacy of the cell-wall disruption procedure should be checked by light microscopy and by a protein quantitation assay, and inhibition of protease activity should be verified, e.g., by one-dimensional SDS-PAGE.

After cell disruption, proteins are solubilized either **(1)** in urea lysis solution, **(2)** urea/thiourea lysis solution, or **(3)** in (hot) SDS lysis solution (*see* Note 7). In the latter case, the extract must be diluted with at least a five-fold excess of urea or urea/thiourea lysis solution before IEF to displace the SDS from the proteins (*see* Note 2). In any case, all extracts should be centrifuged extensively (60 min, 40,000 g, 15 °C) to remove any insoluble material, because solid particles may block the pores of the electrophoresis gel. The resulting protein extracts are either used immediately, but can also be stored in aliquots at –70 °C for up to several months. Do not expose to repeated thawing and freezing.

3.1. Urea Lysis Solution

Contains 9.5 M urea, 2% (w/v) CHAPS, 1% (w/v) DTT, 2% carrier ampholytes. To prepare 50 ml of urea lysis solution, dissolve 30.0 g of urea in approximately 25 ml deionized water (*see* Note 8) and complete to 50 ml. Add 0.5 g of mixed ion-exchange resin (Serdolit MB-1), stir for 15 min and filter. Then add 1.0 g CHAPS, 500 mg DTT, and 1.0 ml Pharmalyte™ to 48 ml of the filtered urea solution (*see* Note 9). Store deep-frozen in aliquots at –70 °C for up to several months (*see* Note 10). Immediately before use, add 1.0 mg Pefabloc® proteinase inhibitor per milliliter of lysis solution (*see* Note 11).

3.2. Urea/thiourea Lysis Solution

Contains 7 M urea, 2 M thiourea, 4% CHAPS, 2% DTT, 2% carrier ampholytes. To prepare 50 ml of thiourea/urea lysis solution, dissolve 22.0 g of urea in approximately 25 ml of deionized water, add 8.0 g of thiourea and adjust the volume to 50 ml with deionized water. Add 0.5 g of Serdolit MB-1 mixed ion-exchange resin (Serdolit MB-1), stir for 15 min and filter. Add 2.0 g CHAPS, 1.0 ml Pharmalyte™ and 0.5 g DTT to 48 ml of the filtrate and store deep-frozen in aliquots –70 °C for up to several months (*see* Notes 8–10). Immediately before use, add 1.0 mg Pefabloc® proteinase inhibitor per milliliter of lysis solution (*see* Note 11).

3.3. SDS Lysis Solution

Contains 1% SDS. To prepare 50 ml of SDS lysis solution, dissolve 500 mg of SDS in 50 ml deionized water and filter. Store in the refrigerator.

4. Notes

1. The major problem associated with urea in aqueous buffers is that urea exists in equilibrium with ammonium (iso)cyanate, which can react with the α-amino groups of the *N*-terminus and the ε-amino groups of lysine residues, thereby forming artifacts such as blocking the *N*-terminus and introducing charge heterogeneities (i.e., altered isoelectric points). To prevent this so-called carbamylation reaction (Fig. 4), it is strongly recommended to avoid temperatures above 37 °C, and to include carrier ampholytes (2% v/v), which act as cyanate scavengers, but which also enhance protein solubility in the urea solution. Given that these precautions have been complied with, protein carbamylation is negligible for a period at least 24 h (*10*), which is sufficiently long for almost all protein extraction and solubilization protocols.
2. The cationic detergent CTAB and, in particular, the anionic detergent SDS are among the most efficient surfactants. Hence, solubilization of proteins in (boiling) SDS solution has often been recommended for complete protein denaturation and solubilization. However, ionic detergents such as SDS are not

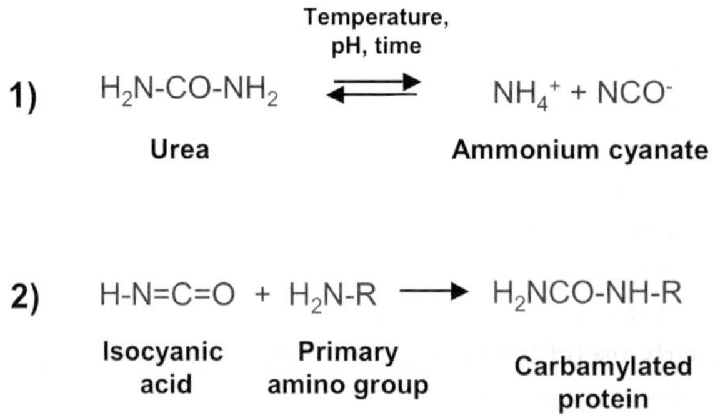

1) $H_2N\text{-}CO\text{-}NH_2$ ⇌ (Temperature, pH, time) $NH_4^+ + NCO^-$

Urea   Ammonium cyanate

2) $H\text{-}N\text{=}C\text{=}O + H_2N\text{-}R \longrightarrow H_2NCO\text{-}NH\text{-}R$

Isocyanic acid   Primary amino group   Carbamylated protein

Fig. 4. Decomposition of urea and reaction of isocyanate with protein amino groups (carbamylation).

compatible with IEF, and horizontal streaks in the 2-D pattern are observed if samples initially solubilized in 1% SDS are not diluted with at least five-fold excess of (thiourea/urea) lysis buffer before isoelectric focusing, to displace the anionic detergent SDS from the proteins and to replace it with a nonionic or zwitterionic detergent to decrease the amount of SDS below a critical concentration (<0.2%).
3. However, TBP reagent has also its disadvantages, the major of which are its low solubility in water and its short half-life. Moreover, TBP (and its solvent dimethyl-formamide, respectively) is toxic, volatile, and has a rather irritating odor. Tris(2-carboxyethyl)phosphine (TCEP) performs better and has been recommended for the saturation labeling procedure in fluorescent difference gel electrophoresis (DIGE).
4. Certain protease inhibitors may also modify proteins and cause charge artifacts.
5. Alternative methods for inhibition of enzymes are disrupting the sample directly in strong denaturants, e.g., by boiling the sample in SDS-buffer (without urea!), or transferring the sample immediately after cell lysis (preferably after grinding the sample in liquid nitrogen) in ice-cold (–20°C) 20% trichloroacetic acid (TCA) in acetone. TCA/acetone precipitation is very efficient for minimizing protein degradation and removing interfering compounds, such as salt, or polyphenols. Commercial sample clean-up kits based on TCA/acetone precipitation are available from several suppliers. However, attention has to be paid to loss of protein because of incomplete precipitation and resolubilization of proteins, and one should be aware that a different set of proteins may be obtained compared to direct extraction with sample solubilization buffer *(11,12)*.
6. Chemicals of the highest purity attainable (ACS or reagent-grade) should be used throughout. The chemicals enclosed in our list have been found to give satisfactory results. However, this does not imply that chemicals from other sources cannot be used with equal or better satisfaction.
7. Heating of the sample in the presence of the detergent SDS can improve solubilization, but should only be done before the addition of urea. Heating of samples containing urea must be avoided under all circumstances to prevent charge heterogeneities caused by carbamylation of the proteins by isocyanate formed in the decomposition of urea at elevated temperatures.
8. All lysis solutions should be prepared in water that has a resistance of 18.2 $M\Omega^{-cm}$. This standard is referred to as "deionized water" in this text.
9. Other detergents (Triton X-100, NP-40, or other nonionic or zwitterionic detergents) can be used instead of (or in combination with) CHAPS. Sulfobetaines such as SB 3–10 or ASB-14 in combination with the urea/thiourea chaotrope are particularly useful for the solubilization of hydrophobic proteins.
10. Urea and thiourea lysis solution should be prepared fresh immediately before use. Alternatively, individual aliquots (1 ml) can be made and stored at –70° C for up to several months. Do not refreeze urea-containing lysis solution after thawing. Use high-purity or de-ionized urea only.

11. Protease inhibitors such as PMSF or Pefabloc may be less active in the presence of thiol reagents (e.g., DTT). To overcome this obstacle, cells may be disrupted in presence of lysis solution containing protease inhibitors, but lacking DTT, which is then added at a later stage, *i.e.* after proteolytic enzymes have been inactivated.

References

1. Dunn, M.J. (ed.) (1993) *Gel electrophoresis: Proteins.* Bios, Oxford, UK.
2. Rabilloud, T (1999) Solubilization of proteins in 2-D electrophoresis, in *2-D Proteome Analysis Protocols* (Link, A.J., ed.), Humana, Totowa, NJ, pp. 9–17.
3. Shaw, M.M, and Riederer, B.M. (2003) Sample preparation for two-dimensional gel electrophoresis. *Proteomics* **3**, 1408–1417.
4. O'Farrell, P.J. (1975) High resolution two-dimensional electrophoresis of proteins. *J. Biol. Chem.* **250**, 4007–4021.
5. Rabilloud, T. (1998) Use of thiourea to increase the solubility of membrane proteins in two-dimensional electrophoresis. *Electrophoresis* **19**, 758–760.
6. Chevallet M., Santoni V., Poinas A., Rouquie D., Fuchs A., Kieffer S., Rossignol M., Lunardi J., Garin J. and Rabilloud T. (1998) New zwitterionic detergents improve the analysis of membrane proteins by two-dimensional electrophoresis. *Electrophoresis* **19**, 1901–1909.
7. Molloy, M.P. (2000) Two-dimensional electrophoresis of membrane proteins using immobilized pH gradients. *Anal. Biochem.* **280**, 1–10.
8. Herbert, B.R., Molloy, M..P, Gooley, A.A., Walsh, B.J., Bryson, W.G. and Williams, K.L. (1998) Improved protein solubility in two-dimensional electrophoresis using tributyl phosphine as reducing agent. *Electrophoresis* **19**, 845–851.
9. Harder, A., Wildgruber, R., Nawrocki, A., Fey, S.J., Larsen, P.M., and Görg, A. (1999) Comparison of yeast cell protein solubilization procedures for two-dimensional electrophoresis. *Electrophoresis* **20**, 826–829.
10. Thoenes, L., Drews, O., Görg, A. and Weiss, W. (2003) Protein Carbamylation – Actually Problem in 2-D Electrophoresis?, In: *Proteomic Forum 03* (Görg, A., ed.), PSP, Freising, Germany, p.93. (Also available at: http://www.wzw.tum.de/proteomik/forum2003/index.htm)
11. Görg, A., Boguth, G., Obermaier, C., and Weiss, W. (1998) Two-dimensional electrophoresis of proteins in an immobilized pH 4–12 gradient. *Electrophoresis* **19**, 1516–1519.
12. Görg, A., Weiss, W., and Dunn. M.J. (2004) Current two-dimensional electrophoresis technology for proteomics. *Proteomics* **4**, 3665–3685.

4

Quantitation of Protein in Samples Prepared for 2-D Electrophoresis

Tom Berkelman

Summary

The concentration of protein in a sample prepared for two dimensional (2-D) electrophoretic analysis is usually determined by protein assay. Reasons for this include the following. (1) Protein quantitation ensures that the amount of protein to be separated is appropriate for the gel size and visualization method. (2) Protein quantitation facilitates comparison among similar samples, as image-based analysis is simplified when equivalent quantities of proteins have been loaded on the gels to be compared. (3) Quantitation is necessary in cases where the protein sample is labeled with dye before separation *(1,2)*. The labeling chemistry is affected by the dye to protein ratio so it is essential to know the protein concentration before setting up the labeling reaction.

A primary consideration with quantitating protein in samples prepared for 2-D electrophoresis is interference by nonprotein substances that may be present in the sample. These samples generally contain chaotropic solubilizing agents, detergents, reductants, buffers or carrier ampholytes, all of which potentially interfere with protein quantitation.

The most commonly used protein assays in proteomics research are colorimetric assays in which the presence of protein causes a color change that can be measured spectrophotometrically *(3)*. All protein assays utilize standards, a dilution series of a known concentration of a known protein, to create a standard curve. Two methods will be considered that circumvent some of the problems associated with interfering substances and are well suited for samples prepared for 2-D electrophoresis. The first method (4.1.1) relies on a color change that occurs upon binding of a dye to protein and the second (4.1.2) relies on binding and reduction of cupric ion (Cu^{2+}) ion to cuprous ion (Cu^{+}) by proteins.

Key Words: Bradford; Lowry; interfering substances; protein assay; 2-D sample preparation.

1. Introduction

1.1. Modified Bradford Protein Assay

The Bradford protein assay is based on the binding of Coomassie Brilliant Blue G-250 dye to proteins (4). Under the conditions used in the assay, a spectral change that accompanies protein binding can be used to determine the protein concentration. All dye binding assays suffer from the limitation of potential interference from any nonprotein substance that can also form a complex with the dye, or otherwise modify the binding interaction between dye and protein. Detergents and chaotropes such as urea can interfere in this manner. Because detergents and chaotropes are universally employed in sample preparation for two-dimensional (2-D) electrophoresis, this has limited the use of dye binding protein assays for this application. Refinements to the original Bradford assay, however, have rendered it relatively insensitive to these additives.

A modified Bradford assay can be used to quantitate 2-D samples as long as the sample is diluted sufficiently to limit interference from 3-([3-cholamidopropyl]dimethylammonio)-1-propanesulfonate (CHAPS) and urea, and as long as no other detergent (e.g., Triton X-100, sodium dodecyl sulfate [SDS]) is present in the sample. This limits the versatility of this assay, and because the sample must be diluted, its applicability is limited to relatively concentrated samples. It is, however, a very simple assay to perform, and despite the disadvantages mentioned, it is probably the most widely used quantitation method for 2-D samples. Commercial Bradford-based protein assay kits are widely available. The Quick Start™ Bradford protein assay (Bio-Rad Laboratories) is a very simple assay to perform and can be adapted to 2-D samples prepared with CHAPS, urea, thiourea, and carrier ampholytes.

1.2. Lowry Protein Assay with Interfering Substances Removed by Precipitation

The so-called Lowry assay is based on the binding of Cu^{2+} by the polypeptide backbone and its reduction to Cu^{+} by certain amino acid side chains, followed by colorimetric determination of Cu^{+} (5). This assay has been widely used since its introduction because the response of protein to the assay is relatively insensitive to amino acid composition. There is thus little protein to protein variability in the response of the assay. The primary disadvantage of the Lowry assay has been is its incompatibility both with reducing agents that can also reduce Cu^{2+} to Cu^{+}, and agents that can form complexes with copper ions (e.g., ethylenediamine tetraacetic acid [EDTA] and thiourea). A solution to this problem involves removal of interfering contaminants by selectively precipitating sample protein, discarding the impurities, and resuspending the protein assay solution (6). The *RC DC*™ protein assay (Bio-Rad Laboratories) is a refinement of the original technique that has been optimized for quantitative removal of interfering impurities without loss of protein. This assay can be used to quantitate protein in virtually any sample solution and is therefore more reliable than the modified Bradford protein assay for dilute, complex or uncharacterized samples. It is, however, relatively complicated and time-consuming.

2. Materials
2.1. Modified Bradford Protein Assay

1. Dye reagent: Quick Start Bradford 1× Dye Reagent (Bio-Rad Laboratories). Store at 4 °C.
2. Protein standard stock solution: 2 mg/ml bovine gamma globulin (Bio-Rad Laboratories). Store at 4 °C.
3. Disposable plastic 1-ml cuvettes.

2.2. Lowry Protein Assay with Interfering Substances Removed by Precipitation

1. *RC DC* Protein Assay Kit I (Bio-Rad Laboratories). This includes: *DC* Reagent S, *DC* Reagent A, *RC* Reagent I, *RC* Reagent II, Protein Standard Solution (2 mg/ml bovine serum albumin). Store *DC* and *RC* reagents at room temperature. Store Protein Standard Solution at 4 °C.
2. Disposable plastic 1-ml cuvettes.

3. Methods
3.1. Modified Bradford Protein Assay

1. Remove the dye reagent from 4 °C storage and allow it to warm to ambient temperature (at least 1 h). Mix the dye reagent by inversion a few times before use.
2. Label eight 0.5-ml or 1.5-ml microcentrifuge tubes with the numerals 1 through 8. Using Table 1 as a guide, prepare dilutions of the 2 mg/ml bovine gamma globulin standard stock solution for the standard curve (*see* Notes 1, 2)
3. Prepare dilutions of each unknown sample according to Table 2 (*see* Notes 3–5).
4. Pipette 20 µl of each standard and diluted sample solution into separate disposable 1-ml cuvettes. For best results, run replicates of each standard and unknown sample.
5. Pipette 1 ml of dye reagent into each cuvette.
6. Mix the contents of each cuvette by pipetting up and down or by placing a small square of Parafilm over each cuvette, pressing with a thumb to seal, and inverting.
7. Incubate at room temperature for at least 5 min but no longer than 1 h.
8. Set the spectrophotometer to read absorbance at 595 nm. Zero the instrument with the blank sample (Tube 8, 0.0 µg/ml). Measure the absorbance of the standards and unknown samples.

Table 1
Dilution of Protein Standard Solution for protein assay standard curve

Tube #	Standard volume (μL)	Source of standard	Diluent volume (μL)*	Final [protein] (μg/mL)
1	70	2 mg/ml stock	0	2,000
2	75	2 mg/ml stock	25	1,500
3	70	2 mg/mL stock	70	1,000
4	35	Tube 2	35	750
5	70	Tube 3	70	500
6	70	Tube 5	70	250
7	70	Tube 6	70	125
8	–	–	70	0

[a] Distilled or de-ionized water may be used as the diluent, but more accurate results will be obtained if the composition of the diluent reflects the final composition of the sample (*see* **Note 2**).

Table 2
Dilution of unknown sample

Dilution factor	Volume of sample (μL)	Volume of water (μL)
4	12.5	37.5
10	5	45

9. Create a standard curve by plotting the values for absorbance at 595 nm (*y*-axis) versus their concentration in μg/ml (*x*-axis). Determine each unknown sample concentration using the standard curve. Adjust the final concentration of the unknown sample by multiplying by the dilution factor.

3.2. Lowry Protein Assay with Interfering Substances Removed by Precipitation

1. Add 5 μl of *DC* Reagent S to each 250 μl of *DC* Reagent A that will be needed for the assay. This solution will be referred to as Reagent A': Each standard or sample assayed will require 127 μl of Reagent A'.
2. Prepare dilutions of the Protein Standard Solution for the standard curve as described in Step 2 under Subheading **3.1**. Use distilled or de-ionized water as the diluent.
3. Prepare dilutions of each unknown sample as described in Step 3 under Subheading **3.1**. Use distilled or de-ionized water as the diluent.
4. Pipette 25 μl of each standard, sample and diluted sample into 1.5-ml microcentrifuge tubes. For best results, run replicates of each standard and sample.

5. Add 125 µl of *RC* Reagent I to each tube. Vortex the tubes and incubate for 1 min at room temperature.
6. Add 125 µl *RC* Reagent II to each tube. Vortex the tubes and centrifuge at 15,000 g for 5 min. Position the microcentrifuge tubes in the microcentrifuge with the cap hinge facing outward (*see* Notes 6,7).
7. Remove the tubes from the microcentrifuge as soon as centrifugation is complete. A small pellet should be visible on the cap-hinge side of the tube. Decant the supernatants. Proceed rapidly to the next step.
8. Carefully reposition the tubes in the microcentrifuge as before, with the cap-hinge facing outward. Centrifuge the tubes again to bring any remaining liquid to the bottom of the tube. A brief pulse is sufficient. Use a micropipette to remove the remaining supernatant. There should be no visible liquid remaining in the tubes (*see* Note 8).
9. Add 125 µl of *RC* Reagent I and 125 µl of *RC* Reagent II to each tube. Vortex the tubes and centrifuge at 15,000 g for 5 min. Position the microcentrifuge tubes in the microcentrifuge with the cap hinge facing outward (*see* Note 9).
10. Remove the tubes from the microcentrifuge as soon as centrifugation is complete. A small pellet should be visible on the cap-hinge side of the tube. Decant the supernatants. Proceed rapidly to the next step.
11. Carefully reposition the tubes in the microcentrifuge as before, with the cap-hinge facing outward. Centrifuge the tubes again to bring any remaining liquid to the bottom of the tube. A brief pulse is sufficient. Use a micropipette to remove the remaining supernatant. There should be no visible liquid remaining in the tubes.
12. Add 127 µl of Reagent A' to each microcentrifuge tube. Vortex the tubes and incubate at room temperature for 5 min. Vortex again before proceeding to the next step.
13. Add 1 ml of *DC* Reagent B to each tube and vortex immediately. Incubate at room temperature for at least 15 min but no longer than 1 h.
14. Set the spectrophotometer to read absorbance at 750 nm. Zero the instrument with the blank sample (Tube 8, 0 µg/ml). Measure the absorbance of the standards and unknown samples.
15. Create a standard curve by plotting the values for absorbance at 750 nm (*y*-axis) versus their concentration in µg/ml (*x*-axis). Determine each unknown sample concentration using the standard curve. If the samples were diluted, adjust the final concentration of the unknown sample by multiplying by the dilution factor.

4. Notes

1. Bovine gamma globulin is recommended as the standard for this assay rather than the more commonly used bovine serum albumin. Serum albumins react anomalously with the Bradford protein assay reagent and give significantly greater color development than most other proteins. Gamma- globulin gives a more typical response. Bovine serum albumin may be used as a standard when the

primary protein in the sample is likely to be serum albumin (as in, for example, body fluids such as serum or cerebrospinal fluid).
2. The response of the modified Bradford protein assay to protein is only slightly affected by concentrations of urea, thiourea, and CHAPS of up to 1.75 M, 0.5 M and 1% (w/v), respectively (*see* Fig. 1). Thus, a 4-fold dilution of a typical 2-D sample solution consisting of 7 M urea, 2 M thiourea, 4% CHAPS may be assayed using a standard curve prepared in water. If greater accuracy is desired, the standard curve may be prepared in a solution reflecting the composition of the diluted sample.
3. Samples prepared for 2-D electrophoresis must be diluted at least fourfold with water before applying the modified Bradford protein assay as described above. The primary reason for this is the high concentration of urea present in 2-D samples which interferes with the assay. The need for dilution limits application of the modified Bradford protein assay to samples with relatively high protein concentration (at least 0.5 mg/ml for a fourfold dilution of the sample to give a reading within the useful range of the standard curve).
4. The modified Bradford protein assay is relatively tolerant of CHAPS, the most commonly used detergent for 2-D samples. Samples containing up to 4% (w/v) CHAPS may be assayed if diluted as described above. If other detergents are present, it is advisable to use a protein assay less subject to interference (such as the modified Lowry protein assay described in this chapter), or to prepare a standard curve containing the detergent in question in order to assess its interference with the modified Bradford protein assay.

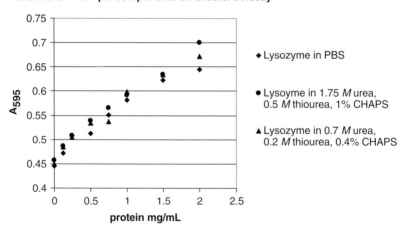

Fig. 1. Dilutions of a standard solution of chicken lysozyme were prepared either in phosphate-buffered saline (PBS); 1.75 M urea, 0.5 M urea, 1% (w/v) CHAPS; or 0.7 M urea, 0.2 M thiourea, 0.4% (w/v) CHAPS. The modified Bradford protein assay was applied as described above with the spectrophotometer zeroed to water.

5. Samples prepared for 2-D electrophoresis typically have protein concentrations in a range between 0.5 mg/ml and 20 mg/ml. It is therefore suggested that both a 4-fold and 10-fold dilution of the sample be assayed to ensure that the sample generates a reading within the standard curve.
6. Tubes should be placed consistently in the microcentrifuge with the cap hinge facing outwards. This ensures that the position of the pellet is consistent so that the supernatant can be withdrawn without disturbing the pellet.
7. Avoid protein losses by not disturbing the pellet and proceeding quickly to the next step following centrifugation so that the pellet does not disperse.
8. The most common problem source of problems with this assay is carry-over of interfering substances through the assay procedure. The procedure therefore incorporates measures to minimize such carry-over. These include brief re-centrifugation of the tubes to facilitate complete removal of the supernatant with a pipette.
9. Steps 9–11 of Section **3.2** are a wash procedure to remove any trace of potentially interfering substance from the sample. It is not necessary with all sample types and these steps may be omitted once this is determined.

References

1. Urwin, V. E. and Jackson, P. (1993) Two-dimensional polyacrylamide gel electrophoresis of proteins labeled with the fluorophore monobromobimane before first-dimensional isoelectric focusing. *Anal. Biochem.* **209**, 57–62.
2. Ünlü, M., Morgan, M. E. and Minden J. S. (1997) Difference gel electrophoresis: A single gel method for detecting changes in protein extracts. *Electrophoresis* **18**, 2071–2077.
3. Sapan, C. V., Lundblad, R. L. and Price, N. C. (1999). Colorimetric protein assay techniques. *Biotechnol. Appl. Biochem.* **29**, 99–108.
4. Bradford, M. M. (1976) A Rapid and sensitive method for the quantitation of microgram quantities of protein utilizing the principle of protein-dye binding. *Anal. Biochem.* **72**, 248–254.
5. Lowry, O. H., Rosebrough, N. J., Farr, A. L. and Randall, R. J. (1951) Protein measurement with the Folin phenol reagent. *J. Biol. Chem.* **193**, 265–275.
6. Bensadoun, A. and Weinstein, D. (1976) Assay of proteins in the presence of interfering materials. *Anal. Biochem.* **70**, 241–250.

5

Removal of Interfering Substances in Samples Prepared for Two-Dimensional (2-D) Electrophoresis

Tom Berkelman

Summary

Biological samples may contain contaminants that interfere with analysis by two-dimensional (2-D) electrophoresis. Lysates or biological fluids are complex mixtures that contain a wide variety of nonprotein substances in addition to the proteins to be analyzed. These substances often interfere with the resolution of the electrophoretic separation or the visualization of the result. Macromolecules (e.g., polysaccharides and DNA) can interfere with electrophoretic separation by clogging gel pores. Small ionic molecules can impair isoelectric focusing (IEF) separation by rendering the sample too conductive. Other substances (e.g., phenolics and lipids) can bind to proteins, influencing their electrophoretic properties or solubility. In many cases, measures to remove interfering substances can result in significantly clearer 2-D patterns with more visible spots and better resolution. It should be borne in mind, however, that analysis of samples by 2-D electrophoresis is usually most successful and informative when performed with minimally processed samples, so it is important that any steps taken to remove interfering substance be appropriate to the sample and only performed when necessary. Procedures for the removal of interfering substances therefore represent a compromise between removing nonprotein contaminants, and minimizing interference with the integrity and relative abundances of the sample proteins. This chapter presents a number of illustrative examples of optimized sample preparation methods in which specific interfering substances are removed by a variety of different strategies.

Key Words: Nuclease treatment; phenol extraction; protein precipitation; 2-D sample preparation.

1. Introduction

1.1. Removal of Nucleic Acids (Preparation of Bacillus subtilis Proteins for Two-Dimensional Electrophoresis)

The presence of nucleic acid, particularly DNA, can interfere with isoelectric focusing (IEF) in the acidic region of the IEF gel. DNA can also impart high viscosity to the sample, which can limit the effectiveness of cell lysis and render the lysate difficult to transfer accurately. Because of this, nucleic acids are often removed by treating the sample with nuclease. This breaks down larger nucleic acids into mono-and oligonucleotides, which no longer interfere with IEF or contribute to viscosity. One needs to be mindful of three factors when using nucleases during sample preparation for two-dimensional (2-D) electrophoresis. (1) Nucleases may be inactive under the strongly denaturing conditions used in sample preparation for 2-D electrophoresis. (2) DNase requires free Mg^{2+} ions for activity. (3) Nucleases are proteins and can show up in the 2-D pattern as extra spots.

Bacterial lysates have a high nucleic acid to protein ratio and generally benefit from nuclease treatment. The procedure presented here uses Benzonase®, a nuclease with specificity for both DNA and RNA that is active in the presence of urea. The amount required for treatment is usually not visible in a 2-D gel against the background of endogenous proteins. Mg^{2+} ion is present during nuclease treatment, but is sequestered with ethylenediamine tetraacetic acid (EDTA) following treatment to inhibit proteases that may also require Mg^{2+} for activity.

1.2. Removal of Polysaccharides and Phenolics (Preparation of Potato Proteins for 2-D Electrophoresis)

Plant tissues represent a particular challenge in sample preparation for 2-D electrophoresis. In addition to having high levels of interfering substances, plant tissues are often rather dilute sources of protein, and sample preparation techniques for plant tissues usually need to incorporate a method for concentrating sample protein. Interfering substances encountered in plant tissues include polysaccharides and phenolic compounds. Polysaccharides can clog gel pores and prevent effective IEF. They can also render a sample viscous and difficult to work with. Phenolic compounds can modify proteins and render them insoluble or generate artifactual heterogeneity. The method presented below circumvents many of these problems using a procedure that involves selectively extracting the proteins from the lysate with phenol leaving polysaccharides and phenolics in the aqueous phase. Sample proteins are then precipitated from the phenol phase with ammonium acetate. They can then be resuspended in 2-D sample solution at high concentration. The procedure for preparation of potato protein presented below is an example developed from a procedure originally described in *(1)*, incorporating refinements described in *(2)*. Potato tubers are particularly rich in polysaccharide and are not amenable to sample preparation by other methods.

1.3. Removal of Salts Endogenous to Body Fluids (Preparation of Cerebrospinal Fluid for 2-D Electrophoresis)

Effective isoelectric focusing requires low sample ionic strength, yet many sample types contain salts and ions that need to be removed if the sample is to be analyzed by 2-D electrophoresis. This is true of body fluids such as cerebrospinal fluid (CSF) and urine. Such materials may also have low protein content and require concentration before analysis. The procedure presented below for preparation of CSF for 2-D electrophoresis is an example of simultaneous salt removal and protein concentration. This is accomplished by precipitating the protein with ethanol and resuspending in a small volume of sample solution for analysis. Among the procedures that have been reported for preparation of CSF for 2-D electrophoresis *(3)*, this procedure is particularly effective and simple to perform.

1.4. Removal of Salts Introduced during Cell Culture (Preparation of HeLa Cell Proteins for 2-D Electrophoresis)

Cultured cells are commonly analyzed by 2-D electrophoresis and it is important to ensure that salt and other components of the culture medium are not carried over in the sample. This is achieved by thorough washing of the cells. Although phosphate-buffered saline (PBS) is most commonly used for washing cultured cells, this high-salt buffer can carry over into the sample and result in less than optimal IEF separation. It is preferable to use a low-salt wash solution. The procedure presented below uses a wash with a Tris-buffered sucrose solution to prepare HeLa cell proteins for 2-D electrophoresis.

1.5. Removal of Lipids (Preparation of Rat Brain Proteins for 2-D Electrophoresis)

Lipids can interfere with the ability of proteins to be fully solubilized by forming insoluble complexes with proteins. Lipids and can also form complexes with detergents and reduce their effectiveness for protein solubilization. One strategy employed has been to use an organic solvent mixture in which lipids are soluble but proteins are not *(4)*. The procedure described below uses a commercial "2-D Cleanup" kit to prepare rat brain extract for 2-D electrophoresis. This procedure incorporates selective protein precipitation and an organic solvent extraction step for effective sample preparation from this lipid-rich tissue.

2. Materials

2.1. Removal of Nucleic Acids (Preparation of B. subtilis Proteins for 2-D Electrophoresis)

1. Benzonase (250 units/µl, Novagen or Sigma-Aldrich). Store at –20°C.
2. 1.2 M dithiothreitol (DTT, 185 mg/ml). Store at –20°C.
3. Lysis solution: 7 M urea, 2 M thiourea, 4% (w/v) 3-([3-cholamidopropyl] dimethylammonio)-1-propanesulfonate (CHAPS), 20 mM Tris-HCl pH 8.0, 1 mM $MgCl_2$. Prepare fresh or store aliquoted at –80°C.
4. 200 mM Na_2EDTA. Store at room temperature.
5. Phenylmethylsulfonyl fluoride (PMSF) solution: 100 mM PMSF in isopropanol. Store at 4°C.
6. Leupeptin solution: 10 mM leupeptin in water (Sigma). Store in aliquots at –20°C.
7. Bestatin solution: 40 mM bestatin in water (Sigma). Store in aliquots at –20°C.
8. E64 solution: 10 mM E64 in water (Sigma). Store in aliquots at –20°C.
11. Probe type sonicator.
12. *B. subtilis* cells grown in medium of choice. A subtilisin-minus strain such as GX4937 should be used.

2.2. Removal of Polysaccharides and Phenolic Compounds (Preparation of Potato Proteins for 2-D Electrophoresis)

1. Liquid nitrogen.
2. Clean sand.
3. Plant tissue lysis solution: 40% (w/v) sucrose, 2.7% (w/v) sodium dodecyl sulfate (SDS), 133 mM Tris-HCl pH 8.0, 6.7% (w/v) β-mercaptoethanol. Prepare fresh.
4. Phenol equilibrated with 10 mM Tris-HCl pH 8.0 (available from Sigma-Aldrich). Store at 4°C.
5. Precipitation solution: 0.1 M ammonium acetate in methanol. Store at –20°C
6. 80% (v/v) acetone. Store at –20°C.
7. 2-D sample solution: 7 M urea, 2 M thiourea, 4% (w/v) CHAPS, 0.2% (w/v) Carrier Ampholyte (Bio-Lyte 3-10, Bio-Rad Laboratories), 40 mM DTT. Prepare fresh.
8. Mortar and pestle.

2.3. Removal of Salts Endogenous to Body Fluids (Preparation of Cerebrospinal Fluid for 2-D Electrophoresis)

1. Absolute ethanol (not denatured).
2. 2-D sample solution: 7 M urea, 2 M thiourea 4% (w/v) CHAPS, 0.2% (w/v) Carrier Ampholyte (Bio-Lyte 3-10, Bio-Rad Laboratories), 40 mM DTT. Prepare fresh.
3. Probe or cup type sonicator.
4. Cerebrospinal fluid (CSF). Store at –80°C.

2.4. Removal of Salts Introduced during Cell Culture (Preparation of HeLa Cell Proteins for 2-D Electrophoresis)

1. Tris-buffered sucrose: 250 mM sucrose, 10 mM Tris-HCl pH 7.4. Store at 4 °C. Allow to come to room temperature before use.
2. 2-D sample solution: 7 M urea, 2 M thiourea 4% (w/v) CHAPS, 0.2% (w/v) Carrier Ampholyte (Bio-Lyte 3-10, Bio-Rad Laboratories), 40 mM DTT. Prepare fresh.
3. Cell scraper.
4. Hemocytometer.
5. HeLa Cells grown in 100-mm culture plate.

2.5. Removal of Lipids (Preparation of Rat Brain Proteins for 2-D Electrophoresis)

1. Liquid nitrogen.
2. Rat brain lysis buffer: 7 M urea, 2 M thiourea, 4% (w/v) CHAPS, 40 mM Tris base, 2 mM tributylphosphine. Prepare fresh.
3. ReadyPrep™ 2-D Cleanup Kit. This includes: Precipitating Agent 1, Precipitating Agent 2, Wash Reagent 1, Wash Reagent 2, Wash 2 Additive. Store Wash Reagent 2 at −20 °C. Store other kit components at room temperature.
4. 2-D sample solution: 7 M urea, 2 M thiourea 4% (w/v) CHAPS, 0.2% (w/v) Carrier Ampholyte (Bio-Lyte 3-10, Bio-Rad Laboratories), 40 mM DTT. Prepare fresh.
5. Mortar and pestle.
6. Probe-type sonicator.
7. Rat brain: May be quick-frozen with liquid nitrogen and stored at −80 °C if not used immediately after harvest.

3. Methods

3.1. Removal of Nucleic Acids (Preparation of B. subtilis Proteins for 2-D Electrophoresis)

1. Pellet the *B. subtilis* cells by centrifugation at 5000 *g* for 15 min at 4 °C.
2. Decant and discard the culture supernatant. Resuspend the cells in 100 ml of water per liter of original culture. Transfer to tared centrifuge tubes. Centrifuge at 10,000 *g* for 10 min at 4 °C.
3. Decant and discard the supernatant. Determine the weight of the cell pellet by weighing the tube and subtracting the tare weight.
4. Suspend the washed cells in lysis solution. Use 1 ml lysis per 100 mg of cells. Transfer the cell suspension to a capped plastic tube (if not in one already). Add 2 µl (500 units) of Benzonase, 10 µl of 100 mM PMSF and 1 µl of the other protease inhibitor stocks (leupeptin, bestatin, E64) per 1 ml of lysis solution. Mix by inversion. Place the tube in a beaker of ice. Proceed rapidly to the next step (*see* Notes 1–3).

5. Sonicate the cells by immersing the tip of the sonicator probe in the cell suspension. Set the sonicator to the maximum power output allowed for the tip. Sonicate in with five 15–30 s bursts allowing the suspension to cool down in a beaker of ice between bursts (*see* Note 4).
6. Leave the tube on ice for 15 min.
7. Add 50 μl of 1.2 M DTT and 10 μl of 200 mM Na_2EDTA per 1 ml of lysis solution used in Step 4. Mix by inversion (*see* Note 5).
8. Transfer the lysate to a centrifuge tube (if not in one already). Centrifuge at 20,000 g for 15 min at 4 °C.
9. Determine the protein concentration of the sample (see chapter on Quantitation of Protein in Samples Prepared for 2-D Electrophoresis in this book).
10. Store the sample at aliquoted at –80 °C if not used immediately.
11. Analyze the sample by 2-D electrophoresis. Figure 1 is an example showing the effect of the inclusion or omission of Benzonase.

3.2. Removal of Polysaccharides and Phenolic Compounds (Preparation of Potato Proteins for 2-D Electrophoresis)

1. Cut and weigh out 200 mg of potato tuber tissue. Place the tissue in a small mortar and freeze by pouring liquid nitrogen into the mortar. Break the tissue up with the pestle. Add clean sand and continue grinding until the material is finely powdered. Add more liquid nitrogen during grinding if necessary to prevent thawing.
2. Transfer material to a 2-ml centrifuge tube. Add 0.6 ml of Plant tissue lysis solution. Vortex for 30 s.

Plus Benzonase   Minus Benzonase

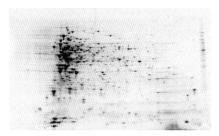

Fig. 1. Effect of Benzonase. The panel on the left shows a 2-D gel of *B. subtilis* proteins prepared as described above. The panel on the right shows a 2-D gel of *B. subtilis* proteins prepared with the same procedure with the Benzonase omitted. Sample load: 50 μg. First dimension: 11 cm pH 3–10 NL IPG strips (Bio-Rad Laboratories). Second dimension: 8–16% Criterion Tris-Cl gels. The gels were stained with Flamingo™ Fluorescent Gel Stain (Bio-Rad Laboratories) and imaged with the Molecular Imager FX™ (Bio-Rad Laboratories).

3. Place the tube in a 100°C heat block for 5 min. Remove and allow to cool (*see* Note 6).
4. Add 0.8 ml of phenol equilibrated with 10 mM Tris-Cl pH 8.0. Vortex for 1 min.
5. Centrifuge at at least 10,000 g for 3 min (room temperature). Carefully transfer the upper phase to a 15-ml solvent-resistant centrifuge tube. Reserve the lower phase.
6. Re-extract the lower phase. Add an additional 0.8 ml of equilibrated phenol. Vortex for 1 min. Centrifuge as in Step 5. Carefully transfer the upper phase and add to the upper phase from the previous extraction.
7. Add 10 ml of cold precipitation solution to the pooled upper phases. Vortex and place at –20°C for at least 1 h.
8. Centrifuge at $\geq$10,000 g for 5 min at 4°C or lower.
9. Carefully decant the supernatant. Add 10 ml of cold 80% acetone. Vortex briefly and centrifuge as in Step 8 with the tube in the same orientation.
10. Carefully decant the supernatant. Allow to air dry for 10–20 min. Do not allow the pellet to become completely dry (*see* Note 7).
11. Resuspend the pellet in 100 µl of 2-D sample solution.
12. Centrifuge the sample at $\geq$10,000 g for 5 min at room temperature (*see* Note 8).
13. Transfer the supernatant and discard the pellet. Determine the protein concentration of the sample (see chapter on Quantitation of Protein in Samples Prepared for 2-D Electrophoresis in this book).
14. Store the sample at –80°C if not used immediately.
15. Analyze the sample by 2-D electrophoresis. An example result is shown in Fig. 2.

3.3. Removal of Salts Endogenous to Body Fluids (Preparation of Cerebrospinal Fluid for 2-D Electrophoresis)

1. Thaw the CSF at room temperature. Vortex the thawed sample (*see* Note 9).
2. Transfer 1 ml of CSF to a centrifuge tube. Add 9 ml of ethanol.
3. Cap the tube and vortex. Store at –20°C for at least 12 h.
4. Centrifuge at 5,000 g for 5 min at 4°C.
5. Decant the supernatant and use a micropipet to remove as much residual supernatant as possible. Do not disturb the pellet.
6. Allow to air dry for 10–20 min. Do not allow the pellet to become completely dry (*see* Note 10).
7. Resuspend the pellet in 100 µl of 2-D sample solution. Complete resuspension will be slow and can be aided by allowing the tubes to stand for 30 min at room temperature followed by 15 s bursts with the sonicator, allowing the solution to cool down in a beaker of ice between bursts.
8. Centrifuge the sample at $\geq$10,000 g for 5 min at room temperature.
9. Transfer the supernatant and discard the pellet. Determine the protein concentration of the sample (see chapter on Quantitation of Protein in Samples Prepared for 2-D Electrophoresis in this book).

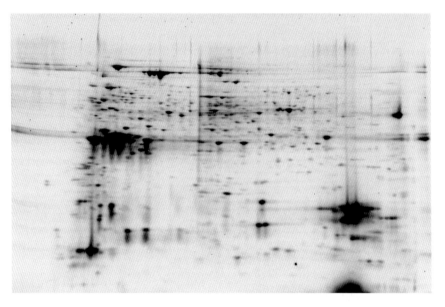

Fig. 2. Potato proteins separated by 2-D electrophoresis. Sample load: 25 µg. First dimension: 11 cm pH 3–10 NL IPG strips (Bio-Rad Laboratories). Second dimension: 8–16% Criterion Tris-Cl gels. The gels were stained with Flamingo™ Fluorescent Gel Stain (Bio-Rad Laboratories) and imaged with the Molecular Imager FX™ (Bio Rad Laboratories)

10. Store the sample at –80 °C if not used immediately.
11. Analyze the sample by 2-D electrophoresis.

3.4. Removal of Salts Introduced during Cell Culture (Preparation of HeLa Cell Proteins for 2-D Electrophoresis)

1. Aspirate culture medium from plate with cells.
2. Gently wash cells *in situ* by adding 10 ml of Tris-buffered sucrose to the plate. Swirl gently and aspirate off the buffer. Repeat.
3. Add 1 ml of Tris-buffered sucrose to the plate. Use a cell scraper to detach the cells. Transfer the cell suspension to a conical 15-ml tube. Dilute the cells with an additional 4 ml of Tris-buffered sucrose.
4. Determine the number of cells in the sample by counting the cells with a hemocytometer
5. Centrifuge the cells at 500 g for 5 min at room temperature.
6. Aspirate off the supernatant and allow the tube to drain well.
7. Add 40 µl of 2-D sample solution per 10^6 cells to the tube. Vortex for 30 s (*see* Note 11).
8. Transfer the lysate to a microcentrifuge tube. Centrifuge at 20,000 g for 1 h at room temperature (*see* Note 12).

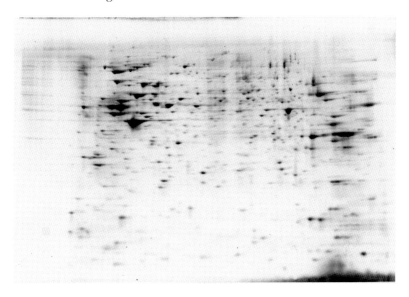

Fig. 3. HeLa Cell Proteins separated by 2-D electrophoresis. Sample load: 50 µg. First dimension: 11 cm pH 3–10 NL IPG strips (Bio-Rad Laboratories). Second dimension: 8–16% Criterion Tris-Cl gels. The gels were stained with Flamingo™ Fluorescent Gel Stain (Bio-Rad Laboratories) and imaged with the Molecular Imager FX™ (Bio-Rad Laboratories).

9. Transfer the supernatant and discard the pellet. Determine the protein concentration of the sample (see chapter on Quantitation of Protein in Samples Prepared for 2-D Electrophoresis in this book).
10. Store the sample at –80 °C if not used immediately.
11. Analyze the sample by 2-D electrophoresis. An example result is shown in Fig. 3.

3.5. Removal of Lipids (Preparation of Rat Brain Proteins for 2-D Electrophoresis)

1. Place brain tissue in a small mortar and freeze the tissue by pouring liquid nitrogen into the mortar. Break the tissue up into small pieces with the pestle.
2. Weigh out 100 mg of brain tissue, working quickly so that the tissue remains frozen.
3. Transfer 100 mg of brain tissue to a 1.5- or 2-ml centrifuge tube. Add 1 ml rat brain lysis buffer.
4. Sonicate the sample by immersing the tip of the sonicator probe in the solution with the broken-up brain tissue. Set the sonicator to the maximum power output allowed for the tip. Sonicate in 15 s bursts allowing the suspension to cool down in a beaker of ice between bursts. Repeat four times (*see* Note 13).

5. Centrifuge at ≥12,000 g for 30 min at room temperature. Transfer the supernatant to a new tube. This crude lysate will contain roughly 10 mg protein per ml. Subsequent steps describe removal of interfering substances from 50 µl of this material, but the procedure may be scaled up or down at will.
6. Transfer 50 µL of the supernatant from Step 5 to a 1.5- or 2-ml centrifuge tube. Add 50 µl of water and 300 µl of Precipitating Agent 1. Vortex and incubate on ice for 15 min.
7. Add 300 µl of Precipitating Agent 2 and vortex (*see* Note 14).
8. Position the microcentrifuge tube in the microcentrifuge with the cap hinge facing outward Centrifuge at ≥12,000 g for 5 min at room temperature. Remove the tube as soon as centrifugation is complete. A small pellet should be visible on the cap-hinge side of the tube (*see* Notes 15,16).
9. Carefully decant the supernatant. Reposition the tube in the microcentrifuge as before, with the cap-hinge facing outward. Centrifuge the tubes again to bring any remaining liquid to the bottom of the tube. A brief pulse is sufficient. Use a micropipet to remove the remaining supernatant. There should be no visible liquid remaining in the tubes.
10. Add 40 µl of Wash Reagent 1 on top of the pellet. Vortex briefly.
11. Reposition the tubes in the microcentrifuge as before, with the cap-hinge facing outward. Centrifuge at ≥12,000 g for 5 min at room temperature. Remove the tube as soon as centrifugation is complete.
12. Remove the Wash Reagent supernatant with a micropipet.
13. Add 25 µl distilled or de-ionized water on top of the pellet. Vortex the tube 10–20 s. The pellet may disperse but will not dissolve.
14. Add 1 ml of cold Wash Reagent 2 and 5 µl of Wash 2 Additive.
15. Incubate the tube at –20°C for 30 min. Vortex the tube for 30 s every 10 min during the incubation period.
16. Reposition the tube in the microcentrifuge as before, with the cap-hinge facing outward. Centrifuge at ≥12,000 g for 5 min at room temperature.
17. Carefully decant the supernatant. Reposition the tube in the microcentrifuge as before, with the cap-hinge facing outward. Centrifuge the tubes again to bring any remaining liquid to the bottom of the tube. A brief pulse is sufficient. Use a micropipet to remove the remaining supernatant. There should be no visible liquid remaining in the tubes.
18. Allow the pellet to air dry 5 min at room temperature. Do not allow the pellet to become completely dry. (*see* Note 17).
19. Resuspend the pellet in 100 µl of 2-D sample solution.
20. Centrifuge the sample at ≥12,000 g for 5 min at room temperature.
21. Transfer the supernatant and discard the pellet. Determine the protein concentration of the sample (see chapter on Quantitation of Protein in Samples Prepared for 2-D Electrophoresis in this book).
22. Store the sample at –80°C if not used immediately.
23. An example result is shown in Fig. 2.

4. Notes

4.1. Removal of Nucleic Acids (Preparation of B. subtilis Proteins for 2-D Electrophoresis)

1. PMSF is a very effective protease inhibitor that is inactivated by sulfhydryl reagents such as DTT. There are therefore no sulfhydryl reagents present in the lysis solution.
2. Magnesium chloride is present in the lysis solution because Mg^{2+} is required for Benzonase activity.
3. Benzonase is inactivated rapidly in the strongly denaturing lysis solution. It should therefore be added only directly prior to lysis.
4. It is very important not to allow the lysate to heat up during sonication.
5. Sulfhydryl reductant (DTT) and EDTA are added in Step 7. DTT was not present during lysis because it prevents protease inhibition by PMSF. PMSF inhibits proteases irreversibly, so DTT can be added following PMSF treatment. EDTA is added to chelate the Mg^{2+} that was present during lysis for Benzonase activity and to inhibit Mg^{2+}-dependent proteases.

4.2. Removal of Polysaccharides and Phenolic Compounds (Preparation of Potato Proteins for 2-D Electrophoresis)

6. The Plant tissue lysis solution contains SDS and the procedure incorporates a 100°C heat treatment. This assures very effective protein solubilization. SDS has limited compatibility with IEF, but the subsequent phenol extraction and precipitation steps remove this potentially interfering detergent.
7. Protein pellets can be very difficult to resuspend and dissolve if completely dry.
8. This step removes any insoluble material.

4.3. Removal of Salts Endogenous to Body Fluids (Preparation of Cerebrospinal Fluid for 2-D Electrophoresis)

9. CSF is potentially biohazardous. Treat all pipet tips, gloves, tubes and supernatants as biohazardous waste. Clean contaminated surfaces with bleach.
10. Protein pellets can be very difficult to resuspend and dissolve if allowed to completely dry.

4.4. Removal of Salts Introduced during Cell Culture (Preparation of HeLa Cell Proteins for 2-D Electrophoresis)

11. Most cultured mammalian cells will lyse spontaneously in 2-D sample solution. There is no need for any treatment more vigorous than vortexing.
12. The purpose of the final centrifugation is to remove nucleic acid.

4.5. Removal of Lipids (Preparation of Rat Brain Proteins for 2-D Electrophoresis)

13. It is very important not to allow the lysate to heat up during sonication.
14. Precipitating Agent 1 selectively precipitates sample protein. Precipitating Agent 2 is a co-precipitatant that ensures quantitative protein recovery.
15. Avoid protein losses by not disturbing the pellet and proceeding quickly to the next step following centrifugation so that the pellet does not disperse.
16. Tubes should be placed consistently in the microcentrifuge with the cap hinge facing outwards. This ensures that the position of the pellet is consistent so that the supernatant can be withdrawn without disturbing the pellet.
17. Protein pellets can be very difficult to resuspend and dissolve if completely dry.

Acknowledgments

The author would like to thank Dr. Dennis Yee for providing *B. subtilis* cells, Dr. Teresa Rubio for providing the HeLa cells, and Dr. Ning Liu for providing the rat brain sample preparation procedure.

References

1. Hurkman, W. J. and Tanaka, C. K. (1986), Solubilization of plant membrane proteins for analysis by two-dimensional gel electrophoresis. *Plant Physiol.* **81,** 802–806.
2. Wang, W., Scali, M., Vignani, R,. Spadafora, A., Sensi, E., Mazzuca, S., and Cresti, M. (2003) Protein extraction for two-dimensional electrophoresis from olive leaf, a plant tissue containing high levels of interfering compounds. *Electrophoresis* **24,** 2369–2375.
3. Yuan, X. and Desiderio, D. M. (2005) Proteomics analysis of human cerebrospinal fluid. *J. Chromatogr. B* **815,** 179–189.
4. Wessel, D. and Flügge, U. I. (1984) A method for the quantitative recovery of protein in dilute solution in the presence of detergents and lipids. *Anal. Biochem.* **138,** 141–143.

6

Protein Concentration by Hydrophilic Interaction Chromatography Combined with Solid Phase Extraction

Ulrich Schneider

Summary

Hydrophilic interaction chromatography (HILIC) is a variant of normal phase chromatography, in which analyte molecules attach to a solid support (e.g., poly [2-hydroxyethyl] aspartamide silica) by the action of a mobile phase containing a high amount of organic modifier such as acetonitrile or propanol. Elution of analyte molecules is achieved, when the resin is washed with a solution devoid of organic solvent. The method and its basic principles have been extensively described by Alpert et al. *(1)*. Applications of HILIC include the isolation of membrane proteins, electroeluted from SDS-PAGE gels *(2)*, glycopeptides *(3)*, and post-translationally modified protein variants *(4)*.

Here, an extended application of hydrophilic interaction chromatography is described, which allows nonselective enrichment of proteins from various dilute sources before two-dimensional gel electrophoresis (2-D PAGE). The use of this approach is demonstrated by processing protein containing samples from high resolution, preparative isoelectric focusing (IEF) separations achieved by carrier free electrophoresis (free flow electrophoresis [FFE]), described in *(5)*. Furthermore the concept of *compatible recovery* is shown which allows buffer exchange, concentration and recovery of bound proteins in one step directly into a sample buffer required for 2-D PAGE.

Key Words: Electrophoresis; hydrophilic interaction chromatography; proteomics; sample preparation; solid phase extraction.

1. Introduction

High resolution protein separation technologies such as two-dimensional gel electrophoresis (2-D PAGE) require concentrated protein samples that are free from interfering contaminants like salts and other small size constituents that impair either isoelectric focusing (IEF) and/or SDS-PAGE. Usually those contaminants are found in protein preparations that have undergone previous extraction and/or prefractionation steps, e.g., by chromatography or preparative, electrophoretic technologies. Many buffering substances used by prefractionation regimens inherently preclude successful downstream analytics such as 2-D PAGE or chromatographic techniques and ask for their efficient removal. Conventional technologies to concentrate and desalt samples for 2-D PAGE usually employ centrifugal ultrafiltration. Those removal procedures are disadvantageous as, because of the time scale of their application, they may lead to enhanced protein degradation and selective depletion of protein species mostly because of precipitation or irreversible adsorption to the membrane surfaces. Prefractionation by preparative isoelectric focusing technologies also suffers from the relatively dilute nature of the resulting samples, which further impairs successful 2-D PAGE; hence, a universal method to furnish samples for 2-D PAGE must meet a number of criteria: (1) Efficient capturing of proteins in dilute solutions, (2) nonselectivity towards protein classes to assure a complete binding of all protein species present in a sample, (3) minimal loss of protein during the course of enrichment, and (4) simple and fast operation.

2. Materials

1. Either Poly (2-hydroxyethyl) aspartamide silica, 12-μm spherical particles, 300Å pore size (PolyLC Inc., Columbia, MD, USA) packed into 96-well SPE plates (custom manufactured) or bulk ware for individually manufactured cartridges (70-mg bulk packing per well).
2. SPE cartridges in 96-well format.
3. Binding buffer: 90% (v/v) isopropanol, 200 mM formic acid, 5 mM NaCl.
4. Elution buffer for 2-D PAGE: 7 M urea, 2 M thiourea, 4% CHAPS, 1% dithiothreitol (DTT), 2% carrier ampholytes, pH 3.10.
5. pI fractionated samples from a preparative IEF-technology (e.g., free flow electrophoresis and MicroRotofor, see other chapters in this book).
6. Cytochrome C (Sigma): 1mg/ml in water.
7. Vacuum manifold for 96-well SPE plates.
8. Vacuum pump.

3. Methods

3.1. Manufacture of HILIC Cartridges

1. Ninety-six-well cartridges for HILIC operation are manufactured by obtaining commercially available SPE cartridges or by obtaining prefilled cartridges, which can be disassembled (see Note 1). Usually, commercially available cartridges are equipped with both a lower and upper frit for proper bed shape. Cartridges receive

a lower frit (porous polypropylene) to hold back the HILIC resin. The frit has to be pushed carefully but firmly into the bottom of the cartridge.
2. The appropriate amount of dry resin is weighed (approximately 70 mg for small scale studies) and suspended with HILIC binding buffer to generate a slurry that can be transferred into the cartridge by manual pipeting.
3. The appropriate volume of slurry is then transferred into the cartridge, allowed to settle and gently compressed by the second, upper frit (see Note 2).
4. The cartridge is then flushed with a minimal amount of 10-bed volumes of binding buffer, followed by at least 10-bed volumes of 200 mM formic acid. Before sample application the cartridge has to be equilibrated with 10-bed volumes of binding buffer again (see Note 3).
5. Some binding buffer should be left above the resin to ensure proper wetting of the resin.

3.2. Binding Properties of HILIC

Before first time usage of HILIC cartridges, a test run using cytochrome C is strongly recommended.

1. The cartridge is prepared and equilibrated as described.
2. 500 μg or 1 mg of a cytochrome c solution is diluted into binding buffer and allowed to bind to a HILIC/SPE cartridge by pipetting the solution on top of the resin and application of the sample by vacuum forces.
3. Because of its reddish color, cytochrome c can be readily traced by visual inspection (see Notes 4–7). Figure 1 shows the result of 4 independent experiments: Cytochrome c appears as a sharp ring at the top end of the HILIC beads packed into the cartridge, indicating a strong and immediate binding. When the

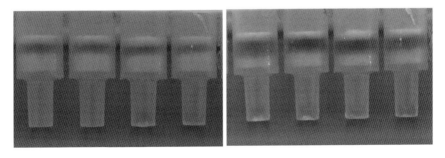

Fig. 1. Cytochrome C loaded onto poly-HEA resin packed into cartridges (four independent experiments each panel). Left panel: 500 mg protein loaded per cartridge. Right panel: Additional 500 mg (1 mg total) protein loaded per cartridge. The reddish color of the protein allows tracing of the proteins whereabouts. Note the presence of the protein on the resin and not on the upper frit.

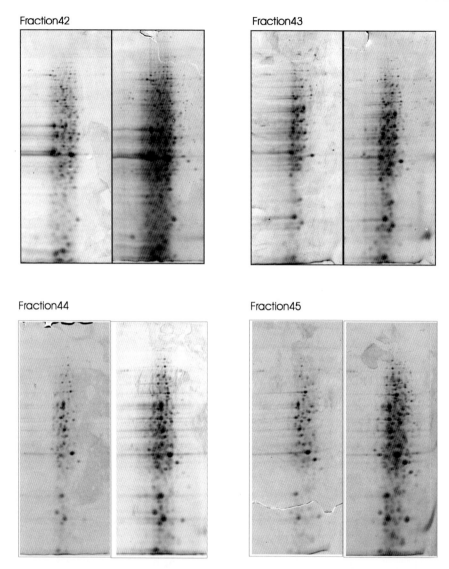

Fig. 2. Direct comparison of four FFE fractions concentrated and desalted by HILIC: Pair wise order of the gels. The left gel of each fraction represents the crude FFE fraction separated by 2DE, whereas the right one represents the sample after it has undergone the HILIC regimen described in the text.

cartridge is flushed with dilute formic acid or any other aqueous solution (see Note 8), the cytochrome c can be immediately recovered from the resin (not shown). Up to 1 mg protein was successfully bound to the resin without significant appearance of protein in the cartridge void.

3.3. Sample Processing by HILIC before 2-D Electrophoresis

1. Conditioned, binding competent cartridges (*see* Section 3.1) are filled with a protein sample that has been diluted 10-fold with binding buffer and slowly applied to the preconditioned HILIC resin. This can be accomplished either by vacuum or by positive pressure, e.g., an inert gas. Even a positive displacement pipette pushed into the cartridge works well.
2. After sample binding and washing, the sample is recovered by step elution in 300-μl sample buffer for 2-D PAGE.

Typical examples of pI-fractionated and HILIC treated protein samples are given in Fig. 2. Four individual fractions of a pI-fractionation procedure using free flow electrophoresis (sample: total yeast extract) were subjected to HILIC concentration and desalting. Crude fractions (# 42 to 45), as well as the HILIC processed fractions were analyzed by 2-D electrophoresis and the gels were stained with a fluorescent dye (SyproRuby™). To ensure comparability of the two experiments, the same sample volumes were subject to the 2-D electrophoresis analyses. Image analysis results of the gels shown in Fig. 2 are summarized in Table 1. The total spot count of the gel pairs, processed versus unprocessed, show an average 3.2-fold increase of the processed sample, which in turn demonstrates a significant concentration effect by HILIC.

Table 1
Quantification of the protein spots visualized after 2-D electrophoresis either before ("raw") and after ("HILIC") processing by HILIC sample processing. The differences between the absolute spot count and the spot quantities are as yet unexplained

Fraction	N° spots	factor	total spotquantity	factor
42, raw	88		3, 149	
42, HILIC	344	3.9	19, 361	6.1
43, raw	66		1, 681	
43, HILIC	176	2.7	7, 528	4.5
44, raw	53		2, 325	
44, HILIC	187	3.5	10, 140	4.4
45, raw	82		2, 062	
45, HILIC	237	2.9	12, 664	6.1

3.4. Resin Regeneration

1. Apply 10-bed volumes of de-ionized water (conductivity <1mS) to remove the previous buffer.
2. Apply 10-bed volumes 200 mM formic acid to clear off remaining protein.
3. Apply 10-bed volumes binding buffer.
4. If the cartridge will be stored for an extended time, a solution of 20% ethanol should be used instead of binding buffer and the cartridge sealed on top and bottom end with Parafilm™.

4. Notes

1. Remove all previous residual resin.
2. As the cartridges are used for step elution procedures only, the usual chromatographic quality criteria (plates/meter, asymmetry) do not really apply here. However, "good" column packing practice should be observed here to avoid artifacts because of voids and channels.
3. For this method it is of utmost importance to have the HILIC resin thoroughly equilibrated with binding buffer.
4. Failure of cytochrome C to bind as described is usually attributable to *(1)* incorrectly equilibrated resin, *(2)* too little organic modifier in the sample.
5. HPLC studies have shown that the concentration of isopropanol in the sample must not be below 80%(v/v)
6. Some publications recommend acetonitrile instead of isopropanol. However, acetonitrile leads to extended protein precipitation in the sample and is therefore no recommended for proteome studies.
7. For initial work a study with well defined test proteins is highly recommended.
8. Complete elution can only be achieved in buffers devoid of organic modifier.
9. Briefly, the prefractionation by free flow electrophoresis "cuts" the protein mixture into discrete fractions representing distinct pH-ranges; hence the proteins are grouped according to their isoelectric points.

Acknowledgments

The author would like to thank Anton Posch for assistance in 2-D gel electrophoresis and image analysis and Gerhard Weber for the opportunity to use the free flow electrophoresis instrument.

References

1. Alpert, A.J., (1990), Hydrophilic-interaction chromatography for the separation of peptides, nucleic acids and other polar compounds. *J Chromatogr*, **499**: 177–196.
2. Jeno, P., et al., (1993), Desalting electroeluted proteins with hydrophilic interaction chromatography. *Anal Biochem*, **215**: 292–298.

3. Zhang, J. and Wang, D.I., (1998), Quantitative analysis and process monitoring of site-specific glycosylation microheterogeneity in recombinant human interferon-gamma from Chinese hamster ovary cell culture by hydrophilic interaction chromatography. *J Chromatogr B Biomed Sci Appl*, **712**: 73–82.
4. Lindner, H., et al., (1996), Separation of acetylated core histones by hydrophilic-interaction liquid chromatography. *J Chromatogr A*, **743**: 137–144.
5. Burggraf, D., Weber, G., and Lottspeich, F., (1995), Free flow-isoelectric focusing of human cellular lysates as sample preparation for protein analysis. *Electrophoresis*, **16**: 1010–1015.

II

Protein Labeling Techniques

7

Difference Gel Electrophoresis Based on Lys/Cys Tagging

Reiner Westermeier and Burghardt Scheibe

Summary

Before separation, proteins of different biological samples are labeled with different fluorescent dyes, the CyDye™ DIGE Fluors. Currently three dyes with spectrally different excitation and emission wavelengths are available. This allows labeling up to three different samples, and coseparating them in one gel. The dyes can either be attached to the ε-amino side group of the lysine without derivatization of the polypeptides or to the cysteines after reduction of the disulfide bonds. For lysine labeling a so called minimal labeling approach is performed: only a low-ratio dye: protein is applied in order to prevent multiple labels per protein. Although only 3% of the proteins are tagged, the sensitivity of detection is comparable with the sensitivity of a good quality silver staining. The dyes are matched for size and charge to obtain migration of differently labeled identical proteins to the same spot positions. The spot pattern achieved with minimal labeling is similar to the pattern obtained with poststained gels. When cysteine tagging is applied, all cysteine moieties are labeled. This modification of the method affords extraordinarily high sensitivity of detection. However, because of multiple labeling, the resulting pattern will look different from nonlabeled or minimal labeled samples.

The labeled samples are mixed together before they are applied on the gel of the first dimension. After separation the gels are scanned with the multifluorescent imager at the different wavelengths. Up to three images of comigrated protein mixtures are compared and evaluated from each gel. This multiplexing technique allows the application of an internal standard for each protein in a complex mixture: One of the labels is applied on a mixture of the pooled aliquots of all samples of an experiment. By coseparating this mixture with each gel an internal standard is created for reliable and reproducible detection and assessment of changes of protein expression levels. Image analysis is performed with special software, which allows codetection of protein spots across the different samples and the internal standard.

From: *Methods in Molecular Biology, vol. 424: 2D PAGE: Sample Preparation and Fractionation, Volume 1*
Edited by: A. Posch © Humana Press, Totowa, NJ

Key Words: Charge-matched labels; CyDye; DIGE; fluorescence imaging; fluorescence labeling; internal standard; multiplexing; size-matched labels; tagging.

1. Introduction

Among various separation methods high resolution two-dimensional (2-D) electrophoresis *(1)* of proteins affords the highest contribution to successful results in the analysis of proteomes. The method is, however, laborious and time consuming. The technique as such demands certain skills from the operator. The evaluation of the spot patterns is highly intricate, slow, and the results are frequently dependent on the operator—in spite of sophisticated image analysis software. Because of gel to gel variations as a major source of shortcomings in 2-D electrophoresis, at least three replicates need to be run for each separation, making the technique even more complex.

Many of these issues can be solved by difference gel electrophoresis, employing labeling the proteins of biological samples with different fluorophores before the separation *(2)*. The samples are then mixed and run in the same 2-D electrophoresis gel. By scanning the gels sequentially at different wave lengths, the images of the emitted light signals of the respective samples are collected. Because the attached fluorescent tags are matched for charge and size, the same proteins belonging to different samples are comigrating to the same position of the isoelectric points and molecular weights. In this way gel-to-gel variation between the sample patterns within one gel is excluded, and the individual spots can be codetected across the samples for the calculation of protein abundance ratios. When one channel in each gel is reserved for a comigrating internal standard, qualitative and quantitative determinations of protein abundances can be performed across different gels with high accuracy. The internal standard is the mixture of aliquots taken from each sample of the experiment, usually labeled with one of the fluorophores. Thus every protein in the population appears on each gel. This makes gel-to-gel matching much easier: the patterns of the internal standard are matched to each other, which are of the identical sample origin. It had been demonstrated for a protein model system using preweighed proteins spiked into *Escherichia coli* lysate, that *only* when protein spot abundances are related to the abundance of the respective internal standard, the quantitative assessments are correct *(3)*. These experimental data have meanwhile been confirmed by a number of real biologic studies, for instance by the work of Friedman et al. *(4)*. Inherent gel-to-gel variations between gels are compensated through the application of the pooled internal standard. Therefore with difference gel electrophoresis, gel replicates are no longer needed, and pattern evaluation is performed almost automatically.

Difference Gel Electrophoresis Based on Lys/Cys Tagging

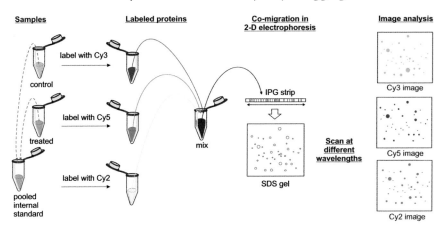

Fig. 1. Schematic drawing of the concept of difference gel electrophoresis using protein tagging with fluorescent dyes. The sample proteins are prelabeled with Cy3 and Cy5. An internal standard is created by applying a mixture of all samples labeled with the third dye Cy2.

Figure 1 shows a simplified schematic drawing of the difference gel electrophoresis workflow with minimal labeling.

Protein labeling with fluorophores offers the possibility of differential detection. The second advantage is the wide linear dynamic range of four orders of magnitude, which allows superior quantification of protein expression changes than with organic dyes like Coomassie Brilliant Blue or silver staining.

An important prerequisite for achieving correct results, however, is appropriate sample preparation and the adherence to exact labeling conditions. It is highly recommended to carry out a sample cleanup based on precipitation of the proteins in order to get rid of endogenous peptides in the sample. These peptides, lysines or cysteines, would be labeled with a CyDye as well, but only be seen in the front. Labeling efficiency is much higher after such a clean-up procedure. Lysine has an average abundance of about 6% of all amino acids and is present in almost all proteins—with some very few exceptions. The ε-amino side group is readily accessible and allows direct labeling without prior derivatization. Usually the pooled internal standard is tagged with Cy2, the other dyes Cy3 and Cy5 are used for the samples. During labeling, the NHS ester group of the CyDye DIGE Fluor minimal dyes gets covalently bound to the ε-amino group of the lysine via an amide linkage (see Fig. 2). The labeling procedure is carried out for 30 min on ice. For this reaction it is very important to maintain a pH value between 8 and 9 in the sample. At this stage the sample must not contain any IPG buffers or carrier ampholytes, and reductants. These

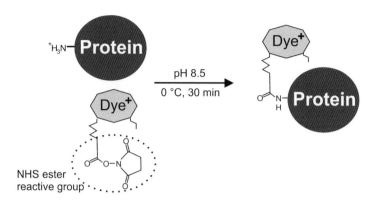

Fig. 2. Simplified representation of minimal labeling of the lysine. The ε-amino group of the lysine is readily accessible, no derivatization is required.

compounds are added after labeling has been completed. In practice this is performed by diluting the sample with 2× sample solution, which contains double concentrations of IPG buffer and reductant compared to a conventional sample buffer.

The dyes have a basic buffering group with a similar pK value like the ε-amino side group of the lysine, in this way the attached dye does not alter the isoelectric point of the protein. Every one of the three dyes add between 450 and 460 Da to the protein. Thus all proteins migrate to the same point in the second dimension. In the medium to high molecular weight range SDS polyacrylamide gel electrophoresis does not resolve nontagged protein from proteins with this additional molecular weight. However, in the lower molecular weight range an offset of nonlabeled from labeled proteins can be observed. Therefore it is important to poststain the gel for spot picking, when further analysis with mass spectrometry is required.

For lysine tagging minimal labeling is applied: that means that only a small portion, 3–5%, of proteins in the sample are labeled. This is achieved by limiting the amount of dye per protein during the labeling procedure. The recommended dye to protein ratio is 400 pM dye per 50 µg of protein. Possible multiple labeled proteins do not show up, because their amount is statistically below detection level. Therefore the resulting spot patterns are the same as those achieved with poststained proteins. The sensitivity is comparable to the most sensitive silver staining. A limit of detection down to 25 pg of a single protein has been determined. Minimal labeling also prevents the proteins from becoming hydrophobic, as already stated in the first paper on difference gel electrophoresis *(2)*.

Cysteines are less abundant among the proteins than lysine. However, tagging of cysteines allows labeling quantitatively all cysteines in all proteins. This offers a very high sensitivity of detection. After the disulfide bridges of the proteins have been cleaved with a reductant like Triscarboxyethylphosphine (TCEP) or dithiothreitol (DTT), the CyDye DIGE Fluor saturation dye is added to the sample. The maleimide reactive group of the dye forms a covalent bond with the thiol group of the cysteine via a thioether linkage (see Fig. 3).

The labeling is performed for 1 h at a pH value of 8.0 and a temperature of 37 °C. Like with minimal labeling, the sample must not contain any IPG buffers or carrier ampholytes. These compounds are added before application on the IPGstrip. For cysteine tagging only two dyes are available: Cy3 and Cy5. The internal standard is usually labeled with Cy3, the sample with Cy5. The dyes are charge-neutral and will not change the isoelectric points of the proteins. Both dyes add a similar mass of about 680 Da per label to the proteins, which results in comigration in the second dimension.

However, in contrast to minimal labeling, multiple labeling will occur: proteins containing more cysteines will carry and gain considerably more molecular weight addition and will deliver a stronger signal than proteins with one or a few cysteines. Therefore the spot maps look different from those resulting from

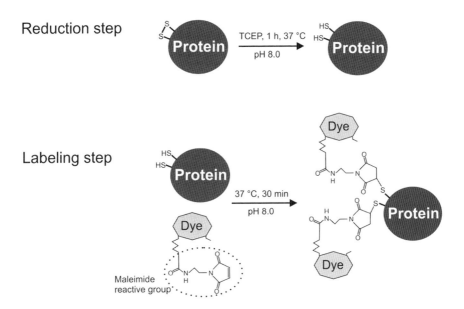

Fig. 3. Simplified representation of saturation labeling of the cysteines. The more cysteinyls are available after reduction, the more dye molecules will become attached to a protein molecule.

nonlabeled and minimal labeled samples *(5)*. Because all cysteine-containing proteins are labeled, the sensitivity of this approach is 20–30 times higher than minimal labeling; proteins with a high number of cysteines show a strongly increased signal; proteins without any cysteines will not show up. The technique is mainly used in cases when only very scarce sample material is available, like analysis of tissue samples acquired with laser capture microdissection or needle dissection. Sample amounts as low as 1,000 mammalian cells, corresponding to 2.5 μg protein, are enough to obtain excellent 2-D maps *(6)*.

2. Materials

2.1. Sample Lysis

Lysis solution: 2 M thiourea, 7 M urea, 4% (w/v) CHAPS, 30 mM Tris; or alternatively: 9 M urea, 4% (w/v) CHAPS, 30 mM Tris.

2.2. Sample Cleanup and Quantification

1. 2-D Cleanup Kit (GE Healthcare Life Sciences, Little Chalfont, UK).
2. 2-D Quant Kit (GE Healthcare Life Sciences, Little Chalfont, UK).

2.3. Protein Labeling

1. CyDye DIGE Fluor dyes, kept in the freezer.
2. Dimethylformamide, a new bottle opened every 3 months.
3. pH indicator paper
4. 50 mM NaOH solution (for pH adjustment).
5. 10 mM lysine solution.

2.3.1. Lysine Tagging (Minimal Labeling)

1. Dye stock solution: CyDye DIGE Fluor minimal dyes solid compounds are reconstituted in Dimethylformamide (DMF) giving a concentration of 1 mM (25 μl DMF to 25 nM of dye). The stock solution of Cy2 will have a deep yellow, Cy3 a deep red, and Cy5 a deep blue color. It is stable at −20 °C for several months.
2. Dye working solution: prior to labeling dilute aliquot of each dye with DMF to 400 pM: Add 3 μl of DMF to 2 μl dye stock solution.

2.3.2. Cysteine Tagging (Saturation Labeling)

1. Dye stock solution: CyDye DIGE Fluor saturation dyes solid compounds are reconstituted in Dimethylformamide (DMF) giving a concentration of 20 mM (20 μl DMF to 400 pM of dye). The stock solution of Cy2 will have a deep yellow, Cy3 a deep red, and Cy5 a deep blue color. It is stable at −20 °C for several months.

2. Dye working solution: prior to labeling dilute aliquot of each dye with DMF to 400 pM: Add 3 µl of DMF to 2 µl dye stock solution.
3. 2 mM TCEP(triscarboxethylphosphine)

2.4. Isoelectric Focusing

1. 2× sample solution: 2 M thiourea, 7 M urea, 4% (w/v) CHAPS, 30 mM Tris, 2% (v/v) Pharmalytes pH 3-10 (or IPG buffer respective to the pH gradient), 2% (w/v) dithiothreitol (DTT), 0.01% (w/v) Bromophenol Blue.

2.5. SDS Electrophoresis

1. Low fluorescence glass cassettes (GE Healthcare Life Sciences, Little Chalfont, UK).
2. Bind-silane (GE Healthcare Life Sciences, Little Chalfont, UK).

3. Methods

3.1. Sample Acquisition and Preparation

Ensure, that you prepare biological replicates (see Note 1). The applied sample acquisition and preparation method is dependent on the sample and the planned experiment; see other chapters of this book and Note 2. For a new sample type it is advised to run some pre-experiments to achieve optimum sample preparation and labeling conditions (see Note 3). Prepare a chart for planned randomization of the samples and the internal standard among the gels (see Note 4).

The proteins are solubilized in lysis solution, which contains either 2 M thiourea, 7 M urea, 4% (w/v) CHAPS, and 30 mM Tris or, alternatively, 9 *M* urea, 4% (w/v) CHAPS, 30 mM Tris (see Note 5).

3.2. Sample Labeling

3.2.1. Lysine tagging

1. Prepare the sample in a way that the protein concentration in the sample solution lies between 2.5–10 mg/ml (see note 6).
2. Check pH value by spotting a small volume of sample on a pH indicator paper. If the pH is lower than 8.0 (a pH of 8.5 would be the optimum), increase the pH by carefully adding 50 mM sodium hydroxide (see Note 7).
3. Prepare the pooled internal standard by taking the same amount of aliquot from each biological sample of the experiment and mixing the aliquots together. A minimum of an equivalent of 50 µg protein of the internal standard per gel is required.

4. Prepare a dye working solution from the dye stock solution (see Note 8).
5. Sample labeling: Add 1 μl dye working solution to a protein sample equivalent of 50 μg (see Notes 9 and 10).
6. Internal standard labeling: Add n μl Cy2 working solution to a protein equivalent of $n \times 50$ μg of the pooled internal standard. n is the number of gels in the experiment (see Note 4).
7. Centrifuge briefly to collect the solution at the bottom of the tube. Leave on ice for 30 min in the dark.
8. Add 1 μl of 10 mM lysine to stop the reaction (add respective amount to internal standard. Mix and spin briefly in a microcentrifuge). Leave for 10 min on ice, in the dark.
9. Mix samples and internal standard according to the experiment design (see Note 4).
10. The labeled samples can be directly applied on the IPG strips or stored for at least 3 months at −70 °C.

3.2.2. Cysteine tagging

1. Prepare the sample in a way that the protein concentration in the sample solution lies between 5–10 mg/ml (see Note 6).
2. Check pH value by spotting a small volume of sample on a pH indicator paper. If the pH is lower than 8.0, increase the pH by carefully adding 50 mM sodium hydroxide (see Note 7).
3. Add a volume of protein sample equivalent to 5 μg to a microfuge tube.
4. Make up to 9 μl with lysis solution.
5. Add 1 μl 2 mM TCEP. (The TCEP:Dye concentration must be 1:2 for efficient labeling)
6. Mix vigorously by pipeting and spin. (See Note 11).
7. Incubate at 37 °C for 1 h in the dark.
8. Add 2 μl 2 mM CyDye DIGE Fluor saturation dye working solution (Note: Label the pooled protein internal standard sample with Cy3 and the experimental protein sample with Cy5).
9. Mix vigorously by pipeting and spin.
10. Incubate at 37 °C for 30 min in the dark.
11. Stop the reaction by adding an equal volume of 2 × sample buffer.
12. Mix vigorously by pipeting and spin.

Depending on the cysteine content of a sample, it can be necessary to optimize the amount of reductant-dye per protein by a pre-experiment. For instance, when a sample contains proteins with high cysteine content, more reductant and dye has to be added—while still maintaining the ratio 1:2 for TCEP to dye. Under-reducing and labeling results in vertical streaking and offset of Cy3 and Cy5 spots of the same protein in the M_r direction; overreducing and -labeling produces charge trains (by labeling of lysines) and horizontal streaking. For the optimization process see chapter 1 in volume 2, by Kai Stühler et al.

3.3. Isoelectric Focusing (see also Note 12)

Before samples are applied on the isoelectric focusing gel, the sample is diluted with 2 × sample solution to reduce the disulfide bonds and increase solubility of the proteins. The samples can be applied via rehydration—or cup-loading, just like in a standard 2-D electrophoresis protocol. During the separation there should not be too bright light shining on the strips in order to avoid bleaching of the dyes. Place the IPGphor chamber at a mediate light place in the laboratory—not a window place, or cover the transparent lid with a dark plastic sheet.

3.4. SDS Polyacrylamide Gel Electrophoresis

The SDS electrophoresis step is performed in the conventional way. For scanning the resulting patterns inside the cassette, the gels must be cast and run in low fluorescent glass cassettes. When certain protein spots should be further analyzed with mass spectrometry, one of the glass plates is prepared with Bind-silane to covalently bind the gel to the glass surface. In this way the gel cannot shrink or swell. This makes it possible for an automatic spot picker to pick the spots according to their position coordinates on the basis of the scanning data. It is also possible to use gels which are covalently bound to the surface of a nonfluorescent film support. When the separation is completed, the gels should not be fixed with an acidic solution, because acids reduce the fluorescence signal. However, when gels are kept overnight in the cassettes in a cold room or a refrigerator until scanning, there is no noticeable diffusion of spots.

3.5. Image Acquisition

The images are acquired with a multifluorescent point laser scanner or a scanning CCD imager. The gels are imaged while they are still in the cassette. Also this approach—additionally—minimizes methodical variation because of shrinking and swelling of the gel during a time-consuming scanning process. Scanners with confocal optics can be adjusted to focus the excitation beam inside the gel layer between the glass plates. In the laser scanner excitation is performed with a single wave length laser beam. In the scanning CCD camera systems monochromatic light is created with white light and narrow band pass filters. With both approaches fluorescence excitation and light emission measurement are performed separately for each dye channel to prevent "crosstalking." This is usually done automatically. Because the fluorescence intensities are different for the different fluorophores, the scanner software has to normalize the images in order to make results comparable. Later, a second normalization of each sample spot volume to the volume of the respective internal standard spot will be applied by the image analysis software.

3.6. Result

Figure 4 shows the images of two samples—extracts from untreated and treated mouse livers—and the related pooled internal standard in false color display. The overlay of all three patterns automatically highlights up- and downregulated proteins: green spots represent proteins only present in sample one, red spots represent proteins only present in sample two, the majority of proteins show white spots, which indicates that they have not changed. Only the dramatic changes are visible. Slight quantitative changes can only be determined with image analysis software.

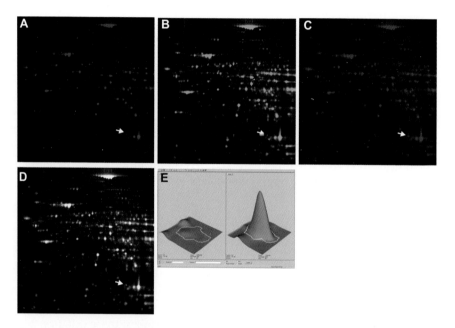

Fig. 4. False color representation of the images of samples and an internal standard, which have been coseparated in one gel by DIGE 2-D electrophoresis. The images have been acquired with a Typhoon™ multifluorescent laser scanner (GE Healthcare BioSciences). **A** Cy2 channel: pooled standard composed of the two samples; **B** Cy3 channel: mouse liver proteins, control; **C** Cy5 channel: mouse liver proteins, treated; **D** overlay of all three images; the arrows point to a protein spot which is up-regulated in the treated sample; **E** three-dimensional view of this spot in the Cy3 channel (left) and Cy5 channel (right). 1st dimension: Isoelectric focusing in IPG pH 3–10, 18 cm; 2nd dimension: SDS polyacrylamide gel electrophoresis in a 12.5% T gel.

3.7. Image Analysis

1. Because multiple samples are separated in the same gel, identical proteins of these samples migrate to the same spot position, making spot matching unnecessary. This feature allows the application of a special algorithm of DeCyder™ software (GE Healthcare), which performs codetection of the same proteins from the different samples within a gel, applying identical spot boundaries on identical protein spots. In this way relative quantification of proteins is much more precise than in one sample per gel experiments. Furthermore image analysis can be carried out fully automatically with the help of a batch processor, thus reducing the user bias to a minimum.
2. With the concept of separating a pooled internal standard in each gel, the comparisons of samples across the gels becomes much easier and accurate: only the patterns of the internal standards are matched, which have been produced from the identical protein mixture. In this way gel-to-gel variations are compensated.
3. Spot volume ratios of the samples are calculated related to the internal standard spot volume. This tool enables detection of highly significant differences with high confidence.

4. Notes

1. The experiment setup should be properly designed: Biological replicates ensure that induced biological changes are not mixed up with inherent biological differences. Gel replicates are not necessary. Optimally all samples of an experiment should be run together in the first and second dimension to minimize methodical variation.
2. Like for conventional 2-D electrophoresis efficient and reproducible sample preparation is very important, which can be achieved by optimization experiments (see all the other chapters of this book).
3. The DIGE technique is very useful for quickly checking the quality of sample preparation. The best strategy is: first perform a same–same sample experiment with two different dyes, run in one gel; second run three sample preparation replicates of the same sample source labeled with three different dyes in a gel. This approach also detects protein individuals with peculiar behavior, which can be contained in some samples.
4. The internal standard must be included in each gel. "Planned randomization" of the samples across the gels avoids systematic errors, and to exclude false positive results caused by preferred labeling. Reverse labeling of wild type and mutant (control and diseased) sample proteins equal out any preferred labeling of one of the dyes. Ideally an Excel sheet is prepared indicating for each gel which combination of label, control and mutant is applied.
5. At the stage of labeling the sample solution must not contain any primary amines, like carrier ampholytes, IPG buffer, or reductants to avoid any interference with the labeling procedure. These additives are applied after labeling has been completed.

6. If the sample is too diluted, labeling will be less efficient.
7. It is essential for labeling, that the sample solution—proteins included—has a pH value between 8 and 9: pH 8.5 is optimal for lysine tagging, pH 8.0 is optimal for cysteine tagging. Samples which have been extracted with TCA acetone or have been cleaned up with precipitation can be rather acidic. Check the pH carefully and adjust the pH value with adding 50 mM NaOH solution.
8. Open a new bottle of DMF every 3 months. The DMF must be high quality anhydrous (ultrapure), and every effort should be taken to ensure it is not contaminated with water. Use of molecular sieves will help keep DMF in an anhydrous condition. DMF, once opened, will start to degrade generating amine compounds, which will react with the CyDye DIGE Fluor minimal dyes reducing the concentration of dye available for protein labeling. Take a small volume of DMF from its original container and dispense into a reagent cup. From this cup remove the specified volume of the aliquoted DMF and add to each new vial of dye. Take the CyDye DIGE Fluor minimal dye from the –20°C freezer and leave to warm for 5 min at room temperature without opening. This will prevent exposure of the dye to water condensation which may cause hydrolysis.
9. Ensure efficient mixing of samples and dyes by rigorously pipeting. Insufficient mixing could lead to preferential labeling of some proteins.
10. Take care, that the sample does not get frozen, because this would cause preferential labeling of some proteins.
11. Because cell lysates are viscous, it is important to mix samples thoroughly in this and all following mixing steps to avoid nonuniform labeling. However, vortexing is not recommended).
12. The multiplex approach is also advantageous for the separation conditions during the isoelectric focusing step: mixing of several samples reduces the conductivity differences between the strips, thus leading to better reproducible spot pattern across the gels.

References

1. O'Farrell, P.H. (1975) High-resolution two-dimensional electrophoresis of proteins. *J Biol Chem.* **250**, 4007–4021.
2. Ünlü, M., Morgan, M.E., and Minden, J.S. (1997) Difference gel electrophoresis: A single gel method for detecting changes in protein extracts. *Electrophoresis* **18**, 2071–2077.
3. Alban, A., David, S., Bjorkesten, L., Andersson, C., Sloge, E., Lewis, S., and Currie, I. (2003) A novel experimental design for comparative two-dimensional gel analysis: Two-dimensional difference gel electrophoresis incorporating a pooled internal standard. *Proteomics* **3**, 36–44.
4. Friedman, D.B., Hill, S., Keller, J.W., Merchant, N.B., Levy, S.E, Coffey, R.J., and Caprioli, R.M. (2004) Proteome analysis of human colon cancer by two-dimensional difference gel electrophoresis and mass spectrometry. *Proteomics* **4**, 793–811.

5. Shaw, J., Rowlinson, R., Nickson, J., Stone, T., Sweet, A., Williams, K., and Tonge, R. (2003) Evaluation of saturation labelling two-dimensional difference gel electrophoresis fluorescent dyes. *Proteomics* **3,** 1181–1195.
6. Sitek, B., Lüttges, J., Marcus, K., Klöppel, G., Schmiegel, W., Meyer, H.E., Hahn, S.A., and Stühler, K. (2005) Application of fluorescence difference gel electrophoresis saturation labelling for the analysis of microdissected precursor lesions of pancreatic ductal adenocarcinoma. *Proteomics* **5,** 2665–2679.

8

Isotope-Coded Two-Dimensional Maps: Tagging with Deuterated Acrylamide and 2-Vinylpyridine

Pier Giorgio Righetti, Roberto Sebastiano, and Attilio Citterio

Summary

Isotope-coded two-dimensional maps, with either D_0/D_3-acrylamide or D_0/D_4 2-vinyl pyridine, are described in detail. They have the advantage of running the two samples under investigation within a single slab gel, thus minimizing errors because of spot matching with software packages when samples are run in parallel maps. Labeling with deuterated acrylamide is very simple and inexpensive, because this chemical is commercially available. The experiment has to be carried out at alkaline pH values (pH 8.5–9.0) and with high molarities of alkylating agent (50–100 mM) to ensure good conversion efficiency. On the contrary, labeling with 2-vinyl pyridine (2-VP) can be performed in much lower alkylant molarities (20 mM) and at neutral pH values, thus ensuring essentially 100% conversion efficiency coupled with 100% specificity, because the reaction is sustained by the partial positive and negative charges on the 2-VP and –SH group, respectively. However, deuterated 2-VP is not commercially available and it has to be synthesized ad hoc.

Key Words: Deuterated acrylamide; deuterated 2-vinyl pyridine; isotope-coded tags; rat sera; two-dimensional maps.

1. Introduction

A key issue in all proteomic approaches (whether by two-dimensional [2-D] electrophoresis or chromatographic means) is quantitation of all spots/peaks resolved, so as to enable a description of the evolution of physiological/pathological events in terms of up- and downregulation of any possible expressed phenotype *(1)*. Such a differential display would allow selecting key proteins involved in such events as potential drug targets in an ever expanding pharmaceutical market seeking development of new, powerful and properly engineered drugs. In the conventional 2-D approach, the ability to display several thousand spots in a single gel slab was achieved with the aid of powerful computer algorithms, developed over the years, such as the Melanie and PD-Quest. These were able to create standard maps of, e.g., physiological versus pathological cell populations and compare them via an overlapping procedure and also detect all spots exhibiting higher or lower stain intensity (typically Coomassie Blue), on the assumption that such densitometric differences would reflect differential protein expression *(2)*. Although this procedure has been amply demonstrated in innumerable publications up to the present, it suffers from three shortcomings: first, the extremely laborious and time-consuming set-up, requiring the generation of at least three maps for each state (control versus disease), in a second instance, the fact that it could not possibly apply to all chromatographic approaches recently described. Third, the large experimental error, forcing to reject all spot intensity variations below a 100% (i.e., twofold) color change (but even below 200–300%, according to a few authors).

As an alternative to the above protocol, DIGE (differential in gel electrophoresis) has been proposed *(3,4)*. It is based on differential labeling with *N*-hydroxy-succinimide ester-modified cyanine fluors, the most popular couple being named Cy3 and Cy5. Cy3 is excited at 540 nm and has an emission maximum at 590 nm, whereas Cy5 is excited at 620 nm and emits at 680 nm. The two samples to be compared are separately labeled with either Cy3 or Cy5, which covalently modify Lys residues in proteins. These dyes have positive charges to replace the loss of charge on the ε-amino group of Lys, and the masses of the dyes are similar to each other (434 and 464 Da, respectively). The reaction is carried out so as to label only a few Lys residues per macromolecule (ideally, in fact, just one). As long as the extent of the reaction is similar between the samples to be compared, the mass shift will be uniform and the pI should be essentially unaltered. Given the distinguishable spectra of the two fluorophores, the two samples can then be combined and run in a single 2-D gel. The differences between the quantities of the individual proteins from each sample can then be determined using specialized 2-D image analysis software. Because both samples to be compared are separated in a single gel, this eliminates gel-to-gel variation, resulting in improved spot matching. As a corollary, the number of parallel and replicate gels required for obtaining reliable results is greatly reduced. Furthermore, fluorescence imparts the ability of detecting proteins over a much broader linear dynamic range of concentrations than visible gel stains *(3,4)*.

As a further addition to DIGE, another protocol for differential Cy3/Cy5 labeling, based on the reaction of a similar set of dyes not any longer on Lys, but on Cys residues, has been reported *(5,6)*. This technique is based on the opposite principle as compared to the original DIGE idea: not any longer "minimal", but "maximal" labeling, i.e., saturation of all possible Cys reacting sites. This would fulfil two goals at once: on the one hand, it would automatically enhance the stain sensitivity; on the other hand, it would block further reactivity of reduced Cys residues. The reacting end of these Cys-dyes is a maleimide residue, permitting an addition of the –SH group to the double bond of the maleimide moiety, thus forming a thioether link (although the structure of the dyes has not been disclosed as yet, their mass has been reported to be 673 and 685, respectively). However, "saturation labeling" seems to be besieged by problems. This bulky reagent seems to suppress the MS signal (by quenching the ionization of peptides) and also to interfere with the trypsin digestion, thus producing fewer cuts than expected. There are other matters of concern, of course. Among them, the extent of reaction: does the Cys blocking procedure achieve 100%, or is it considerably less, considering their bulky structure? Other serious problems are apparent: first of all, the massive shift of all proteins spots towards higher apparent M_r values, because of the bulky size of the cyanine dye. Secondly, the fact that quite a few of the spots in the 2-D maps reported *(5,6)* appear blurred and out of focus, as though they have a tendency to precipitate along the migration path. In addition, the fact that the total number of spots is considerably less than in silvering protocols, notwithstanding the much higher fluorescent signal of the saturation label, makes one wonder if, during the labeling protocol, a number of barely soluble proteins might precipitate out of solution because of increased hydrophobicity brought about by the cyanine dyes, thus disappearing from the map just at the onset of the 2-D mapping procedure.

Aware of the advantages of performing differential analysis on a single 2-D gel, but also of the disadvantages of DIGE methods, as just discussed, we investigated the possibility of carrying out a novel type of DIGE, based on the use of deuterated compounds *(7,8)*. The idea of stable isotope labeling was first proposed by Aebersold's group, with a labeling method named ICAT (isotope-coded affinity tag) *(9,10)*. The technique was so well accepted, that a host of methods were reported, based on a variety of isotope coded labels and even mass-coded labels, as reviewed in *(11–15)*. It must be emphasized, though, that most of these methods are only applicable to tryptic peptides, as customarily done in 2-D chromatography. Thus, our protocol of applying isotope-coded tags to differential analysis in 2-D maps appears to be novel and quite revolutionary. In particular, the deuterated acrylamide method was simultaneously described by us *(7)* and by Sechi *(16)* and fully validated experimentally by Cahill et al. *(17)*.

2. Materials
2.1. Acrylamide Alkylation
1. Electrophoresis-grade acrylamide from Bio Rad (Hercules, CA, USA).
2. 98% enriched 2,3,3'-D3-acrylamide (deuterium-labeled acrylamide) from Cambridge Isotope Labs., Woburn, MA, USA.

2.2. 2-Vinyl Pyridine Alkylation
1. 2- and 4-vinyl pyridine (VP) from Aldrich Chemical CO (Dorset, UK).
2. Tetra-deuterated 2-VP: by synthesis, see below.

2.3. Reagents Common to Both Methods
Urea, thiourea, tributylphosphine (TBP), glycine, sodium dodecyl sulfate (SDS), and 3-[3-cholamidopropyl dimetilammonio]-1-propansulfonate (CHAPS) were obtained from Fluka Chemie (Buchs, Switzerland). Acrylamide, N',N'-methylenebisacrylamide, ammonium persulfate, TEMED, SYPRO ruby as well as the linear Immobiline dry strips pH gradient 3–10 (7 cm long), were from Bio-Rad Laboratories (Hercules, CA, USA). Ethanol, methanol, glycerol, sodium hydroxide, hydrochloric acid, acetone, and acetic acid were from Merck (Darmstadt, Germany). Sigma-Aldrich Chemie GmbH (Steinheim, Germany) provided Tris, mineral oil, phosphoric acid, Coomassie Brilliant Blue G-250, DL-dithiothreitol (DTT), and bovine α-lactalbumin (LCA) (accession number in SwissProt database P00711).

3. Methods
3.1. Acrylamide Alkylation
1. One hundred μl of rat serum (containing 6 mg total protein) are added to 50 mM Tris-HCl buffer, pH 8.5, containing 5 mM TBP, 2 M thiourea, and 7 M urea (solubilizing buffer) and left for 2 h at room temperature.
2. The resulting mixture is divided in two parts; the first is alkylated with 100 mM D_0-acrylamide, whereas the second fraction is alkylated with 100 mM D_3-acrylamide.
3. The resulting two fractions are then mixed in appropriate ratios (typically 50/50), dialyzed overnight (against solubilizing buffer minus TBP and in 10 mM Tris-HCl, pH 8.5) and subjected to 2-D gel analysis.

3.2. 2-Vinylpyridine Alkylation
1. Alkylation of protein standards with either light or heavy VPs is implemented under the following conditions: 20 mM Tris-acetate, pH 7.0, containing 5 mM

tributyl phosphine (TBP), 2 M thiourea and 7 M urea, with the simultaneous addition of 20 mM 2-VP.
2. The reaction is quenched after 1 h by addition of an equimolar (20 mM) amount of DTT, so as to destroy any excess of VP. The resulting D_0- or D_4-VP reacted species are subjected to digestion before MS analysis.

3.3. Synthesis of Tetra-Deuterated 2-Vinyl Pyridine

1. Vinylmagnesium bromide (20 ml of a 1-M solution in tetrahydrofuran) is cooled to −5°C under nitrogen. To it, dropwise, is added, during 20 min, penta-deutero pyridine (2.08 g, 24 mmol). The mixture becomes turbid and yellow, and, after 30 min, is heated at 35–40°C for 1 h.
2. The solution is rapidly filtered on silica gel (5.0 g) with ethyl ether (40 ml) as eluent. The first 20 ml were flash chromatographed on silica gel (pentane:ethyl ether 3:2) allowing to separate fractions which contain the tetradeuterated 2-vinyl pyridine, detected by GC-MS. Careful distillation at atmospheric pressure of the pure fractions allows to obtain a pure sample of tetradeuterated 2-vinyl pyridine as an oil (235 mg, 19% yield).

3.4. Results

3.4.1. Two-Dimensional Maps of Rat Serum Tagged with D_0/D_3 Acrylamide

The main steps of the present approach, in analysing 2-D maps of tissue extracts or proteins in biological fluids, are depicted in Fig. 1. It can be appreciated that this isotope coded labeling is quite similar to the DIGE technique, in that the two samples under analysis are labeled with either light or heavy tags and then run simultaneously in a single gel, after mixing in appropriate ratios. Thus the quantitation of protein expression in a control versus "pathological" sample (or sample obtained under a variety of experimental conditions, such as drug treatment and the like) is done via analysis of the isotopic ratios of the two different protein populations contained in a single gel spot, just as performed in DIGE via ratio of fluorescent signals of Cy3 versus Cy5 dyes contained in a single gel spot. It should be pointed out that the analysis of the intact protein (as shown in the path in the lower right corner) is an optional step that may be used as an early screening for possible gel-induced and/or post-translational modifications. In the vast majority of cases, proteins are identified by searching for the exact mass and number of the peptides (and, when available, short sequences of some selected peptides) obtained by digestion of each single spot in the 2-D map against suitable databases, such as SWISS-PROT, TrEMBL, and NCBInr, by using the ProteinLynx program (Micromass) or NCBInr database by using the ProFound program (http://129.85.19.192/profound_bin/WebProFound.exe).

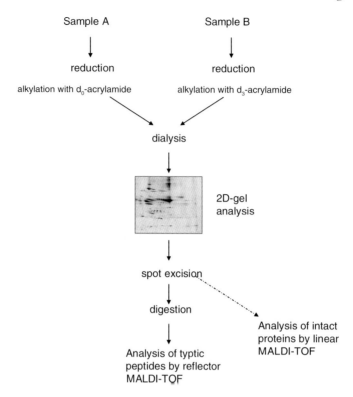

Fig. 1. Protocol for alkylating samples, before 2-D map analysis with D_0/D_3-acrylamide, followed by dialysis, mixing of the two separate samples, running a single gel, excision of bands and MALD-TOF MS analysis.

In the particular experiment depicted in Fig. 1, the two alkylated rat sera were mixed in a ratio of 30:70 of D_0/D_3-acrylamide before 2-D map analysis; a number of separated proteins were digested in situ and subjected to reflector MALDI-TOF analysis. A representative spectrum pertaining to albumin (the most intense band in Fig. 1) is given in Fig. 2(**a**), whereas short scan intervals taken from the same spectrum are presented in Figs. 2 (**b**, **c**), pertaining to peptides with m/z = 1899.7 and m/z = 1679.5, respectively. The observed isotopic distributions in both spectra are attributed to the indicated sequences, each of which contains a single cysteine residue. Furthermore, the more intense peaks in both monoisotopic spectra are displaced by 3 Da from their weaker counterparts. Such displacement coincides with a single alkylation channel associated with the D_3-acrylamide. Considering the relative peak heights in both isotopic distributions we have calculated a ratio of 34:66, which is in good agreement with the labeling ratio 30:70 before 2-D separation.

Isotope-Coded Two-Dimensional Maps 93

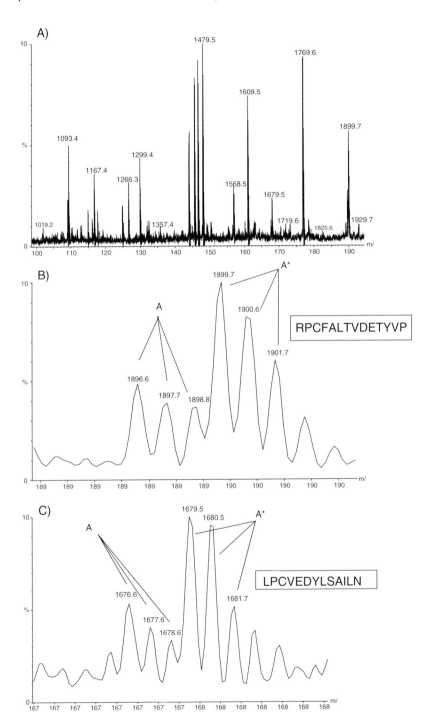

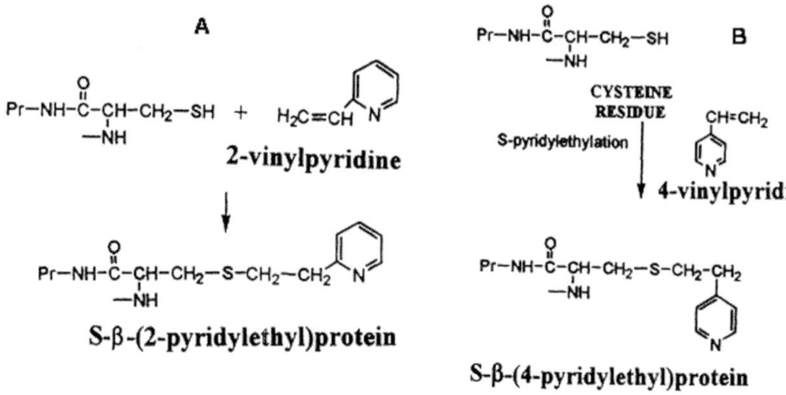

Fig. 3. Alkylation process with 2-VP (**A**) and 4-VP (**B**), respectively. This reaction produces S-β-pyridylethyl protein derivatives.

3.4.2. Analysis of Marker Proteins Labeled with D_0/D_4 2-Vinyl Pyridine

Figure 3 shows the alkylation process in the case of 2-VP (**A**) and 4-VP (**B**), respectively. The process gives rise to S-β-pyridylethyl protein derivatives. To prove the principle of light/heavy 2-VP labeling, a marker protein (here bovine α-lactalbumin) was tagged with each of the two –Cys labels, digested and separately analyzed by MALDI TOF MS. A representative spectrum of the two tryptic digests is shown in Fig. 4. Figure 5 offers a zoom of the peptide m/z 1252,7 (D_0) as compared with the same one labeled with D_4-2-VP (lower tracing): It can be appreciated that both peptides are fully resolved down to four monoisotopic peaks and that they are spaced apart by 4 Da, as expected, considering that this fragment contains a single Cys residue. The same resolution is obtained when the light/heavy labeled peptides are injected in a mixture (not shown). Figure 6 shows a corresponding zoom in the area of the peptide m/z 1989,1; as this fragment contains two Cys residues, it can be appreciated that all the monoisotopic distributions are spaced apart by, precisely, 8 Da. It is thus shown that this type of labeling can be fully exploited for quantitative proteomics.

◀───

Fig. 2. *(Continued)* (**a**) Reflector MALDI mass spectrum of an in-situ digest of albumin (the most intense spot) in Fig. 1. (**b**) and (**c**) are two short intervals taken from spectrum in (**a**) associated with the indicated peptide sequences. Note the 3 Da difference between the monoisotopic distributions of the D_0 versus the D_3-acrylamide derivatives.

Isotope-Coded Two-Dimensional Maps

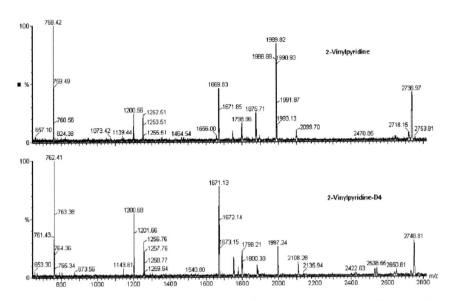

Fig. 4. MALDI-TOF mass spectra of tryptic digests of bovine α-lactalbumin labeled with either D_0- (upper tracing) or with D_4 (lower tracing) 2-VP.

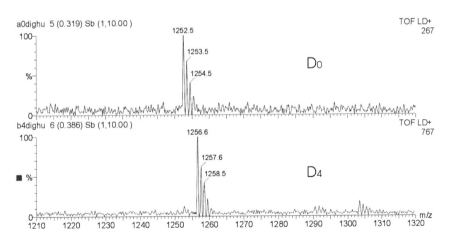

Fig. 5. Expanded view of monoisotopic distributions of the m/z 1252,5 peptide labeled with D_0 2-VP and the corresponding Mz 1256,6 peptide labeled with D_4 2-VP. Note the 4 Da differences between the two monoisotopic series, because of alkylation of a single Cys residue.

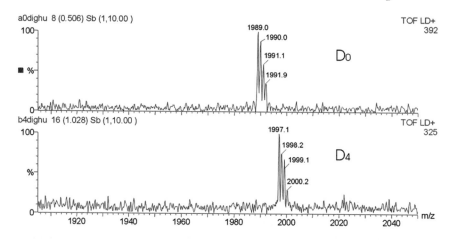

Fig. 6. Expanded view of monoisotopic distributions of the m/z 1989,0 peptide labeled with D_0 2-VP and the corresponding Mz 1997,1 peptide labeled with D_4 2-VP. Note that this peptide contains 2 Cys residues, thus the 8 Da difference between the two series of peaks.

4. Notes

1. Elimination of excess of all types of alkylating agents. This is a must, before entering the electric field (e.g., first dimension of a 2-D map, SDS-PAGE in monodimensional runs). The reaction conditions here reported have been optimized so as to drive the alkylation solely on the –SH groups of Cys. Prolonged incubation times in a large excess of alkylating agent will drive the reaction also on ε-amino groups of Lys and the $-NH_2$ termini of polypeptide chains. Such over-alkylation will be disastrous in quantitative proteomics, because it will give false results and often misclassification of peptides, based on wrong mass data. In addition, especially during the focusing process in the first dimension of 2-D maps, which can last for long periods of time (typically overnight) the excess of alkylant (present all throughout the gel strip for the neutral acrylamide, in the neutral to alkaline region for the charged 2-VP) will continue to alkylate amino acid residues, also aided by the electric field, in a process that we have described as "electrically engendered chemical reactions" *(18)*.
2. Alkylation with acrylamide. In this case, high molarities of alkylant are required, typically of the order of 50–100 mM acrylamide. The reason is that, even at pH 8.5–9.0, where the –SH group of Cys (pK 8.3) is highly ionized, and thus highly reactive, acrylamide is a slow reacting species. Thus, to shift the equilibrium towards the formation of reacted species, high levels of acrylamide are required.
3. Alkylation with 2-VP. Here the reaction conditions are quite different. First of all, because of the much higher reactivity of 2-VP, its concentration can be reduced to 20 mM. It should also be noted that the reaction here is conducted at neutral pH values (pH 7.0). This is optimal for the reactivity of both compounds (i.e., 2-VP

and –SH groups of Cys) because it is about half way between the pK values of the –SH groups (8.3) and the pK value of the tertiary amino group of 2-VP (pK 5.4): under these conditions, the reaction is driven by the partial negative and positive charges, respectively, of the two reacting species and continues at a fast rate till full alkyation of all Cys residues. As an additional bonus, this neutral pH during reaction ensures its very high specificity, because at this pH value the ε-amino groups of Lys, being fully protonated, are unreactive. Thus, 2-VP appears to be among the best alkylating agents, because it couples 100% reactivity with 100% specificity, conditions rarely achievable with any other alkylating agent.

4. Alkylation with acrylamide derivatives versus iodinated compounds. It is well known that the typical alkylating agent for Cys residues, exploited in most protocols up to the present, is iodoacetamide. Indeed, our data have shown that alkylation with acrylamide and/or its derivatives are to be preferred, especially when dealing with 2-D maps. The preferred solubilization cocktail for tissue and cell proteins, in use today, contains a mixture of 2 M thiourea and 7 M urea. We have demonstrated that thiourea is a scavenger of iodoacetamide, to the point at which all the amounts added (typically of the order of 50 mM and higher) are consumed within 5 min of incubation *(19,20)*. Thus, under these conditions, alkylation with iodoacetamide is incomplete and hardly quantitative. That is also the reason why the classical ICAT reagent *(9,10)* cannot possibly be used in 2-D mapping: besides the fact that this reagent is terribly expensive, because it has a reacting iodinated tail, it will be immediately destroyed upon incubation of proteins dissolved in the solubilization cocktail, thus impeding proper tagging and proper quantitative analysis.

5. Health hazards. It should be remembered that both acrylamide and 2-VP are neurotoxins, as typical of all compounds containing a reactive acrylic double bond (as reviewed in *(21)*). Thus, great care should be exerted in handling them. For acrylamide, which is less toxic than 2-VP, because it is a solid, one should nevertheless be careful at not accidentally breath in the powder when handling it for preparing a polyacrylamide gel or for alkylating purposes. In the case of 2-VP, although this compound is more toxic because it is more volatile (its physical state is a liquid), nevertheless it is more easily handled because it can be transferred by pipeting. In both cases, though, these compounds should be handled only under a fume hood.

6. 2-VP or 4-VP? In principle, one could use either of these compounds, because they have the same alkylating efficiency and very similar pK values. However, we much prefer using 2-VP, because of its higher stability on storage and under reacting conditions.

7. Why 2-VP? There is an additional advantage in using 2-VP alkylation, especially when the protein sample is further analyzed by MS. In MS, all peptides labeled with 2-VP, because of the positive charge in this molecule, will give a much stronger signal, considering that most peptide analyses are performed in the positive ion mode.

8. More on 2-VP and 4-VP. Although S-pyridylethylation is scarcely used today for blocking -SH groups in proteins, this reaction has been known for more that 35 years and it has been accepted as the best method for the modification of Cys residues in proteins for subsequent analysis and sequence determination. An extra bonus of this reaction is that, whereas free cysteine and cystine residues are unstable under conditions used for acid hydrolysis of peptide bonds, their 2-VP or 4-VP derivatives are fully stable under the same harsh hydrolysis conditions, thus permitting quantitative recovery of these amino acid residues. This reaction has also been proposed for measuring D-Cys, homocysteine, glutathione, tryptophan, dehydroalanine, and furanthiol in food flavors. The excellent review by Friedman *(22)* on S-pyridylethylation of proteins should be mandatory reading for all working in the field of proteome analysis.

Acknowledgments

Supported in part by a FIRB grant No. RBNE01KJHT_002 from MIUR, by Fondazione Cariplo and by PRIN-2006 (MURST,Rome).

References

1. Hamdan, H. and Righetti, P.G. (2005) Proteomics Today, Wiley-Interscience, Hoboken, pp. 1–426.
2. Marengo, E., Robotti, E., Antonucci, F., Cecconi, D., Campostrini, N. and Righetti, P.G. (2005) Numerical approaches for quantitative analysis of two-dimensional maps: a review of commercial software and home-made systems. Proteomics 5, 654–666.
3. Unlu, M., Morgan, M.E. and Minden, J.S. (1997) Difference in gel electrophoresis: a single gel method for detecting changes in protein extracts. Electrophoresis 18, 2071–2077.
4. Tonge, R., Shaw, J., Middleton, B., Rawlinson, R., Rayner, S., Young, J., Pognan, F., Hawkins, E., Currie, I. and Davison, M. (2001) Validation and development of fluorescence two-dimensional differential gel electrophoresis proteomics technology. Proteomics 1, 377–396.
5. Shaw, J., Rowlinson, R., Nickson, J., Stone, T., Sweet, A., Williams, K., Tonge, R. (2003) Evaluation of saturation labelling two-dimensional difference gel electrophoresis fluorescent dyes. Proteomics 3, 1181–1195.
6. Kondo, T., Seike, M., Mori, Y., Fujii, K., Yamada, T., Hirohashi, S. (2003) Application of sensitive fluorescent dyes in linkage of laser microdissection and two-dimensional gel electrophoresis as a cancer proteomic study tool. Proteomics 3, 1758–1766.
7. Gehanne, S., Cecconi, D., Carboni, L., Righetti, P.G., Domenici, E. and Hamdan, M. (2002) Quantitative analysis of two-dimensional gel-separated proteins using isotopically marked alkylating agents and matrix-assisted laser desorption/ionization mass spectrometry. Rapid Commun. Mass Spectrom. 16, 1692–1698.

8. Sebastiano, R, Citterio, A., Lapadula, M. and Righetti, P.G. (2003) A new deuterated alkylating agent for quantitative proteomics. Rapid Commun. Mass Spectrom. 17, 2380–2386.
9. Gygi, S.P., Rist, B., Geber, S.A., Turecek, F., Gelb, M.H. and Aebersold, R. (1999) Quantitative analysis of complex protein mixtures using isotope-coded affinity tags. Nature Biotech. 17, 994–999.
10. Tao, V.A. and Aebersold, R. (2003) Advances in quantitative proteomics via stable isotope tagging and mass spectrometry. Curr. Opin. Biotechnol. 14, 110–118.
11. Hamdan, M. and Righetti, P.G. (2002) Modern strategies for protein quantification in proteome analysis: advantages and limitations. Mass Spectrom. Reviews 21, 287–302.
12. Righetti, P.G., Campostrini, N., Pascali, J., Hamdan, M. and Astner, H. (2004) Quantitative proteomics: a review of different methodologies. Eur. J. Mass Spectrom. 10, 335–348.
13. Righetti, P.G., Castagna, A., Antonucci, F., Piubelli, C., Cecconi, D., Campostrini, N., Antonioli, P., Astner, H. and Hamdan, M. (2004) Critical survey of quantitative proteomics in two-dimensional electrophoretic approaches. J. Chromatogr. A 1051, 3–17.
14. Julka, S. and Regnier, F. (2004) Quantification in proteomics through stable isotope coding: a review. J. Proteome Res. 3, 350–363.
15. Moritz, B. and Meyer, H.E. (2003) Approaches for the quantification of protein concentration ratios. Proteomics 3, 2208–2220.
16. Sechi, S. (2002) A method for identifying and simultaneously determining the relative quantities of proteins isolated by gel electrophoresis. Rapid Commun. Mass Spectrom. 16, 1416–1424.
17. Cahill, M.A., Wozny, W., Schwall, G., Schroer, K., Hoelzer, K., Poznanovic, S., Hunzinger, C., Vogt, J.A., Stegmann, W., Matthies, H. and Schrattenholz, A. (2003) Analysis of relative isotopologue abundance for quantitative profiling of complex protein mixtures labelled with the acrylamide/D_3-acrylamide alkylation tag system. Rapid Commun. Mass Spectrom. 17, 1283–1290.
18. Righetti, P.G., Gelfi, C., Sebastiano, R., and Citterio, A. (2004) Surfing silica surfaces superciliously. J. Chomatogr. A 1053, 15–26.
19. Galvani, M., Hamdan, M., Herbert, B. and Righetti, P.G. (2001) Alkylation kinetics of proteins in preparation for two-dimensional maps: a MALDI-TOF mass spectrometry investigation. Electrophoresis 22, 2058–2065.
20. Galvani, M., Rovatti, L., Hamdan, M., Herbert, B. and Righetti, P.G. (2001) Proteins alkylation in presence/absence of thiourea in proteome analysis: a MALDI-TOF mass spectrometry investigation. Electrophoresis 22, 2066–2074.
21. Hamdan, M., Galvani, M., Bordini, E. and Righetti, P.G. (2001) Protein alkylation by acrylamide, its N-substituted derivatives and cross-linkers and its relevance to proteomics: A matrix assisted laser desorption/ionization-time of flight-mass spectrometry study" Electrophoresis 22, 1633–1644.
22. Friedman, M. (2001) Application of the S-pyridylethylation reaction to the elucidation of the structures and functions of proteins. J. Protein Chemistry 20, 431–453.

9

Stable Isotope Labeling by Amino Acids in Cell Culture (SILAC)

Albrecht Gruhler and Irina Kratchmarova

Summary

Quantitative proteomics has become a pivotal tool that has been applied to the investigation of many different biological processes such diverse as the detection of biomarkers in tissue samples, the regulation of cell signaling, and the characterization of protein interactions. Stable isotope labeling techniques have facilitated the precise quantitation of changes in protein abundance by mass spectrometry. Among different choices, Stable Isotope Labeling by Amino acids in Cell culture (SILAC) is an easy and reliable method for unbiased comparative proteomic experiments, which has been employed to study posttranslational modifications such as protein phosphorylation and methylation, to characterize signaling pathways and to determine specific protein interactions. Here we describe detailed procedures for SILAC experiments in mammalian and yeast cells.

Key Words: Cell culture; mass spectrometry; protein quantitation; proteomics; stable isotope labeling.

1. Introduction

The precise quantitation of protein abundance by proteomic technologies has become an important tool to study the dynamics of cellular processes. Examples include the study of cell signaling pathways or the phosphorylation on cell surface receptors *(1,2)*. In addition, quantitative data have been used to determine specific protein–peptide interactions and to localize proteins to complexes and subcellular compartments *(3–6)* There are a number of different methods to acquire quantitative protein data. Some of them are gel-based such as the densitometric measurement of stained proteins, the use of protein specific fluorescent stains, Western blotting, or protein labeling with radioactive isotopes *(7)*. In recent years, mass spectrometry has become an increasingly more important tool in quantitative proteomics. The direct determination of protein and peptide amounts by mass spectrometry is difficult, because the measured signal intensity depends besides the sample concentration also on other ill-defined factors such as ionization efficiency, sample composition, and gas-phase stability of the ions. However, comparative measurements between the same peptide in different experiments and a modified/unmodified peptide pair in the same experiment can be used to determine relative peptide and protein amounts. Particularly the introduction of stable isotope labeling facilitated the reliable measurement of relative and absolute protein amounts with unprecedented exactness *(8)*. There exist different possibilities to incorporate stable isotopes into individual proteins or to isotope encode complete proteomes. It can be achieved by chemical modification of extracted proteins *(9)* or by *in-vivo* labeling of cells or organisms.

In-vivo stable isotope labeling can be accomplished by growing the cells in the presence of a ^{15}N containing nitrogen source. This method has been reported for yeast and bacterial cultures and for higher organisms as *C. elegans* and Drosophila *(10–12)*. In a similar way Stable Isotope Labeling by Amino acids in Cell culture (SILAC) also relies on the metabolic encoding of the proteome *(13,14)*. SILAC has been successfully used in mammalian, yeast, and plant cells *(2,13,15)* and applications include the study of protein phosphorylation regulation, phosphopeptide-interaction screens, and the study of EGF receptor stimulation and signaling. Recently, experiments with two different variants of modified arginine allowed the temporal resolution of early signaling events upon stimulation of the EGF receptor *(16)*. In addition, comparison of two signaling pathways using SILAC led to determination of a control point during mesenchymal stem cell differentiation *(17)*.

An outline of the SILAC procedure is given in Fig. 1. Cells are either grown in normal medium or in medium that contains one or more stable isotope labeled amino acid(s) instead of the unlabeled amino acid(s). The modified amino acids are incorporated into the proteome, leading to a defined mass shift of labeled ("heavy") proteins and peptides depending on the number and type of the heavy amino acid. Commonly used amino acids are ^{2}H$_{3}$-leucine, ^{13}C$_{6}$-arginine, ^{13}C$_{6}^{15}$N$_{4}$-arginine, and ^{13}C$_{6}$-lysine with a mass difference of 3, 6, 10, and 6 Da respectively, compared to the unlabeled ("light") amino acids. Equal amounts of labeled and unlabeled cells or protein extracts are combined followed by other sample preparation steps as required by the experiment, generation of peptides by proteolysis, and analysis by mass spectrometry. The most commonly used protease in proteomic experiments is trypsin, which cleaves carboxy-terminal to lysine and arginine residues. Double labeling with ^{13}C$_{6}$-lysine and

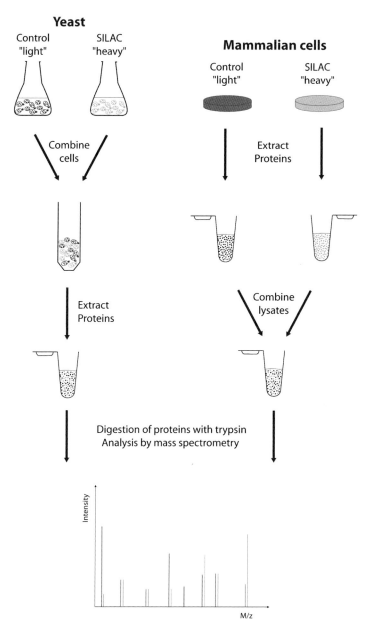

Fig. 1. Basic outline of SILAC labeling. Cells are grown in either "light" (control) or "heavy" SILAC medium containing amino acids with stable isotopes (^{13}C, ^{15}N, or ^{2}H). After completion of labeling and differential treatment of SILAC cultures, cells or protein extracts are combined. Peptides are generated by incubation of proteins with trypsin and analyzed by mass spectrometry. Intensities of unlabeled and stable

$^{13}C_6$-arginine ensures that every tryptic peptide (except C-terminal peptides) contains a heavy amino acid and can be used for protein quantitation. Peak intensities of light and heavy peptide pairs can now be directly recorded and compared in a single mass spectrum. Differences in the intensities reflect a change in protein abundance between the two samples, which is typically expressed as the ratio between the intensity of the heavy peptide peak divided by the intensity of the light peptide peak.

Complete and uniform incorporation of the heavy amino acids is essential for successful SILAC experiments. Therefore, the cells need to be grown for at least five generations in the labeling medium to facilitate the turn-over of unlabeled proteins (Fig. 2). In cell lines, where the heavy amino acids are essential, 100% labeling efficiency is reached, such as for lysine labeling in mammalian cells or in auxotrophic yeast cells. For nonessential amino acids, the incorporation efficiency has to be determined in control experiments. In a recent study with prototrophic Arabidopsis cells, incorporation of $^{13}C_6$-arginine reached an equilibrium at approximately 80%, even at prolonged labeling times and high arginine concentrations *(15)*. However, when the incomplete labeling was taken into account during protein quantitation, quantitative proteomic experiments were still possible.

In the following sections we describe detailed procedures for the labeling of mammalian and yeast cells.

2. Materials

2.1. Mammalian Cell Culture

1. Dulbecco's Modified Eagle Medium (DMEM); supplemented with 10% fetal bovine serum (FBS) (Gibco, Invitrogen, Carlsbad, CA).
2. 200 mM glutamine stock solution in 0.85% NaCl solution, 10,000 U penicillin/10,000 µg streptomycin stock solution and solution of trypsin EDTA: 200 mg/L trypsin and 500 mg/L Versene-EDTA (Gibco, Invitrogen, Carlsbad, CA).
3. Sterile phosphate buffered saline (PBS) (Cambrex Bio Science Copenhagen ApS, Denmark).
4. DMEM deficient in lysine, arginine, and methionine (a custom media preparation), supplemented with 10% dialyzed FBS (Gibco, Invitrogen, Carlsbad, CA) (see Notes 1 and 2).

Fig. 1. *(Continued)* isotope labeled peptide pairs in the mass spectrum can be used to calculate relative abundances in light and heavy SILAC samples. This basic scheme can be altered by introducing additional steps during sample purification, e.g., affinity purification of selected proteins, chromatographic fractionation, or gel electrophoresis.

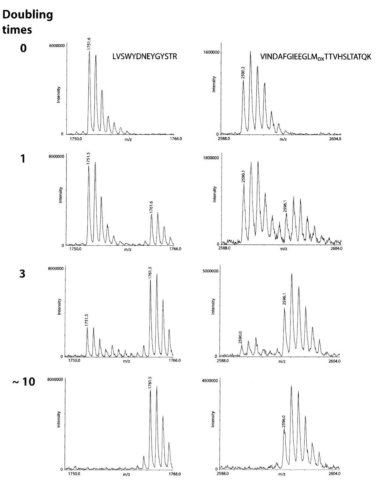

Fig. 2. Time course of incorporation of $^{13}C_6^{15}N_4$-arginine and $^{13}C_6$-lysine in YAL6B yeast cells. YAL6B yeast cells were grown in SILAC media containing $^{13}C_6^{15}N_4$-arginine and $^{13}C_6$-lysine. At the indicated doubling times aliquots were taken from the cell suspension and proteins extracted by alkaline lysis. Proteins were subjected to SDS-PAGE, stained with Coomassie blue and bands of the same molecular weight range were excised from samples corresponding to the different time points. Peptides were generated by in-gel digestion with trypsin and analyzed by MALDI-MS on a Bruker Ultraflex instrument. Peptide sequences were identified by MALDI-MSMS. The figure displays two peptides from Tdh3p (LVSWYDNEYGYSTR, MH+: 1752.56 Da) containing a C-terminal arginine and (VINDAFGIEEGLM$_{ox}$TTVHSLTATQK, MH+: 2591.15 Da) with a C-terminal lysine. The incorporation of $^{13}C_6^{15}N_4$-arginine and $^{13}C_6$-lysine leads to the appearance of peptide peaks at 1762.1 Da and 2597.1 Da with a mass difference of 10 and 6 Da, respectively. After 10 doubling times, no light peaks are observed any longer, demonstrating complete labeling of the yeast cells.

5. Amino acids: L-lysine, L-arginine:HCl, and L-methionine (Sigma Chemicals, Copenhagen, Denmark).
6. Filter units, MF75TM series (Nalge Nunc International, NY, USA) (see Note 3).
7. Epithelial adherent HeLa cells (ATCC® Number: CCL-2), (American Type Culture Collection (ATCC), Manassas, VA 20108 USA).

2.2. Yeast Cell Culture

1. Yeast nitrogen base w/o amino acids (Difco™, BD).
2. D(+)-Glucose monohydrate (Merck KG, Darmstadt, Germany).
3. Uracil (Sigma).
4. Adenine hemisulfate salt (Sigma).
5. Tyrosine (Serva Electrophoresis GmbH; Heidelberg, Germany).
6. Amino acids: L-arginine:HCl; L-lysine:HCl; L-methionine; L-tryptophane; L-histidine:HCl; L-phenylalanine; L-leucine (Serva Electrophoresis GmbH; Heidelberg, Germany).
7. Yeast strain: YAL6B (Mat **a**; his3Δ1; leu2Δ 0; met15Δ 0; ura3Δ 0; YHR018c::kanMX4, YIR034c::kanMX4; not tested for met15Δ 0, lys2Δ 0) *(2)*.

2.3. Modified Amino Acids

L-arginine:HCl (U-$^{13}C_6$, 98%) ($^{13}C_6$-arginine); L-lysine:2HCl (U-$^{13}C_6$, 98%) ($^{13}C_6$-lysine); L-arginine:HCl (U-$^{13}C_6$,$^{15}N_4$, 98 %) ($^{13}C_6^{15}N_4$-arginine) (all from Cambridge Isotope Laboratories, Andover, MA) (*see* Note 4).

3. Methods

3.1. SILAC in Mammalian Cells

SILAC can be applied to a variety of mammalian cell cultures. The only essential requirement is that the cells should be able to undergo five to six population doublings that are necessary for full incorporation of the labeled amino acids into their proteome. The described protocol is optimized for the labeling of HeLa cells in Dulbecco's Modified Eagle Medium (DMEM) using modified arginine. After five to six population doublings the entire proteome of the cells is normally encoded with the labeled amino acid to 90–100% efficiency.

3.1.1. Preparation of SILAC Media

1. Unlabeled L-lysine, L-methionine and L-arginine are dissolved in sterile PBS at concentrations of 146 g/L, 30 g/L, and 84 g/L, respectively, and sterile filtered using a 0.2 μm filter. These stock solutions can be stored for up to 3 months at 4 °C.
2. The modified amino acids $^{13}C_6$-arginine and $^{13}C_6^{15}N_4$-arginine are dissolved in sterile PBS at a concentration of 84 g/L.

Stable Isotope Labeling by Amino Acids in Cell Culture (SILAC)

3. "Light" (control) SILAC medium: To 1 L of DMEM (deficient in L-lysine, L-methionine, and L-arginine), 1 mL of unlabeled lysine and methionine stock solutions and 330 µl of the unlabeled arginine stock solution are added (see Notes 5 and 6).
4. "Heavy" SILAC media: To 1 L of DMEM (deficient in L-lysine, L-methionine, and l-arginine), 1 mL of L-lysine and L-methionine stock solutions, and 330 µL of either $^{13}C_6$-arginine or $^{13}C_6^{15}N_4$-arginine stock solutions are added.
5. All media are filtered through 1 L filter units from Nalge Nunc.
6. Light and heavy SILAC media are supplemented with 10% dialyzed FBS, 2 mM Glutamine, 100 U/L penicillin, and 100 µg/L streptomycin.
7. The media are now ready for use and can be stored at 4°C (see Note 7).

3.1.2. Cell growth

1. HeLa cells are grown in DMEM in culture dishes with addition of 10% FBS for two to three population doublings in an incubator at 37°C, 5% CO_2, and 85% humidity.
2. When the cells are approximately 80% confluent, the media is aspirated, the cells are washed with PBS and then trypsinized for 5 min.
3. 1/3 of the cells are seeded into the light (control) SILAC medium, the remaining cells are divided between heavy SILAC media containing either $^{13}C_6$-arginine- or $^{13}C_6^{15}N_4$-arginine, respectively (see Notes 8 and 9).
4. The cells are propagated and expanded in the SILAC media for 5–6 population doublings.
5. After the 5–6 population doublings the cells can be used for different types of experiments such as investigation of signaling cascades or the determination of changes that occur in response to variety of growth factors, etc. (see Note 10).

3.2. SILAC in Saccharomyces cerevisiae Yeast Cells

Complete incorporation of the modified amino acids might require auxotrophic yeast strains, where the biosynthesis of the respective endogenous amino acids is inhibited. Otherwise, the labeling efficiency might not reach 100% and the concentration of the stable-isotope labeled amino acids that is required for optimal labeling needs to be determined. The following protocol is given for the strain YAL6B, which has been successfully used for SILAC experiments *(2)*.

3.2.1. Preparation of Minimal Medium

1. 10×YNB stock solution: 67 g/L YNB, 0.55 g/L adenine, and 0.55 g/L tyrosine are dissolved in H_2O and autoclaved for 15 min at 121°C.
2. 20% glucose: 200 g/L glucose are dissolved in H_2O and autoclaved for 15 min at 121°C.

3. 100× amino acid mix stock solution: 1 g/L histidine:HCl, 6 g/L leucine, 1 g/L methionine, 6 g/L phenylalanine, and 4 g/L tryptophane are dissolved in H_2O and filtered through a 0.2 μm filter.
4. 50× uracil stock solution: 2.25 g/L uracil are dissolved in H_2O and sterile filtered. The solution is kept at 25 °C.
5. SILAC amino acid solutions: $^{13}C_6$-arginine and $^{13}C_6$-lysine are dissolved in sterile H_2O at a concentration of 100 mg/ml.
6. Unlabeled arginine and lysine stock solutions: 100 mg/ml lysine and 100 mg/ml arginine are dissolved separately in H_2O and filtered through a 0.2 μm filter.
7. All stock solutions (except uracil) should be kept refrigerated and handled under sterile conditions.
8. For 100 ml of minimal medium combine 10 ml 10× YNB, 10 ml 20% glucose, 1 ml 100× amino acid mix, 2 mL 50× uracil, and 77 ml H_2O. Thirty mg/L arginine and 30 mg/L lysine are added to the light (control) medium from the stock solutions. The heavy SILAC medium is supplemented with 30 mg/L $^{13}C_6$-arginine and 30 mg/L $^{13}C_6$-lysine from the stock solutions.

3.2.2. Cell Growth

1. Yeast cells are grown in Erlenmayer flasks on a shaker at 30 °C with 200 rpm (see Note 11).
2. A preculture is prepared by growing YAL6B yeast cells over night in unlabeled minimal medium or full medium.
3. 100 ml of light and heavy SILAC minimal media are inoculated with yeast cells from the preculture at an optical density $OD_{600} \leq 0.01$. The cells are grown until they reach mid-log phase $OD_{600} \approx 0.5–1$.
4. The cultures are now ready for differential treatments such as stimulation of cell signaling, heat shock, etc.
5. The OD_{600} of the two cell cultures is measured in order to combine equal numbers of cells from the unlabeled control and heavy SILAC cultures. If it is important to stop biological processes in the cells quickly, the cell suspensions from both cultures should be poured into a centrifuge bottle containing ice to cool them rapidly.
6. The combined cells from both cultures are harvested by centrifugation for 5 min at $1,600g$ at 4 °C and washed once with H_2O.
7. The cell pellet is now ready for cell lysis, protein extraction and analysis by mass spectrometry.

4. Notes

1. Other cell lines may require addition of other compounds such as sodium bicarbonate or glucose in order to assure normal growth. Especially some stem and primary cells and some blood cell types require higher concentrations of serum or the addition of growth factors, hormones and vitamins in order to assure

normal cell growth. These substances can be added prior to sterile filtration of the light and heavy SILAC media. For SILAC-based experiments any growth media can be used provided it is depleted of the specific labeled amino acid used (e.g., Minimum essential Medium (MEM), RPMI 1640, etc.).
2. Dialysis of the serum is necessary to remove amino acids. Other dialyzed serum types such as calf or horse serum can also be used.
3. It is recommended to use the filter units supplied by Nalge Nunc because they do not absorb any of the essential components of the media.
4. Double labeling with pairs of modified lysines and arginines can be employed for kinetic studies with several time points. In this case, L-lysine:HCl-4,4,5,5-d_4 (98 atom % D) and L-lysine:HCl ($^{13}C_6, ^{15}N_2$, 98%) (Isotec, Sigma-Aldrich, Miamisburg, OH, USA) can be used together with $^{13}C_6$-arginine and $^{13}C_6^{15}N_4$-arginine *(18)*. SILAC labeling using tyrosine (L-tyrosine (U-$^{13}C_9$, 98%) (Cambridge Isotope Laboratories, Andover, MA) for phosphorylation analysis has been reported as well *(19)*.
5. The arginine concentration has been reduced as compared to the media formulation (28 mg/L versus 84 mg/L in the original formula of DMEM). This has been done to avoid bioconversion of excess of labeled arginine into proline *(20)* and to save on reagent costs. For each cell line, the optimal concentration of labeled amino acids, where full incorporation is achieved without impairing the growth of the cells should be established prior to performing SILAC experiments. The efficiency of the incorporation of the employed labeled amino acids during the expansion should be monitored for each cell line before performing a SILAC experiment for the first time. In our laboratory we use different concentrations of the labeled amino acid (for example 16.8 mg/L, 21 mg/L, 28 mg/L, 42 mg/L, and 84 mg/L labeled arginine is tested for each cell line that is maintained in DMEM media) to grow and expand the cells and perform mass spectrometry-based analysis in order to estimate if there is a problem with bioconversion of $^{13}C_6$-arginine to $^{13}C_5$-proline or synthesis of endogenous arginine. Once the optimal concentration of arginine is determined we proceed with the planned experiments.
6. For media different than DMEM, the final concentrations of the amino acids should be adjusted according to the media formulation. For example the final concentration of arginine in RPMI media should be 240 mg/L.
7. It is recommended to prepare fresh light and heavy SILAC media containing dialyzed serum for each new experiment, because repeated heating of the serum could lead to the inactivation of the present growth factors and hormones.
8. The growth capabilities and growth rate of different cell lines in dialyzed serum must be tested and compared to the growth rate in media containing undialyzed serum.
9. For optimal labeling conditions, light and heavy SILAC cultures should be seeded from the same pool of cells.
10. Subsequent experiments might require a different degree of cell expansion and the culturing conditions (dilution of cells, etc.) can be adjusted accordingly.

11. The growth conditions of the yeast cells depend on the biological experiment, which might require high/low temperature incubations, stimulation of cells or the harvesting of cells at a certain growth phase and the protocol might therefore needed to be adapted. Important is, that the yeast cells are grown for at least five doubling times in SILAC medium to ensure complete labeling.

References

1. Blagoev, B., Kratchmarova, I., Ong, S.E., Nielsen, M., Foster, L.J. and Mann, M. (2003) A proteomics strategy to elucidate functional protein-protein interactions applied to EGF signaling. Nat Biotechnol. 21(3), 315–318.
2. Gruhler, A., Olsen, J.V., Shabaz, M., Mortensen, P., Faergman, N., Mann, M. and Jensen, O.N. (2005) Quantitative phosphoproteomics applied to the yeast pheromone signaling pathway. Molecular and Cellular Proteomics. 4(3), 310–27.
3. Ranish, J.A., Yi, E.C., Leslie, D.M., Purvine, S.O., Goodlett, D.R., Eng, J. and Aebersold, R. (2003) The study of macromolecular complexes by quantitative proteomics. Nat Genet. 33(3), 349–55.
4. Schulze, W. and Mann, M. (2004) A novel proteomic screen for peptide-protein interactions. J Biol Chem. 279(11), 10756–10764.
5. Dunkley, T.P., Watson, R., Griffin, J.L., Dupree, P. and Lilley, K.S. (2004) Localization of organelle proteins by isotope tagging (LOPIT). Mol Cell Proteomics. 3(11), 1128–34.
6. Andersen, J.S., Lam, Y.W., Leung, A.K., Ong, S.E., Lyon, C.E., Lamond, A.I. and Mann, M. (2005) Nucleolar proteome dynamics. Nature. 433(7021), 77–83.
7. Westermeier, R. and Marouga, R. (2005) Protein detection methods in proteomics research. Biosci Rep. 25(1–2), 19–32.
8. Ong, S.E. and Mann, M. (2005) Mass spectrometry-based proteomics turns quantitative. Nat Chem Biol. 1(5), 252–62.
9. Leitner, A. and Lindner, W. (2004) Current chemical tagging strategies for proteome analysis by mass spectrometry. J Chromatog B. 813(1–2), 1.
10. Krijgsveld, J., Ketting, R.F., Mahmoudi, T., Johansen, J., Artal-Sanz, M., Verrijzer, C.P., Plasterk, R.H. and Heck, A.J. (2003) Metabolic labeling of C. elegans and D. melanogaster for quantitative proteomics. Nat Biotechnol. 21(8), 927–31.
11. Lahm, H.W. and Langen, H. (2000) Mass spectrometry: a tool for the identification of proteins separated by gels. Electrophoresis. 21(11), 2105–14.
12. Oda, Y., Huang, K., Cross, F.R., Cowburn, D. and Chait, B.T. (1999) Accurate quantitation of protein expression and site-specific phosphorylation. PNAS. 96(12), 6591–6596.
13. Ong, S.E., Blagoev, B., Kratchmarova, I., Kristensen, D.B., Steen, H., Pandey, A. and Mann, M. (2002) Stable Isotope Labeling by Amino Acids in Cell Culture, SILAC, as a Simple and Accurate Approach to Expression Proteomics. Mol Cell Proteomics. 1(5), 376–86.

14. Zhu, H., Pan, S., Gu, S., Bradbury, E.M. and Chen, X. (2002) Amino acid residue specific stable isotope labeling for quantiative proteomics. Rapid Communications in Mass Spectrometry. 16(22), 2115–2123.
15. Gruhler, A., Schulze, W.X., Matthiesen, R., Mann, M. and Jensen, O.N. (2005) Stable Isotope Labeling of Arabidopsis thaliana Cells and Quantitative Proteomics by Mass Spectrometry. Mol Cell Proteomics. 4(11), 1697–1709.
16. Blagoev, B., Ong, S.E., Kratchmarova, I. and Mann, M. (2004) Temporal analysis of phosphotyrosine-dependent signaling networks by quantitative proteomics. Nat Biotechnol. 22(9), 1139–45.
17. Kratchmarova, I., Blagoev, B., Haack-Sorensen, M., Kassem, M. and Mann, M. (2005) Mechanism of divergent growth factor effects in mesenchymal stem cell differentiation. Science. 308(5727), 1472–7.
18. Olsen, J.V., Blagoev, B., Gnad, F., Macek, B., Kumar, C., Mortensen, P. and Mann, M. (2006) Global, in vivo, and site-specific phosphorylation dynamics in signaling networks. Cell, 127, 635–648.
19. Amanchy, R., Kalume, D.E., Iwahori, A., Zhong, J. and Pandey, A. (2005) Phosphoproteome analysis of HeLa cells using stable isotope labeling with amino acids in cell culture (SILAC). J Proteome Res. 4(5), 1661–71.
20. Ong, S.E., Kratchmarova, I. and Mann, M. (2003) Properties of 13C-substituted arginine in stable isotope labeling by amino acids in cell culture (SILAC). J Proteome Res. 2(2), 173–81.

10

ICPL—Isotope-Coded Protein Label

Josef Kellermann

Summary

Stable isotope labeling in combination with mass spectrometry has emerged as a powerful tool to identify and relatively quantify thousands of proteins within complex protein mixtures. Here we describe a method, termed isotope-coded protein label (ICPL), which is capable of high-throughput quantitative proteome profiling on a global scale. Because ICPL is based on stable isotope tagging at the frequent free amino groups of isolated intact proteins, it is applicable to any protein sample, including extracts from tissues or body fluids, and compatible to all separation methods currently employed in proteome studies. The method shows highly accurate and reproducible quantification of proteins and yields high sequence coverage, indispensable for the detection of post-translational modifications and protein isoforms.

Key Words: High-throughput; isotope-coded protein label; mass spectrometry; quantitative proteomics; two-dimensional gel electrophoresis.

1. Introduction

One key focus in proteome research is the determination of changes in protein expression and their modifications. Two-dimensional gel electrophoresis (2-DE) followed by mass-spectrometry (MS) based techniques like peptide mapping or MS-sequencing (MS/MS) is still the most popular approach to relatively quantify and identify proteins within complex mixtures *(1,2)*. Despite its high resolving power, 2-DE suffers from some limitations *(3,4)* that confine high-throughput proteome analysis. This can be primarily ascribed to the difficult automation of the workflow and the rare detection of low-abundant or membrane proteins. Labeling of the proteins with fluorescent dyes, as introduced during the DIGE technology, partly overcomes the limitations of the 2-DE *(5)*. In 1999 Gygi and colleagues *(6)* have introduced a new approach based on stable isotope labeling of proteins using isotope coded affinity tags (ICAT), which has been shown to overcome some of the drawbacks mentioned above and thus represents a powerful alternative to 2-DE. The isotope labeling methods shift the quantification problem from image analysis to MS based quantification. Since that time, many groups have adopted the principle of this strategy, generating different approaches with their own strengths and weaknesses *(7)*. These techniques are based on differential isotopic labeling of proteins or peptides derived from two cell states (e.g., healthy and tumor cells) with either light or heavy tags. In the latter case, hydrogen atoms are exchanged by deuterium increasing the M_r by one Dalton *per* incorporated isotope. More recently, other isotopes like 13C or 15N have been employed as well, avoiding different elution times during reverse-phase LC and assuring more accurate quantification *(8,9)*. After modification, both samples are combined and analyzed using multidimensional LC followed by MS/MS. Because isotopes have identical physico-chemical properties, the light labeled peptides elute with their heavy counterparts after the separation steps simultaneously into the mass spectrometer. Quantitative analysis is then performed by comparing the relative signal intensities of the light and heavy labeled peptide in the MS-spectra. Finally, the labeled peptides are identified by MS/MS analysis followed by protein database searching *(10,11)*. With the ICAT approach *(6,12)*, a very elegant strategy was developed that is based on differential isotope labeling of the rare cysteine residues in proteins with biotin containing tags, which allow the isolation of modified peptides by avidin affinity chromatography. This step reduces complexity by 10-fold and simplifies the separation effort *(12)* but on the other hand reduces the sequence information of proteins by almost the same factor. Recent 2-DE/MS studies demonstrate that single gene products result in an average of 10–15 different spots, mostly derived from posttranslational modification (PTMs) *(13,14)*. To characterize these modifications, which have a great influence on protein activity and biological function, high sequence information of the modified proteins is essential *(15)*. Other isotope labeling techniques termed global internal standard technology (GIST) approaches *(16,17)* allow isotopic tagging of all peptides obtained after separate enzymatic cleavage of the two protein samples. Recently a higher multiplexing GIST technology was introduced. The core of this iTRAQ methodology *(18)* is a multiplexed set of four isobaric reagents that yield amine-derivatized peptides. These peptides are indistinguishable in MS, but exhibit low-mass MS/MS signature ions that support quantitation. This comprises one severe drawback of this technology, because for quantification MS/MS spectra of each peptide are needed. Although almost every peptide is modified by the GIST technologies, the successful separation of complex proteomes obtained from higher organisms may be critical, because all reduction of complexity is limited to the peptide level.

Here we describe an approach *(19)*, termed isotope coded protein label (ICPL), which is based on isotopic labeling of all free amino groups in proteins. This method provides highly accurate and reproducible quantitation, high protein sequence information, including PTMs and isoforms, and is compatibility to all commonly used protein and peptide separation techniques. A *N*-nicotinoyloxy-succinimide (Nic-NHS) (Fig. 1) isotope containing label has been developed. Thereby six 12C are replaced by 13C isotopes. For higher multiplexing a third label can be introduced, where four H are replaced by their deuterium isotope.

The complete workflow is shown in Fig. 2. Two protein mixtures obtained from two distinct cell states or tissues are individually reduced and alkylated to denature the proteins and to ensure easier access to free amino groups that are subsequently derivatized with the 12C (light) or 13C containing (heavy) form, respectively, of the ICPL reagent. After combining both mixtures, any separation method can be adopted to reduce the complexity of the sample on the protein level and, after digestion, on the peptide level followed by high throughput MS and MS/MS. Because peptides of identical sequence derived from the two differentially labeled protein samples differ in mass, they appear as doublets in the acquired MS-spectra. From the ratios of the ion intensities of these sister peptide pairs, the relative abundance of their parent proteins in the original samples can be determined. Finally, proteins are identified either

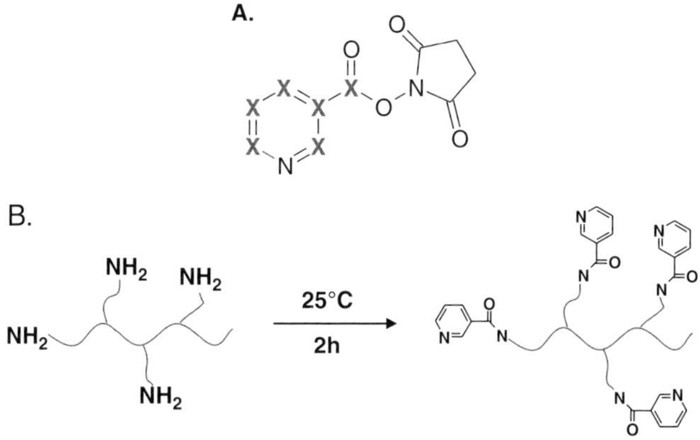

Fig. 1. Molecular structure (**A**) and reaction scheme (**B**) of the ICPL-label. X represents the $^{12}C/^{13}C$ isotopes.

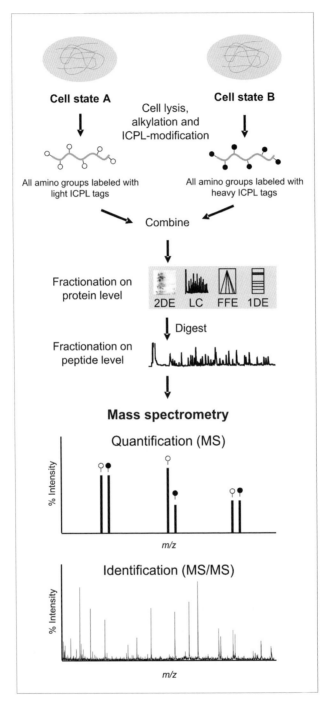

Fig. 2. Workflow of the ICPL reagent technology.

by peptide mass fingerprint (PMF) or collision induced dissociation (CID) of single peptides followed by correlation with sequence databases using highly sophisticated search algorithms.

2. Materials

2.1. Sample Preparation

1. Lysis-Buffer: 6 M guanidinium hydrochloride, 0.1 M HEPES, pH 8.5 (Serva Electrophoresis, Heidelberg, Germany).
2. The protein concentrations of the samples are determined either by Bradford (BioRad Laboratories, Munich, Germany) or the ProteoQuant Method (Serva Electrophoresis, Heidelberg, Germany).

2.2. Reduction and Alkylation of Cysteine Residues (Carbamidomethylation)

1. Reduction solution: 0.2 M TCEP in 0.1 M HEPES, pH 8.5 (Serva Electrophoresis, Heidelberg, Germany).
2. Alkylation solution: 0.4 M iodoacetamide in 0.1 M HEPES, pH 8.5. Prepare fresh every time.
3. Stop solution: 0.5 M *N*-acetyl-cysteine in 0.1 M HEPES, pH 8.5.

2.3. Isotope Labeling of the Protein Samples

1. A 0.15 M solution of each derivative of the *N*-hydroxysuccinimide-ester (Serva Electrophoresis, Heidelberg, Germany) in dimethylsulfoxide (DMSO) is prepared.
2. Stop solution: 1.5 M hydroxylamine-HCl, pH 8.5 (Serva Electrophoresis, Heidelberg, Germany).
3. To adjust the pH of the reaction mixture 2N NaOH and 2N HCl is used, respectively.

2.4. Purification of the Labeled Proteins by Acetone Precipitation

1. Acetone, cooled to -20°C.
2. 80% acetone/20% water, cooled to -20°C.

2.5. Enzymatic Digestion of the Labeled Proteins

1. Buffer for tryptic cleavage: 25 mM Tris-HCl, pH 8.5, 4 M urea.
2. Buffer for cleavage with endoproteinase Glu C: 25 mM Tris-HCl, pH 7.8, 4 M urea.
3. Trypsin (Roche Diagnostics, Penzberg, Germany).
4. Endoproteinase Glu C (Roche Diagnostics, Penzberg, Germany).

3. Methods

3.1. Sample Preparation

1. Dissolve 100 µg of protein in 20 µl lysis buffer (see Note 1) and vortex for 2 min.
2. Incubate sample with gentle agitation for 20 min. at 25°C.
3. Vortex 2 min and sonicate 4 times for 30 sec. in ultrasound bath (cool sample during pauses in ice water for 2 min.).
4. Vortex 1 min and incubate sample with gentle agitation for 15 min at 25°C.
5. Vortex 2 min and spin sample for 30 min. at 100,000g.
6. Use the supernatant directly for protein assay and further analysis.
7. Adjust protein concentration to 5 mg/ml with lysis buffer before you proceed with the labeling protocol (see Note 1).

3.2. Reduction and Alkylation of Cysteine Residues (Carbamidomethylation)

1. Check the pH of the sample buffer and if necessary, adjust to 8.5 ± 0.1 by addition of HCl or NaOH! A micro pH-meter electrode is recommended. Before every measurement, the tip of the electrode is rinsed with distilled water and dried very carefully with a dry, dust-free tissue to avoid sample dilution.
2. The carbamidomethylation protocol is identical for both, sample A and sample B.
3. Add 0.5 µl reduction solution to 20 µl sample solution (equivalent to 100 µg protein) and reduce proteins for 30 min at 60°C.
4. Cool sample to room temperature and spin down condensed solution from the lid.
5. Add 0.5 µl of this freshly prepared alkylation reagent to each sample, wrap samples quickly in aluminium foil for light protection and leave samples for 30 min at 25°C.
6. Stop reaction by adding 0.5 µl stop solution 1 to each sample and incubate for 15 min at 25°C.

3.3. Isotope Labeling of the Protein Samples

1. After carbamidomethylation add 3 µl of ^{12}C-Nic-reagent solution to sample A and 3 µl of ^{13}C-Nic-reagent solution to sample B.
2. Overlay both samples with argon (or equivalent) to exclude oxidation, vortex (10 sec.) and sonicate for 1 min in ultrasound bath. Spin down samples.
3. Incubate samples for 2 h at 25°C.
4. Add 2 µl of stop solution to each sample and shake for 20 min at 25°C to destroy excess reagent.
5. Combine both ICPL labeled samples and vortex thoroughly.
6. Adjust the pH of the mixture to 11.9 ± 0.1 by adding 2N NaOH (about 2 µl for 2 × 20 µl sample volume) to destroy possible esterification products. After 20 min add the same amount of 2N HCl to neutralize sample (usually it is not necessary to check the pH).

3.4. Purification of the Labeled Proteins by Acetone Precipitation

1. Add the equal amount of distilled water to your sample, i.e., about 57 µl (= 2 × (20 µl sample + 1.5 µl reduction + 3 µl reagent + 2 µl stop sol.) + 4 µl pH adjustment).
2. Add fivefold excess (related to total volume of sample and water) of ice-cold acetone to your sample, i.e., 570 µl acetone (= 114 µl × 5) and leave sample at -20 °C overnight.
3. Spin down precipitated proteins at 100,000g for 30 min at 4°C.
4. Discard supernatant.
5. Overlay precipitated proteins with approximately 100–200 µl ice-cold 80% acetone, shake carefully in your hands and spin down again at 100,000g for 5 min at 4°C.
6. Discard supernatant and let the remaining acetone evaporate at room temperature (leave the cup open).
7. The samples can be stored now at -80°C or directly dissolved in appropriate buffers for protein separation (1-D, 2-DE, free flow electrophoresis or chromatography, see Note 2).

3.5. Enzymatic Digestion of the Labeled Proteins for Direct MS-Analysis

Enzymatic digestions of the labeled samples are performed according to common protocols. We recommend using trypsin or endoproteinase Glu C as enzymes.

1. Dissolve sample in 20 µl 25 mM Tris-HCl, pH 8.5, 4 M urea for tryptic cleavage or pH 7.8 for cleavage with endoproteinase Glu C.
2. Add enzyme in a protein/enzyme ratio of 50:1 for trypsin or 30:1 for endoproteinase Glu C. Incubate sample for 4 h at 37°C.
3. After digestion, the samples can be directly analyzed by mass spectrometry, preferentially by LC-MALDI-TOF-/TOF or LC-ESI-MS-/MS.
4. The ratios of the proteins are calculated by the ratio of the peak areas. When using mass spectrometer of low resolution, the calculation should be done by peak-height.
5. After the modification of the lysine (K) residues by ICPL, lysine is protected against proteolytic digestion. Trypsin therefore only cleaves C-terminal of arginine (R). For this reason database searches should be done using endoproteinase Arg C as enzyme entry (Endoproteinase Lys C can not be used at all!).
6. For Mascot searches, the ICPL modified residues have to be added to the modification file (mod_file) on the local Mascot server (see Note 3). As an example MS spectra of a virtual proteome of four different proteins in various amounts are shown in Fig. 3. The protein composition and amount is shown in Table 1.

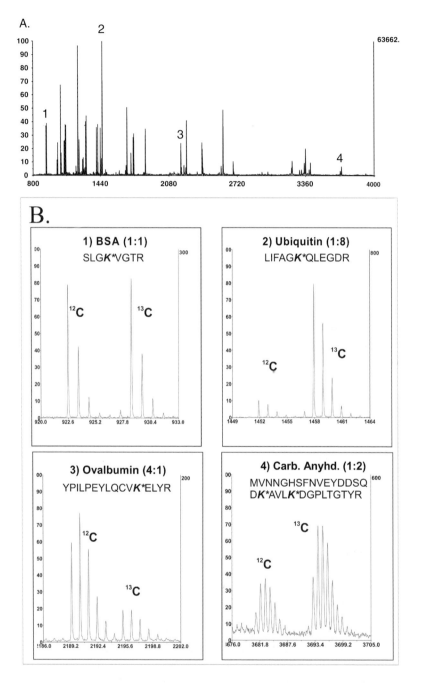

Fig. 3. MS spectra of a tryptic digest of four standard proteins (**A**) present in different amounts in proteinmix A (^{12}C) and proteinmix B (^{13}C). Zoom view (**B**) of four isotopic pairs as example for each protein. For more details see text.

Table 1
Protein composition of proteinmix A and B

Protein	Proteinmix A	Proteinmix B	Relative amount
BSA	13.5 μg	13.5 μg	1:1
Ovalbumin	24.0 μg	6.0 μg	4:1
Carbonic anhydrase	12.0 μg	24.0 μg	1:2
Ubiquitine	0.5 μg	4.0 μg	1:8

4. Notes

1. The protocol is optimized for a protein concentration of 5 mg/ml. However, it works as well with protein concentrations of 2.5 mg/ml. As the recovery rate of the protein precipitation step below depends strongly on the total protein concentration, losses are likely when working with lower protein concentrations. Therefore, it is extremely important, to keep the concentrations of the reagents strictly as recommended. If you want to work with increased sample volumes of 40 μl (for example to facilitate the pH measurement), you rather have to double the sample amount and also have to double volumes of the reagents given in this protocol!

2. The further protocol depends on the complexity of the used sample. Complex proteome samples should be separated to a convenient complexity by any protein fractionation method (1-DE, 2-DE, free flow electrophoresis, chromatography or any combination of these techniques) before MS analysis. In combination with 2-D electrophoresis the ICPL technology provides some improvements over current 2-DE/MS technologies. The multiplexing of several proteome states allows for the simultaneously separation of differentially labeled protein samples in the same gel. Therefore problems of electrophoretic variations between gels are avoided and protein quantification is more accurate and confident. The modification of the basic amino groups with Nic-NHS changes the migration behaviour of labeled basic proteins during 2-DE towards the more acidic side, making extreme basic proteins more accessible for analysis. Therefore pH-ranges for isoelectric focussing from 3-6 should be used.

   The protein mixture also can be cleaved directly by trypsin. For a better solubility, we recommend to dissolve the acetone precipitate using a buffer containing 4 M urea. This urea concentration is tolerated by trypsin as well as endoproteinase Glu C (check the enzyme data sheet).

3. The following four entries ($^{12}C/^{13}C$) have to be made (titles can be chosen arbitrarily): (i) Title: ICPL_heavy; Residues: K 239.13680 239.22440 (ii) Title: ICPL_light; Residues: K 233.11640 233.27000 (iii) Title: ICPL-heavy (N-term); ProteinNterm: 112.04970 112.05830 (iv) Title: ICPL-light (N-term); Protein-Nterm: 106.02930 106.10390. Working with Mascot version 2.1 or below, all four modifications have to be selected as variable modifications.

Working with Mascot version 2.2 the light modifications have to be selected fixed, heavy as variable modifications: (i) Carbamidomethylation has to be defined as fixed modification (ii) The mass difference of ^{12}C- and ^{13}C-labeled peptides is 6.0204 Da per labeled amino group. (iii) The mass difference between labeled (^{12}C/^{13}C) and unlabeled peptides are 105.0215Da/111.0419 Da for each modified amino group.

The H/L ratios of the proteins are calculated as average or median values of the H/L ratios of the individual peptides. In general, the calculation should be done by peak-height of the entire isotopic cluster as enabled in the respective quantification software of the instrument manufacturer. Some instruments may provide superior data if peak areas are used as quantification measure, though.

References

1. Klose,J., Nock,C., Herrmann,M., Stuhler,K., Marcus,K., Bluggel,M., Krause,E., Schalkwyk,L.C., Rastan,S., Brown,S.D., Bussow,K., Himmelbauer,H. & Lehrach,H. (2002) Genetic analysis of the mouse brain proteome. *Nat. Genet.* **30**, 385–393.
2. Wildgruber,R., Reil,G., Drews,O., Parlar,H. & Gorg,A. (2002) Web-based two-dimensional database of Saccharomyces cerevisiae proteins using immobilized pH gradients from pH 6 to pH 12 and matrix-assisted laser desorption/ionization-time of flight mass spectrometry. *Proteomics* **2**, 727–732.
3. Hoving,S., Gerrits,B., Voshol,H., Muller,D., Roberts,R.C. & van Oostrum,J. (2002) Preparative two-dimensional gel electrophoresis at alkaline pH using narrow range immobilized pH gradients. *Proteomics* **2**, 127–134.
4. Patton,W.F., Schulenberg,B. & Steinberg,T.H. (2002) Two-dimensional gel electrophoresis; better than a poke in the ICAT? *Current Opinion in Biotechnology* **13**, 321–328.
5. Unlu, M., Morgan, M. E., and Minden, J. S. (1997) Difference gel electrophoresis: a single gel method fordetecting changes in protein extracts. *Electrophoresis.* 18, 2071–2077.
6. Gygi, S. P., Rist, B., Gerber, S. A., Turecek, F., Gelb,M.H., Aebersold,R., (1999). Quantitative analysis of complex protein mixtures using isotope-coded affinity tags. *Nat. Biotechnol.*, **17**, 994–999.
7. Moritz, B., Meyer, H. E., (2003) Approaches for the quantification of protein concentration ratios. *Proteomics*, *3*, 2208–2220.
8. Krijgsveld,J., Ketting,R.F., Mahmoudi,T., Johansen,J., Artal-Sanz,M., Verrijzer,C.P., Plasterk,R.H. & Heck,A.J. (2003) Metabolic labeling of C. elegans and D. melanogaster for quantitative proteomics. *Nat. Biotechnol.* **21**, 927–931.
9. Zhang,R., Sioma,C.S., Thompson,R.A., Xiong,L. & Regnier,F.E. (2002) Controlling deuterium isotope effects in comparative proteomics. *Anal. Chem.* **74**, 3662–3669.
10. Washburn,M.P., Wolters,D. & Yates,J.R. (2001) Large-scale analysis of the yeast proteome by multidimensional protein identification technology. *Nat. Biotechnol.* **19**, 242–247.

11. Yates,J.R., III, Eng,J.K., McCormack,A.L. & Schieltz,D. (1995) Method to correlate tandem mass spectra of modified peptides to amino acid sequences in the protein database. *Anal. Chem.* **67**, 1426–1436.
12. Gygi,S.P., Rist,B., Griffin,T.J., Eng,J. & Aebersold,R. (2002) Proteome analysis of low-abundance proteins using multidimensional chromatography and isotope-coded affinity tags. *J. Proteome Res.* **1**, 47–54.
13. Fountoulakis,M., Juranville,J.F., Berndt,P., Langen,H. & Suter,L. (2001) Two-dimensional database of mouse liver proteins. An update. *Electrophoresis* **22**, 1747–1763.
14. Fountoulakis,M., Berndt,P., Langen,H. & Suter,L. (2002) The rat liver mitochondrial proteins. *Electrophoresis* **23**, 311–328.
15. Mann,M. & Jensen,O.N. (2003) Proteomic analysis of post-translational modifications. *Nat. Biotechnol.* **21**, 255–261.
16. Chakraborty,A. & Regnier,F.E. (2002) Global internal standard technology for comparative proteomics. *J. Chromatogr. A* **949**, 173–184.
17. Goodlett,D.R., Keller,A., Watts,J.D., Newitt,R., Yi,E.C., Purvine,S., Eng,J.K., von Haller,P., Aebersold,R. & Kolker,E. (2001) Differential stable isotope labeling of peptides for quantitation and de novo sequence derivation. *Rapid Commun. Mass Spectrom.* **15**, 1214–1221.
18. Philip L. Ross,P.L., Huang, Y.N., Marchese, J.N., Williamson, B., Parker, K., Hattan, S., Khainovski, N., Pillai, S., Dey, S., Daniels, S., Purkayastha, S., Juhasz, P., Martin, S., Bartlet-Jones, M., He, F., Jacobson, A. and Pappin, D.J. (2004) Multiplexed protein quantitation in *Saccharomyces cerevisiae* using amine-reactive isobaric tagging reagents. *Mol. Cell. Proteomics* 3, 12, 1154–1169.
19. Schmidt,A., Kellermann,J. & Lottspeich,F. (2005) A novel strategy for quantitative proteomics using isotope-coded protein labels. *Proteomics* **5**, 4–15.

11

Radiolabeling for Two-Dimensional Gel Analysis

Hélian Boucherie, Aurélie Massoni, and Christelle Monribot-Espagne

Summary

Radiolabeling is a highly sensitive method for protein detection, which is easily performed by the incorporation of radioactive amino acids into proteins. This makes radiolabeling a method of choice for visualizing proteins separated on two-dimensional (2-D) gels. This chapter presents protocols to determine *in vivo* labeling conditions and to label proteins for the comparison of protein samples by means of 2-D gel electrophoresis.

Key Words: [^{35}S]-Methionine labeling; protein synthesis; proteome analysis; radiolabeling; two-dimensional gel electrophoresis.

1. Introduction

Radiolabeling is a method of choice to visualize and quantify proteins separated on two-dimensional (2-D) gels. The procedure for radiolabeling proteins is simple: it is obtained by *in vivo* incorporation of radioactive amino acids into proteins and only requires the addition of radioactive amino acids in the culture medium. In addition it is a very sensitive method of detection, the more sensitive method currently available. Finally, combined with the storage phosphor screen technology, radiolabeling makes it possible to obtain quantitative data with a linear dynamic range of four orders of magnitude.

Radiolabeling differs from all the other methods used for detecting proteins on 2-D gels (e.g., Coomassie Blue, silver staining, and fluorescent dyes) by the fact that it is the only method where proteins are labeled *in vivo*. The other methods rely on the staining of proteins after extraction. Hence, whereas the later procedures are limited to the description of cellular protein content, radiolabeling raises the possibility to visualize the proteins synthesized during a short period of time. This is of particular interest for investigating the cellular response to environmental changes or the reprogramming of genome expression. Under these conditions the pattern of the proteins newly synthesized may be markedly different from the pattern of the proteins accumulated before the change (*see* **Fig. 1**). Taken together, three different levels of proteome investigation are accessible by radiolabeling depending on the labeling conditions: protein synthesis (pulse labeling), protein content (long term labeling), and protein degradation (pulse labeling followed by a chase).

The main purpose of radiolabeling is the characterization of differences between protein samples. In conventional methodology, the protein samples to be compared are labeled with the same radioisotope and separated independently on different gels. The protein spots are detected and quantified on each gel; then identical spots are matched between gels for comparison of their intensity in each sample. Because radioisotopes such as ^{3}H and ^{14}C, or ^{3}H and ^{35}S, can be easily differentiated, it is also possible to label the samples to be compared with different radioisotopes. Then the samples can be submitted to comigration on the same gel and the ratio of the two isotopes determined. Two different methods are used to determine the isotope ratio, scintillation counting (*see* for example (*1*)) and differential gel exposure (*2*).

Radiolabeling can be used to investigate the proteome of a wide variety of species. It is particularly suitable to study microorganisms. Hence it has been extensively used for proteomic studies of model organisms such as *Escherichia*

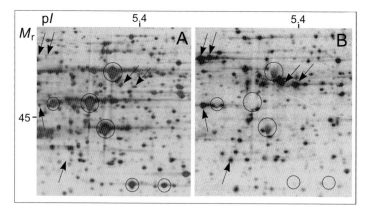

Fig. 1. Comparison of *in vivo* protein content and protein synthesis during the diauxic shift of yeast cells. Yeast cells (S288C strain) were labeled with L-[^{35}S]-methionine (500 Ci/mmol; 300 µCi/mL) 15–30 min after glucose exhaustion from the culture medium. After labeling proteins were extracted and separated on a 2-D gel. The gel was silver stained for detection of *in vivo* protein content, and exposed to phosphor screen for detection of proteins synthesized after glucose exhaustion. (**A**) Silver stained proteins. (**B**) Radiolabeled proteins.

coli *(3)*, *Saccharomyces cerevisiae (4)*, or *Bacillus subtilis (5)*. The interest of radiolabeling is not limited to unicellular organisms. In plants it has been used to study the proteins synthesized during seed germination *(6)* or seedling development *(7)*. In mammals, it can be used to label proteins from various cell lines *(8)* or to label ex vivo biopsies *(9)*.

In this chapter we describe the procedures for determining labeling conditions and for labeling proteins in view of 2-D gel analysis. Once the proteins are labeled they can be extracted and separated on 2-D gels according to any standard protocols. We have limited the description of procedures to those required for the conventional methodology where samples to be compared are labeled with the same radioisotope. Nevertheless these procedures remain valuable for dual-labeling investigations.

2. Materials

See Note 1 for general comments on the Materials section.

2.1. Determining labeling conditions

1. Culture medium: for radiolabeling, cells are grown in a defined minimal medium devoid of the amino acid used for the labeling (*see* Note 2). Amino acids and bases are added as required if using auxotrophic strains.
2. Choice of the labeling amino acid: we generally use L-[^{35}S]-methionine (specific activity 1,000 Ci/mmol, 10 µCi/µL) (*see* Note 3). L-[^{35}S]-methionine is stored as single use aliquots (60 µl) at -80°C.
3. Unlabeled L-methionine stock solutions (3.3×0^{-4} M and 6×10^{-5} M). The solutions are sterilized by filtering through 0.45-µm pore sterile filters (Millipore, Bedford, MA), and kept at 4°C.
4. L-[^{35}S]-methionine solutions for pulse labeling test (1.25×10^{-5} M, 2.5×10^{-5} M, and 5×10^{-5} M; molar specific activity 40 Ci/mmol). To prepare 5×10^{-5} M [^{35}S]-methionine test solution with a specific activity of 40 Ci/mmol mix commercialized [^{35}S]-methionine solution (1,000 Ci/mmol, 10 µCi/µL) with 6×10^{-5} M unlabeled methionine stock solution in a ratio 1:4 (v/v), respectively. The 1.25×10^{-5} M and 2.5×10^{-5} M [^{35}S]-methionine test solutions (specific activity 40 Ci/mmol) are obtained by appropriate dilution of 5×10^{-5} M test solution, using sterile deionized water. The solutions can be stored for a few days at 4°C.
5. L-[^{35}S]-methionine labeling solutions for long term labeling test (0.5×10^{-4} M, 10^{-4} M, and 2.5×10^{-4} M; specific activity 10 Ci/mmol). Prepare 2.5×10^{-4} M [^{35}S]-methionine test solution with a specific radioactivity of 10 Ci/mL by mixing commercialized [^{35}S]-methionine solution with 3.3×10^{-4} M unlabeled methionine stock solution in a ratio 1:3 (v/v), respectively. The 10^{-4} M and 0.5×10^{-4} M [^{35}S]-methionine test solutions (specific activity 10 Ci/mmol) are obtained by appropriate dilution of 5×10^{-5} M test solution, using sterile deionized water.

5. Bovine serum albumin (BSA) solution 3 mg/L.
6. 10-ml glass tubes containing 100 μl of BSA solution. Prepare just before use. Keep on ice.
7. 5% (w/v) TCA containing 1g/L of methionine.
8. Microfiber filters (GF/C, Whatman, Maidstone, UK) incubated in 5% (w/v) TCA containing 1g/L of methionine.
9. 95% ethanol.
10. Liquid scintillation: Ready Value (Beckman Coulter, Fullerton, CA).
11. Counting vials.
12. Equipment: water bath at 90°C, filter apparatus, β counter.

2.2. Labeling Cells for 2-D Gel Analysis

1. Culture medium: see Subheading 11.2.1.1.
2. L-[^{35}S]-methionine labeling solution: depending on the results of the labeling-test experiments described in Subheading 11.3.1., use pure commercialized L-[^{35}S]-methionine solution (1,000 Ci/mmol, 10 μCi/μL) or prepare just before use a [^{35}S]-methionine labeling solution of the molarity and specific activity determined from the labeling test (see Subheading 11.3.1.2. for preparing the labeling solution).
3. 20 ml sterile plastic tubes sealed with cotton plug.
4. 5% (w/v) TCA containing 1g/L of methionine.
5. Microfiber filters (GF/C, Whatman) incubated in 5% (w/v) TCA containing 1g/L of methionine.
6. Liquid scintillation. Ready Value (Beckman Coulter).
7. Counting vials.
8. Bio-Rad Protein Assay (Bio-Rad, Hercules, CA).

2.3. Visualization of Radiolabeled Proteins on 2-D gels

1. Fixing solution 1: 50% (v/v) ethanol, 7.5% (v/v) acetic acid in distilled water.
2. Fixing solution 2: 25% (v/v) ethanol, 2.5% (v/v) acetic acid in distilled water.
3. 3MM paper (Whatman).
4. Saran wrap.
5. Equipment: slab gel dryer, phosphor screens, storage phosphor imaging system.

3. Methods

3.1. Determining Labeling Conditions

The efficiency of radiolabeling is dependent upon two parameters, the concentration of the radiolabeled amino acid and its molar specific activity (see Note 4). Concentration must be high enough to allow linear incorporation of the labeled amino acid during the labeling period (i.e., the labeled amino acid should not be limiting). Specific activity must be sufficient to permit enough radioactivity incorporation for protein detection. These parameters are determined under the culture conditions used for the labeling experiment. We describe here a procedure based on the use of L-[^{35}S]-methionine, which is the most common labeled amino acid used for protein labeling. The procedure is described for labeling proteins from microorganisms. The rationale remains the same when using other radiolabeled amino acids and the procedure can be easily adapted to the labeling of any type of organism.

3.1.1. Determining Concentration of the Labeling Solution

The optimum concentration is determined by following the time course of radioactivity incorporation under different concentrations of labeled [^{35}S] methionine (*see* Note 5). The incorporation is followed during a period of time exceeding the labeling period planned for the experiment. This labeling period may last from a few minutes, when investigating protein synthesis, to hours when investigating protein accumulation.

For each labeling condition investigated:

1. Culture is performed in 500 ml Erlenmeyer flasks containing 50 ml of medium. Growth is monitored at 600 nm.
2. Transfer 1 ml sample solution of appropriate cell density (*see* Note 6) to a 20 ml sterile plastic tube and add 5–20 µl of [^{35}S]-methionine labeling-test solution as defined below.
3. Immediately vortex and harvest 20 µl which are transferred to a glass tube containing 100 µl of BSA. Add 2.5 ml of 5% TCA with methionine (0 time) and vortex. Duplicate the 0 time point.
4. Seal the tube with a cotton plug and put it immediately under the same conditions as the starting culture (temperature, rotary shaking).
5. Harvest 20 µl samples periodically and transfer to a glass tube containing 100 µl of BSA. Add 2.5 ml of 5% cold TCA with methionine and vortex. "Periodically" means every 5 min over a 20 min period for a pulse labeling test or every 20 min over a period covering at least 1.5 generation time for long-term labeling test. Duplicate each time point.
6. Keep on ice for at least 30 min.
7. Incubate the tubes in a water bath at 90°C for 10 min.
8. Keep on ice for at least 30 min.
9. Filter the content of each tube on GF/C filters incubated in 5% (w/v) TCA containing 1g/L of methionine.
10. Rinse the tube twice with 2.5 ml of 5% (w/v) TCA containing 1g/L of methionine.
11. Rinse the filter with 2.5 ml of 95% ethanol.
12. Let filters dry at room temperature for 30 min.
13. Place the dried filters in a counting vial, add 5 ml of liquid scintillation and count in a β-counter.

To determine the optimum conditions for pulse labeling experiments, test first the following final methionine concentrations: 0.5×10^{-7} M, 10^{-7} M, and 1.5×10^{-7} M. They are obtained by adding 5 μl, 10 μl, and 15 μl of 10^{-5} M [^{35}S]-methionine test solution to 1-ml sample of culture, respectively (*see* Note 7). If no linear incorporation of methionine is observed under any of these conditions, continue with the following final methionine concentrations: 2×10^{-7} M, 5×10^{-7} M, and 10^{-6} M. They are obtained by adding 20 μl of 1.25×10^{-5} M, 2.5×10^{-5} M, and 5×10^{-5} M labeling-test solution, respectively, to 1-ml sample of culture.

To determine conditions for long term labeling experiments test the following final methionine concentrations: 10^{-6} M, 2×10^{-6} M, and 5×10^{-6} M. They are obtained by adding 20 μl of 0.5×10^{-4} M, 10^{-4} M, and 2.5×10^{-4} M [^{35}S]-methionine test solutions to 1-ml sample of culture, respectively.

The optimum concentration is the lower concentration for which a linear incorporation of radiolabeled methionine is observed. Examples of [^{35}S]-methionine incorporation by exponentially growing yeast cells under different methionine concentrations are shown in **Fig. 2**.

3.1.2. Determining Molar Specific Activity of the Labeling Solution

If the optimum concentration requires the use of a labeling solution with a molarity higher than the molarity of the commercialized [^{35}S] methionine, it is

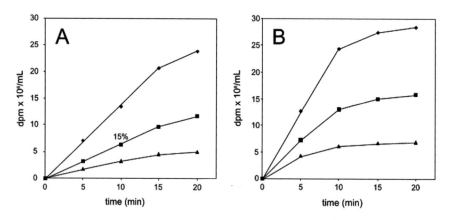

Fig. 2. Time course of L-[^{35}S]-methionine incorporation of yeast cells labeled at two different stages of growth, under different [^{35}S]-methionine concentrations. Yeast cells (strain S288C) were grown in YNB glucose medium at 22 °C and were labeled with [^{35}S]-methionine (40 Ci/mmol) when culture reached an optical density of 1 (**A**) or 3 (**B**). Concentration of [^{35}S]-methionine was 2.5×10^{-7} M (▲), 5×10^{-7} M (■), and 10^{-6} M (♦). The data are the average from two measurements. The indicated percentage corresponds to the percent of the added radioactivity incorporated under condition 5×10^{-7} M after 10 min.

necessary to perform an isotopic dilution of the supplied [^{35}S]-methionine (*see* Note 7 *and* 8). The molar specific activity of the labeling solution to be used is determined as follows:

1. Calculate the percent of radioactivity incorporated during the labeling period under the selected condition.
2. From this percentage, estimate the radioactive concentration (μCi/mL of culture) necessary to incorporate enough radioactivity for protein detection.
3. The specific activity of the labeling solution is calculated by combining the information on methionine concentration and on radioactive concentration.

See Note 9 for calculating the specific activity of the labeling solution and Note 10 for preparing the labeling solution.

3.2. Labeling Cells for 2-D Gel Analysis

1. Cultures are performed in 500 ml Erlenmeyer flasks containing 50 ml of medium.
2. When cultures reach appropriate cell density, corresponding generally to mid-log phase, transfer a 1-ml sample to a 20-ml sterile plastic tube and add 25–50 μl of a [^{35}S]-methionine labeling solution with the molarity and specific activity determined in Subheading 11.3.1 (*see* Note 6 and 10).
3. Sealed the tube with a cotton plug and put it immediately in the same conditions as the starting culture (temperature, rotary shaking).
4. After labeling transfer the sample into a 2-ml microcentrifuge tube previously kept on ice and spin down the cells for 1 min at 11,000g at 4°C.
5. Rinse the pellet twice with the starting volume of ice-cold deionized water.
6. Remove the supernatant and keep the cells frozen at -80°C or immediately extract the proteins.
7. After protein extraction, radioactivity and protein concentration are determined.
8. 2-μl sample are spotted on a microfiber filter for counting the radioactivity. Dry filters at room temperature for 30 min. Soak filters twice for 10 min in 5% (w/v) TCA containing 1g/L of methionine. Then place the dried filters in a counting vial, add 5 ml of liquid scintillation and count in a β-counter.
9. 10 μl are used for determining protein concentration with the Bio-Rad Protein Assay.
10. Store the protein sample in aliquots at -80°C.

3.3. Visualization of Radioactive Proteins on 2-D Gels

Proteins are separated according to standard 2-D gel electrophoresis procedure. Generally about 5×10^6 dpm are loaded on the gel (gel size 24cm×18cm). After 2-D gel electrophoresis, either briefly rinse the gel with fixing solution 1 and dry before exposure to phosphor screen plates or leave the gel in fixing solution 1 overnight and then allow the gel to re-swell to its original size for 2 h in fixing solution 2 (*see* Note 11). A comparison of 2-D pattern of protein synthesis obtained by labeling cells with L-[^{35}S]-methionine is shown in **Fig. 3**.

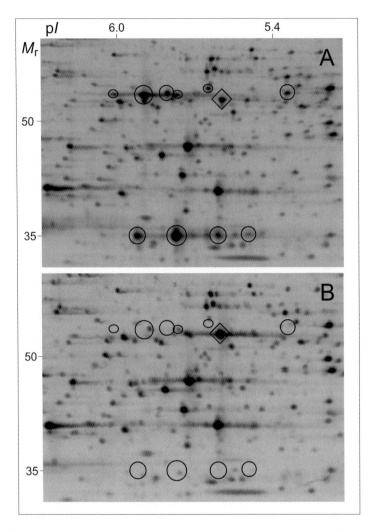

Fig. 3. Comparison of the pattern of proteins synthesized by *Saccharomyces cerevisiae* in the presence (**A**) or absence (**B**) of thiamine in the culture medium (2-D gel detail). Exponentially growing cells were labeled for 10 min with ^{35}S-methionine (500 Ci/mmol; 100 µCi/mL). 5×10^6 dpm were loaded on each gel. Gels were exposed one night to phosphor screen. Circles indicate proteins induced in the absence of thiamine and diamond, a protein that is repressed.

1. Put the gel on a piece of 3MM paper.
2. Cover the gel with a sheet of Saran wrap.
3. Place the gel with the paper-side down.
4. Dry for 1 h at 70 °C.

5. Remove the Saran wrap and expose the dried gels to phosphor screens for the appropriate period of time. Scan the screens with a storage phosphor imaging system (*see* Note 12). For quantitative analysis be careful to avoid saturation of the screen.

4. Notes

1. General comments on the Materials section:
   All solutions should be prepared in deionized water that has a resistivity of 18 MΩ-cm, except culture medium which is prepared with distilled water. Our laboratory is equipped with a Milli-Q Water system (Millipore) for deionized water.
   Manipulating radioactivity requires some specific precautions: gloves must be worn at all times, manipulate radioactivity behind a safety screen, use microcentrifuge tubes with a screw-cap rather than with a snap-cap, use pipet tips with a filter. Always use a bench protection. Working area should be regularly monitored for contamination. Decontaminate materials with a detergent special for radioactivity such as Decon 90 (Decon Laboratories Ltd, Hove, UK).
2. Rich medium should not be used as it contains large amounts of amino acids.
3. Radiolabeled [^{35}S]-methionine is used extensively for labeling proteins as it is commercially available at a high molar specific activity (>1,000 Ci/mmol) and β-emission is easily detectable. Another benefit of [^{35}S]-methionine is that the small intracellular pool of methionine allows reaching isotopic equilibrium very rapidly. However, some proteins may be devoid of methionine (removal of the *N*-terminal methionine). When interested in such proteins radiolabeled [^{14}C]-leucine or a mixture of L-[U-^{14}C]-amino acids can be used. Note that L-[U-^{14}C]-leucine and mixture of L-[U-^{14}C]-amino acids have a markedly lower specific activity than [^{35}S]-methionine (> 300 mCi/mmol and > 50 mCi/mg atom carbon, respectively). This may render difficult obtaining an efficient labeling for protein detection. Avoid L-[^{3}H]-amino acids as the energy emission of ^{3}H isotope is not high enough to be detected through acrylamide gels. Detection of ^{3}H isotope requires transfer on PVDF membrane.
4. Describing radiolabeling experiment requires providing two types of information, the radioactive concentration (Ci/mL or Bq/mL) and the molar specific activity of the labeled amino acid (Ci/mmol or Bq/mmol). The concentration of the radiolabeled amino acid during labeling is deduced from these two parameters.
5. When cells to be compared have significantly different metabolic activity, the concentration of the radiolabeled amino acid should be determined under the condition of the highest metabolic activity. Similarly when different stages of growth are investigated, determine the optimum concentration for the highest cell density.
6. Generally labeling experiments are carried out when culture reach mid-log phase.

7. In order to facilitate determining the labeling conditions, the molarity of the [^{35}S]-methionine test solutions, 10^{-5} M, is corresponding to the molarity of the [^{35}S]-methionine solution (1,000 Ci/mol, 10 µCi/µL) commercially supplied.
8. If isotopic dilution is required, it must be kept in mind that the final volume of labeling solution to be added to 1 ml of culture should not exceed 50 µl in order to avoid excessive dilution of the culture medium that may induce alteration in cell physiology.
9. Example of calculation of specific activity of labeling solution. Protein synthesis will be investigated during a 10-min period. The selected final molarity of methionine for labeling is 5×10^{-7} M. Under this condition 15 % of total radioactivity is incorporated into proteins after a 10-min labeling (see Fig. 2). It is necessary to incorporate 15 µCi per ml of culture for efficient protein detection. Thus the radioactive concentration of the labeling medium will be 100 µCi/mL. The specific activity is calculated through the following equation: Specific activity (Ci/mmol) = radioactive concentration (Ci/ml)/methionine concentration (mmol/mL). In this example specific activity = (10^{-4}Ci/mL)/(5 10^{-7} mmol/mL) = 200 Ci/mmol.
10. To prepare a [^{35}S]-methionine labeling solution such as addition of 25 µl of this solution to 1-ml sample of culture results in predetermined conditions of radioactive concentration (Rad_C), and methionine concentration (Met_C), mix commercialized [^{35}S]-methionine solution (1,000 Ci/mmol; 10 µCi/µL) and unlabeled methionine solution in a ratio v_1/v_2 respectively, where $v_1 = 10^{-1} Rad_C$ (Rad_C expressed in µCi/mL) and $v_2 = 25 - v_1$. The methionine molarity, M, of the unlabeled solution is calculated through the following equation: M = (10^6 $Met_C - 10^{-2}$ v_1)/(v_2) where Met_C is expressed in mmol/mL. In the example reported in Note 9 where Rad_C = 100 µCi/mL and Met_C = 5×10^{-7} mmol/mL, v_1 = 10, v_2 = 15, and M = 2.66 × 10^{-2} M.
11. Leaving the gel overnight in fixing solution 1 reduces the background when gels are exposed to phosphor screen.
12. When 5×10^6 dpm are loaded on the gel, the protein pattern can be visualized after a one night-exposure. For quantitative analysis be careful to avoid saturation of the screen.

References

1. Godon, C., Lagniel, G., Lee J., Buhler, J.M., Kiefer, S., Perrot, M., Boucherie, H., Toledano, M.B. and Labarre, J. (1998) The H_2O_2 stimulon in *Saccharomyces cerevisiae*. *J. Biol. Chem.* **273**, 22480–22489.
2. Monribot-Espagne, C. and Boucherie, H. (2002) Differential gel exposure, a new mehodology for the two-dimensional comparison of protein samples. *Proteomics* **2**, 229–240.
3. VanBogelen, R. A., Abshire, K. Z., Moldover, B., Olson, E. R. and Neidhardt, F. C. (1997) Escherichia coli proteome analysis using the gene-protein database. *Electrophoresis* **18**, 1243–1251.

4. Haurie, V., Sagliocco, F. and Boucherie, H. (2004) Dissecting regulatory networks by means of two-dimensional gel electrophoresis: Application to the study of the diauxic shift in the yeast *Saccharomyces cerevisiae*. *Proteomics* **4**, 364–373.
5. Bernhardt, J., Büttner K., Scharf, C. and Hecker, M. (1999) Dual channel imaging of two-dimensional electropherograms in *Bacillus subtilis*. *Electrophoresis* **20**, 2225–2240.
6. Rajjou, L., Gallardo, K., Debeaujon, I., Vandekerckhove, J., Job, C. and Job, D. (2004) The effect of α-amanitin on the Arabidopsis seed proteome highlights the distinct roles of stored and neosynthesized mRNAs during germination. *Plant Physiol.* **134**, 1598–1613.
7. Rajjou, L., Belghazi, M., Huguet, R., Robin, C., Moreau, A, Job, C. and Job, D. (2006) Proteomic investigation of the effect of salicylic acid on Arabidopsis seed germination and establishment of early defense mechanisms. *Plant Physiol.* **141**, 910–923.
8. Peyrat JP, Hondermarck H (2001). Proteomic detection of changes in protein synthesis induced by fibroblast growth factor-2 in MCF-7 human breast cancer cells. *Exp Cell Res.* **262**, 59–68.
9. Westbrook JA, Yan JX, Wait R. and Dunn MJ (2001) A combined radiolabelling and silver staining technique for improved visualisation, localisation, and identification of proteins separated by two-dimensional gel electrophoresis. *Proteomics* **1**, 370–376.

III

FRACTIONATION OF PROTEINS BY CHEMICAL REAGENTS AND CHROMATOGRAPHY

12

Sequential Extraction of Proteins by Chemical Reagents

Stuart J. Cordwell

Summary

Reproducible techniques for the prefractionation of proteins prior to two-dimensional gel electrophoresis (2-DE) are essential for increasing the number of unique proteins that can be identified and assayed following biological experimentation. A simple and robust technique for separating highly soluble (hydrophilic) cytoplasmic proteins from poorly soluble (hydrophobic) membrane-associated proteins uses differential solubility in a progressive series of extraction buffers, each containing more potent solubilizing chaotropes and detergents. This "sequential extraction" procedure is based on protein solubility in Tris buffer for the initial removal of highly soluble proteins, whereas proteins from the insoluble pellet are then extracted in 2-DE sample buffers containing urea and CHAPS. The final step of the procedure uses thiourea and amidosulfobetaine-14 (ASB-14) to solubilize CHAPS-insoluble proteins. This procedure has been optimized for the analysis of outer membrane porins from Gram negative bacteria, as well as the separation of plasma membrane proteins from mammalian cells grown in culture, and finally for the removal of insoluble cytoskeletal structures from mammalian heart tissue.

Key Words: Differential solubility; hydrophobic proteins; proteomics; sequential extraction; two-dimensional gel electrophoresis.

1. Introduction

Despite many technical advances in two-dimensional gel electrophoresis (2-DE), including improved reproducibility and standardization, the method remains incapable of visualizing several important protein subsets *(1,2)*. These include highly hydrophobic proteins, and/or those with several (generally greater than 3) transmembrane-spanning regions (TMR), proteins present at the

From: *Methods in Molecular Biology, vol. 424: 2D PAGE: Sample Preparation and Fractionation, Volume 1*
Edited by: A. Posch © Humana Press, Totowa, NJ

extremes of isoelectric point (pI) and molecular mass (M_r); and finally, lower abundance proteins that cannot be visualized because of the dynamic range effects of high abundance proteins and the limited resolving power of 2-DE. The significant challenge, therefore, in gel-based proteomics research has been to overcome these technical challenges and maximize the number of proteins that can be separated using this approach. Protein prefractionation has become a widely accepted strategy to improve the resolution of protein separation.

Several of the known limitations of 2-DE may be overcome by taking such a "sub-proteomics" approach *(3)*. Prefractionation may be based on several physical, chemical or functional protein properties *(4)*, allowing a specific subset of proteins to be analyzed in relative purity *(5,6)*. Several prefractionating devices using isoelectric focusing (IEF) or mass-based separations are commercially available and have been applied to the separation of proteins in conjunction with microrange or "zoom" IPG 2-DE *(7,8)* for enhanced protein and peptide separations *(9–12)*. Some of these devices have also been used to further enrich for particular organelles from within a cell lysate *(13)*.

The sequential extraction of proteins from a complex cell or tissue can aid in the visualization of a greater percentage of the proteome *(3,14)*. This technique uses the different solubilities of proteins, allowing them to be extracted in a series of progressively harsher extraction reagents (Fig. 1). In the first step, highly soluble proteins are extracted with 40 mM Tris-HCl buffer (pH 7.8). Proteins solubilized in this extraction are precipitated and resolubilized in a 2-DE-compatible sample buffer. Identification of these proteins generally reveals that they are cytoplasmic in subcellular location and are often key metabolic enzymes involved in critical processes such as glycolysis, the tricarboxylic acid cycle, as well as chaperonins, translation factors, and ribosomal proteins. The insoluble pellet from this step is then washed extensively and resolubilized with a standard 2-DE sample buffer containing 5–8 M urea and 2% (w/v) 3-[(3- Cholamidopropyl)dimethylammonio]-1- propanesulfonate (CHAPS) as the detergent. The remaining insoluble pellet derived after centrifugation of this mixture can then be washed again and subjected to one of two methods—first, solubilization in 2-DE sample buffer supplemented with 2 M thiourea, and the detergents 2% sulfobetaine 3–10 (SB3-10) and 1% (w/v) amidosulfobetaine 14 (ASB-14); or second, in SDS-containing sample buffer followed by SDS-PAGE one-dimensional separation alone. The SB3-10/ASB-14-soluble fraction is generally enriched for more hydrophobic, membrane-associated proteins. In particular, these include porins from Gram-negative bacteria, which although quite hydrophilic are significantly enriched using this procedure *(14)*. For tissue studies, we and others have also shown that this sequential extraction procedure allows for the removal of myofilament and cytoskeletal proteins (Fig. 2; *(15)*), hence providing improved recovery of lower abundance cytoplasmic proteins in the Tris-soluble fraction.

Sequential Extraction of Proteins by Chemical Reagents 141

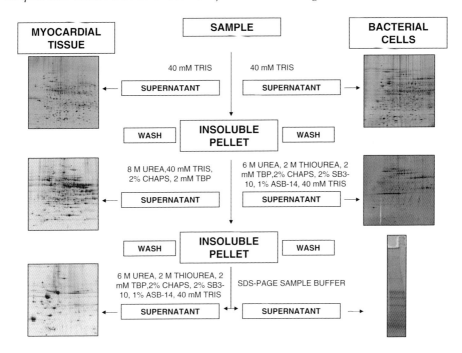

Fig. 1. Schematic diagram showing sequential extraction procedure for cells and tissues.

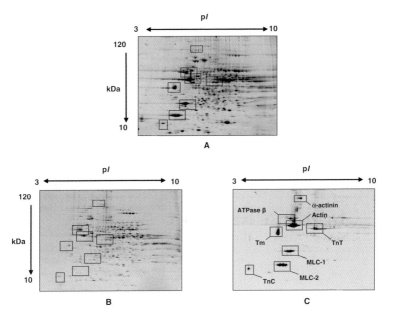

Fig. 2. Sequential extraction of proteins from rabbit myocardium. (**A**) Whole tissue lysate; (**B**) Tris-soluble proteins; (**C**) Detergent-soluble myofilament-associated proteins. Identifications were generated by MALDI-TOF MS, following in-gel tryptic digest.

2. Materials
2.1. Cell Disruption
1. Tip-probe sonicator or French pressure cell for bacterial and cultured cells.
2. Mechanical homogenizer for mammalian tissue samples.

2.2. Sequential Extraction
1. Tris buffer: 40 mM Tris-HCl (pH 7.8).
2. Ice-cold methanol.
3. 2-DE Sample Buffer I: 8 M urea, 2 mM tributylphosphine (TBP; see Note 1), 2% (w/v) CHAPS, 40 mM Tris, 0.2% (v/v) carrier ampholytes, and 0.002% (w/v) bromophenol blue dye (see Notes 2 and 3).
4. 2-DE Sample Buffer II: 6 M urea, 2 M thiourea, 2 mM TBP (See Note 1), 2% (w/v) CHAPS, 2% (w/v) sulfobetaine 3–10 (SB3-10), 1% (w/v) amidosulfobetaine 14 (ASB-14), 40 mM Tris, 0.2% (v/v) carrier ampholytes, and 0.002% (w/v) bromophenol blue dye (see Notes 2 and 3).
5. SDS-PAGE Sample Buffer: 75 mM Tris-HCl (pH 6.8), 2% (w/v) sodium dodecyl sulfate(SDS), 7.5% (v/v) glycerol, 100 mM DTT, 0.002% (w/v) bromophenol blue dye.

3. Methods
3.1. Cell Disruption
3.1.1. Bacterial Cells (See Note 4)
1. Wash the cells in phosphate-buffered saline (PBS) and centrifuge to collect. Take care to remove excess PBS by gently tapping the end of the tube against tissue paper.
2. Add three volumes of ice cold Tris buffer to the washed cell pellet (begin with approximately 10–40 mg dry weight cells) and vortex to resuspend cells. Add 20 U endonuclease and appropriate phosphatase and protease inhibitors, as required.
3. Subject cell mixture to at least triplicate rounds of tip-probe sonication (30–45 sec) as per the manufacturers' instructions (See Note 5).

3.1.2. Mammalian Tissue (See Note 4)
1. Wash the cells in phosphate-buffered saline (PBS) and centrifuge to collect. Take care to remove excess PBS by gently tapping the end of the tube against tissue paper.
2. Add three volumes of ice cold Tris Buffer to the washed tissue (begin with approximately 300 mg wet weight tissue) and vortex vigorously. Add 20 U endonuclease and appropriate phosphatase and protease inhibitors, as required.
3. Homogenize the tissue samples with a hand-held homogenizer using three 2 × 7 sec bursts on ice. Perform a low speed (2500g) centrifugation to remove

nonhomogenized tissue or continue rounds of homogenization until no particulate matter is visible.

3.2. Sequential Extraction

1. Centrifuge the protein lysate at 12,000g for 15 min at 4°C to remove insoluble material.
2. Collect the supernatant and keep the pellet.
3. Place the supernatant in a 50-ml centrifuge tube (*see* Note 6) and add ice-cold methanol to a final volume of 40 ml. Place at −80°C for a minimum of 2 h, but preferentially overnight.
4. Centrifuge at 16,000g for 30 min at 4°C. Carefully remove and discard supernatant and add 1 ml of 2-DE Sample Buffer I to the precipitated proteins. This fraction contains highly soluble, predominantly cytosolic proteins (*see* Note 7).
5. Wash the pellet from step 4 at least twice in Tris buffer to remove any contaminating Tris-soluble proteins and centrifuge at 12,000g for 15 min.
6. Add 1 ml of 2-DE Sample Buffer I, vortex strongly, and tip-probe sonicate/homogenize as described in Section 12.3.1, step 3.
7. Centrifuge in a bench-top centrifuge at 12,000g for 15 min at 4°C.
8. Collect the supernatant and remove to a fresh 1.5-ml centrifuge tube. This is the detergent-soluble fraction (*see* Note 8).
9. Wash the pellet twice in 2-DE Sample Buffer I (*see* Note 9).
10. Add 500 µl of 2-DE Sample Buffer II to the insoluble pellet from step 9, vortex strongly and tip-probe sonicate / homogenize as described in step 3 and 7.
11. Centrifuge in a bench-top centrifuge at 12,000g for 15 min at 4°C.
12. Collect the supernatant and remove to a fresh 1.5-ml centrifuge tube. This is the membrane-associated protein enriched fraction (*see* Note 10). Optional steps (*see* Note 11):
13. Wash the insoluble pellet twice in 2-DE Sample Buffer II (*see* Note 9).
14. Centrifuge at 12,000g for 15 min at 4°C.
15. Add SDS-PAGE Sample Buffer to the remaining insoluble pellet and perform 1-D SDS-PAGE.

4. Notes

1. An alternative to 2 mM TBP is 100 mM dithiothreitol (DTT).
2. If proteases and phosphatases are a problem, buffers should be supplemented with protease inhibitors such as 1 mM phenylmethylsulfonyl fluoride (PMSF), 1 mM EDTA, or protease inhibitor cocktails (20 µg/ml), as well as phosphatase inhibitors including 0.2% (v/v) okadaic acid. Nucleic acids are generally removed using *Serratia marcesans* endonuclease or benzonase at 20 U/ml.
3. Stock solutions of 2-DE Sample Buffer I and II can be made and stored at -20°C or -80°C, however, stock solutions should be made without TBP. TBP is volatile and unstable in aqueous solutions and therefore should be added immediately before use.

4. When using this method, it is important that expectations of the results be realistic. There has been much debate about the limit of 2-DE, even in conjunction with novel solubilizing detergents, for separating highly hydrophobic proteins (generally defined as those with positive Grand Average of Hydropathy [GRAVY] values) and for integral membrane proteins with many TMR *(16)*. Although the sequential extraction approach is undoubtedly effective at enriching for lower abundance proteins in the Tris-soluble fraction, and membrane-associated proteins in the detergent soluble fractions, it is important to recognize that these membrane-associated proteins are generally still hydrophilic in nature and contain only a very few (generally up to 3) TMR *(17)*. Very hydrophobic proteins and those integral membrane proteins with multiple TMR will remain under-represented and are best analyzed using multidimensional chromatography/tandem-MS following tryptic digest to generate soluble peptides *(18)*.
5. After each round of sonication, place the cell lysate on ice for 1–2 min. This step will be different depending on sample types. The described method should fracture approximately 90% of Gram-negative outer membranes. Additional rounds (at higher sonication values) may be required for Gram positive, acid-fast, and/or spore-forming bacteria.
6. Use solvent resistant, high-speed centrifuge tubes.
7. This fraction should contain enough protein to run triplicate large-format (20 × 20 cm) 2-DE gels.
8. This fraction should contain enough protein to run triplicate large-format (20 × 20 cm) 2-DE gels.
9. This procedure is often modified depending on the biological sample under investigation. This is because for many samples, the vast majority of proteins will be extracted in the first two steps of the procedure whereas for others, the first two steps will produce overlapping protein spot patterns on 2-DE gels, and novel proteins may only appear following the final extraction step. For the best possible results, a good knowledge of the methods required for physical disruption of the sample should be taken into consideration. The sequential extraction technique will not provide satisfactory results if near-to-complete cellular lysis is not achieved in the first (Tris buffer) protein extraction buffer. This is because subsequent rounds of extraction will result in the lysis of additional cells and hence, result in near identical 2-DE patterns. Although some overlap is almost completely unavoidable, minimizing it is critical to the success of the technique.
10. This fraction should contain enough protein to run a single large-format (20 × 20 cm) 2-DE gel.
11. The most common variation of the method is the use of SDS-PAGE sample buffer to either replace the final solubilizing step, or to provide a fourth solubility fraction (Fig. 1). Insoluble proteins resuspended in buffer containing 1% SDS can then be separated via 2-DE, following dialysis to lower the final concentration of SDS, or on a 1-D SDS-PAGE gel *(3)*. In our experience, however, very few, if any, additional protein identifications can be achieved using this strategy as a

fourth step in the extraction process, because the solubilizing efficiency of SDS does not appear to be significantly greater than that achieved using a cocktail of CHAPS, SB3-10, and ASB-14.

References

1. Görg A., Weiss, W., and Dunn, M.J. (2004) Current two-dimensional electrophoresis technology for proteomics. Proteomics 4, 3665–3685.
2. Görg A., Obermaier, C., Boguth, G., Harder, A., Schiebe, B., Wildgruber, R., and Weiss, W. (2000) The current state of two-dimensional electrophoresis with immobilized pH gradients. Electrophoresis 21, 1037–1053.
3. Cordwell, S.J., Nouwens, A.S., Verrills, N.M., Basseal, D.J., and Walsh, B.J. (2000) Subproteomics based upon protein cellular location and relative solubilities in conjunction with composite two-dimensional electrophoresis gels. Electrophoresis 21, 1094–1103.
4. Lee, W.-C. and Lee, K.H. Applications of affinity chromatography in proteomics. (2004) Anal. Biochem. 324, 1–10.
5. Righetti, P.G., Castagna, A., Herbert, B., Reymond, F., and Rossier, J.S. (2003) Prefractionation techniques in proteome analysis. Proteomics 3, 1397–1407.
6. Righetti, P.G., Castagna, A., Antonioli, P. and Boschetti, E. (2005) Prefractionation techniques in proteome analysis: the mining tools of the third millennium. Electrophoresis 26, 297–319.
7. Wildgruber, R., Harder, A., Obermaier, C., Boguth, G., Weiss, W., Fey, S.J., Larsen, P.M., and Görg, A. (2000) Towards higher resolution: two-dimensional electrophoresis of *Saccharomyces cerevisiae* proteins using overlapping narrow immobilized pH gradients. Electrophoresis 21, 2610–2616.
8. Westbrook, J.A., Yan, J.X., Wait, R., Welson, S.Y., and Dunn, M.J. (2001) Zooming-in on the proteome: very narrow-range immobilized pH gradients reveal more protein species and isoforms. Electrophoresis 22, 2865–2871.
9. Puchades, M., Westman, A., Blennow, K., and Davidsson, P. (1999) Analysis of intact proteins from cerebrospinal fluid by matrix-assisted laser desorption / ionization mass spectrometry after two-dimensional liquid-phase electrophoresis. Rapid Commun. Mass Spectrom. 13, 2450–2455.
10. Locke, V.L., Gibson, T.S., Thomas, T.M., Corthals, G.L., and Rylatt, D.B. (2002) Gradiflow as a prefractionation tool for two-dimensional electrophoresis. Proteomics 2, 1254–1260.
11. Herbert, B. and Righetti, P.G. (2000) A turning point in proteome analysis: sample prefractionation via multicompartment electrolyzers with isoelectric membranes. Electrophoresis 21, 3639–3648.
12. Zuo, X. and Speicher, D.W. (2002) Comprehensive analysis of complex proteomes using microscale solution isoelectricfocusing prior to narrow pH range two-dimensional electrophoresis. Proteomics 2, 58–68.
13. Zischka, H., Weber, G., Weber, P.J.A., Posch, A., Braun, R.J., Bühringer, D., Schneider, U., Nissum, M., Meitinger, T., Ueffing, M., and Eckerskorn, C. (2003)

Improved proteome analysis of *Saccharomyces cerevisiae* mitochondria by free-flow electrophoresis. Proteomics 3, 906–916.

14. Molloy, M.P., Herbert, B.R., Walsh, B.J., Tyler, M.I., Traini, M., Sanchez, J.-C., Hochstrasser, D.F., Williams, K.L., and Gooley, A.A. (1998) Extraction of membrane proteins by differential solubilization for separation using two-dimensional gel electrophoresis. Electrophoresis 19, 837–844.
15. Abdolzade-Bavil, A. Hayes, S., Goretski, L., Kroger, M., Anders, J., and Hendriks, R. (2004) Convenient and versatile subcellular extraction procedure, that facilitates classical protein expression profiling and functional protein analysis. Proteomics 4, 1397–1405.
16. Rabilloud, T., Blisnick, T., Heller, M., Luche, S., Aebersold, R., Lunardi, J., and Braun-Breton, C. (1999) Analysis of membrane proteins by two-dimensional electrophoresis: comparison of the proteins extracted from normal of *Plasmodium falciparum*-infected erythrocyte ghosts. Electrophoresis 20, 3603–3610.
17. Nouwens, A.S., Cordwell, S.J., Larsen, M.R., Molloy, M.P., Gillings, M., Willcox, M.D.P., and Walsh, B.J. (2000) Complementing genomics with proteomics: the membrane subproteome of *Pseudomonas aeruginosa* PAO1. Electrophoresis 21, 3797–3809.
18. Cordwell, S.J. (2006) Technologies for bacterial surface proteomics. Curr. Opin. Microbiol. 9, 320–329.

13

Reducing Sample Complexity by RP-HPLC: Beyond the Tip of the Protein Expression Iceberg

Gert Van den Bergh and Lutgarde Arckens

Summary

Because the dynamic range of most cell or tissue proteomes is enormous, separation of such complex protein samples by two-dimensional electrophoresis (2-DE) on broad pH gradients often results in the visualization of only the most abundantly expressed proteins. It is, therefore, often beneficial to first subdivide the proteome in smaller, less complex fractions before 2-DE. This enables the analysis of a larger number of proteins. One approach to prefractionate protein samples is by reversed-phase high-performance liquid chromatography (RP-HPLC), separating proteins according to their hydrophobicity. This effectively introduces a third separation dimension, increasing the spatial resolution of the experiment. Here, we will describe a procedure for separating whole protein lysates by RP-HPLC, before their analysis by 2-DE or 2-D difference gel electrophoresis

Key Words: CyDyes; difference gel electrophoresis; prefractionation; reversed-phase high-performance liquid chromatography; two-dimensional electrophoresis.

1. Introduction

Although the use of multidimensional chromatography directly coupled to mass spectrometry is becoming more and more widespread for quantitatively analyzing protein expression level differences in complex proteomes *(1,2)*, two-dimensional electrophoresis (2-DE) remains the protein separation method of choice in a large number of laboratories. This is primarily because of its high separation power and its complementarity to the currently available gel-free approaches in visualizing specific protein subclasses. Moreover, the quantitative aspect of 2-DE has drastically improved over the last few years, with the introduction of two-dimensional difference gel electrophoresis (2-D DIGE) *(3,4)*. 2-D DIGE uses fluorescent dyes in a sophisticated manner, thereby increasing the sensitivity and linearity of the protein detection compared to classical protein stains such as silver or Coomassie blue stains. Moreover, the 2-D DIGE multiplexing approach enables the comparison of multiple samples on a single gel and thus elegantly overcomes otherwise unavoidable gel-to-gel variation. As such, the sensitivity, accuracy and reproducibility of 2-DE based comparative proteomics studies have all been greatly improved *(5–7)*.

Despite these innovations, some important drawbacks of 2-DE still remain. The exclusion of highly hydrophobic proteins or proteins with very large or small isoelectric points or molecular weights limits the thoroughness of the proteome investigations. And most importantly, 2-DE is incapable of visualizing low-copy-number proteins in the presence of highly abundant gene products. Only about 2,000 different protein spots are generally visible on a standard-sized 2-DE gel, compared to the 10,000–30,000 proteins that are typically expressed by a cell or tissue type. This is not really a sensitivity issue but seems more related to the finite resolution of 2-DE. Indeed, using a more sensitive protein detection method will not increase the visible number of spots beyond a certain limit, as high abundant protein spots will mask the underlying faint spots. It is, therefore, advantageous to reduce the complexity of the protein sample that is run on a 2-DE gel, at the same time increasing the protein amount loaded onto the gel, providing higher chances of detecting the low abundant protein spots.

One approach to decrease sample complexity is by fractionating a protein mixture in several subsets, based on subcellular localization *(8,9)* or on biochemical protein characteristics, for example by using affinity chromatography or immunoprecipitation, as for the separation of phosphorylated proteins *(10)*. Another prefractionation technique that has been successfully applied is based on the subdivision of proteins based on their differential solubility in buffers with sequentially increasing solibilization abilities *(11)*, again followed by 2-DE of each of the obtained fractions. The protein prefractionation approach that we will describe in this chapter is reversed-phase high-performance liquid chromatography (RP-HPLC), that separates proteins based on their hydrophobicity, a global protein characteristic that, in combination with 2-DE, introduces essentially a third independent separation dimension next to isoelectric point and molecular weight. The goal of every prefractionation approach is the reduction of the number of proteins in each fraction while at the same time the most abundant proteins are confined to a single subfraction, freeing up physical space for the low abundant proteins at these positions in the gels in which the other protein subfractions are separated. This combination

Reducing Sample Complexity by RP-HPLC

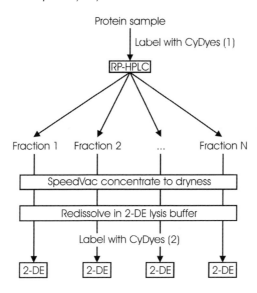

Fig. 1. Schematic overview of the procedure of protein prefractionation with RP-HPLC, before 2-DE. Two possible positions, either before (1) or after (2) RP-HPLC, where protein samples can be labeled with CyDyes are indicated. Depending on the gradient and the number of 2-DE gels one wants to run, the protein sample can be subdivided in a certain number of fractions (N).

of RP-HPLC and subsequent 2-DE separation of the obtained samples was first successfully introduced by Badock et al. *(12)*, using silver staining for protein detection. In our laboratory, we effectively combined this RP-HPLC fractionation method with a subsequent 2-D DIGE analysis *(13)*.

Here, we will describe a procedure for obtaining protein samples from brain tissue, followed by RP-HPLC fractionation of these samples and subsequent quantitative analysis of each of the obtained fractions by 2-DE (*see* Fig. 1). We will also provide notes for amending this technology to exploit the strengths of the 2-D DIGE methodology.

2. Materials

2.1. Sample Preparation

1. 2-DE lysis buffer: 7 M urea, 2 M thiourea, 4% CHAPS, 1% DTT, 40 mM Tris-base. Store in 1 ml aliquots at $-20°C$ or use fresh. Do not allow warming above room temperature because this can result in the degradation of urea to isocyanate leading to the carbamylation of proteins. Add 40 μl complete protease inhibitor (Roche, Basel, Switzerland) to 1 mL aliquot before use (stock solution: 1 tablet in 2 ml HPLC grade water. Store for maximal 1 month at $-20°C$). (*see* Note 1)

2. Bath sonicator (Model 5510, Branson Ultrasonics, Danbury, USA).
3. An ultracentrifuge (e.g., Optima LE80K, Beckman Coulter).
4. PBS buffer: 0.1 M NaH_2PO_4, pH 7.4.
5. Protein Assay kit (Bio-Rad, Hercules, CA, USA).
6. An ELISA plate reader (Labsystems Multiskan RC, Thermo Electron Corporation, Brussels, Belgium).

2.2. Reversed-Phase HPLC Separation

1. A standard reversed phase HPLC system is required. In our experiments, we use a 625LC system from Waters (Milford, MA, USA). The separation system should be connected to a UV monitoring system, recording the chromatogram at an absorption wavelength of 210 nm and/or 280 nm, to visualize the eluted peak and to enable the pooling of eluted fractions after the RP-HPLC run. We apply an LKB 2141 Variable Wavelength Monitor (GE Healthcare, Freiburg, Germany).
2. Depending on the desired protein fractionation characteristics, a column suitable for protein separations can be chosen. We use a Poros C_4 reversed phase column (150 × 4.6 mm, 5 µm, 300 Å) (Applied Biosystems, Foster City, CA, USA), with a compatible guard column (*see* Notes 2–3).
4. HPLC solvent A: 0.1% trifluoroacetic acid (TFA) in HPLC-gradient water.
5. HPLC solvent B: 0.1% TFA in acetonitrile (ACN) (*see* Note 4).
6. An automatic fraction collector (Model 2128 Fraction Collector, Bio-Rad, Hercules, CA, USA).
7. A SpeedVac vacuum centrifuge (A290, Savant, Farmingdale, NY, USA).

3. Methods

3.1. Sample Preparation

1. Collect fresh or frozen cells or tissue. In our experiments *(13,14)* we usually collect protein samples from a small area of mammalian brain, namely the neocortex. To this end, 200 µm thick cryostat sections are cut and an area of gray matter of approximately 160 mm² (+/–32 mm³) is cut from these sections on dry ice to prevent protein degradation. Collected tissue is rapidly transferred to a volume of ice-cold lysis solution. For the above-described amount of tissue, we use a standard volume of 100 µl lysis solution. This volume should be changed according to the volume of the cells or tissue to be lysed (*see* Note 5).
2. Homogenize brain tissue on ice. Avoid excessive foaming of the lysis buffer, as this could result in protein loss.
3. Sonicate in a bath sonicator for approximately 1 min at room temperature.
4. Put protein sample at room temperature for 1 h, to allow for complete solubilization of the proteins in the lysis buffer.
5. Sonicate again for 1 min at room temperature.
6. Clear the protein lysate by ultracentrifugation for 20 min. at 100,000g at 4°C.

7. Determine the protein concentration of the supernatant. We use a modified Bradford assay, as described by Qu et al. *(15)*, but any protein quantification method that is compatible with the contents of the lysis solution (high concentrations of urea, thiourea, CHAPS...) can be used. Make a 2 µg/µl stock solution of ovalbumin and prepare a standard dilution series, adding 400 µl of a solution containing 500, 300, 100, 50, 25, and 12.5 µg ovalbumin in PBS and 300 µl 0.1 M NaOH. Prepare the samples to be measured by mixing 1 µl of protein sample, 30 µl 0.1 M NaOH and 69 µl PBS buffer. Place 20 µl of the samples and ovalbumin standard in triplicate on a flat-bottomed 96-well plate. Add 200 µl of Bio-Rad protein assay solution (diluted 1/10 in HPLC grade water), read the absorbance at 595 nm with an ELISA plate reader and determine protein concentration.

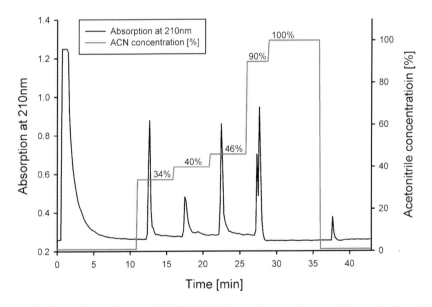

Fig. 2. Illustration of a typical RP-HPLC run for protein sample prefractionation. In gray, the profile of the acetonitrile concentration is illustrated, eluting proteins in four steps of increasing ACN concentrations: 34%, 40%, 46%, and 90%. The column is recycled by washing with 100% ACN and then with 1%. In black, the elution profile of the fractionated proteins (absorption measured at 210nm) is shown. The first large peak is a typical injection peak, whereas the peaks at 34%, 40%, 46%, and 90% ACN represent the different fractions that are obtained by RP-HPLC separation of a mammalian neocortical protein sample.

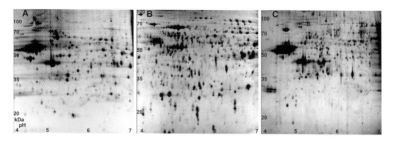

Fig. 3. Example of the effect of RP-HPLC fractionation on the 2-D protein expression patterns of adult cat primary visual cortex. **Panel A** corresponds to the unfractionated protein sample, whereas **panels B** and **C** represent protein spot patterns from the 40% and 46% ACN RP-HPLC fractions, respectively. Proteins were labeled with Cy3. Molecular weight and isoelectric point are indicated on the gel images. Image analysis shows that the protein spot patterns are radically different between the unfractionated pattern and the RP-HPLC fractions (especially the 40% ACN fraction). In B and C, spots are visible that are not detectable on the unfractionated pattern, demonstrating the enrichment of proteins of low abundance.

3.2. Reversed-Phase HPLC Separation

1. Equilibrate the column and guard column by pumping 1% HPLC solvent B in solvent A for 15 min at a flow rate of 3 ml/min (*see* Notes 6–8).
2. Inject approximately 1 mg protein of the cleared supernatant, mixed 1:1 (volume) with HPLC solvent A, onto the column. Wash the column for 11 min with 99% of solvent A and 1% of solvent B to obtain a stable baseline (*see* Note 8).
3. Perform a step gradient elution, with four steps of increasing concentrations of solvent B in solvent A: 5 min. at 34% of solvent B, 5 min. at 40%, 5 min. at 46%, and 3 min at 90%. Recycle the column by washing with 100% solvent B for 7 min, followed by 7 min with 1% solvent B (*see* Fig. 2 and *see* Note 9).
4. Automatically collect fractions every minute.
5. Pool the fractions of each step and dry in a SpeedVac vacuum centrifuge.
6. Dissolve dried sample in 50 µl lysis solution and determine protein concentration.
7. Analyze protein expression profiles of each fraction by 2-D PAGE or 2-D DIGE (*see* Fig. 3 and Notes 10–11).

4. Notes

1. All water used during sample preparation, RP-HPLC and two-dimensional electrophoresis should be HPLC-grade or similar quality, having a resistance of 18.2 MΩ.
2. We successfully used Poros C4 reversed phase material, but any other reversed phase material suitable for protein separation should work as well.

3. Because some proteins could precipitate after mixing the sample 1:1 with HPLC solvent A or during injection of the sample onto the column, the separating column could become clogged by protein aggregates. It is, therefore, of utmost importance to protect the column by placing a guard column in front of it, trapping these aggregates before they can do damage to the separating column. These guard columns should be replaced when the absorbance in blank runs after a separation run does not return to baseline, or an increase in pressure is observed.
4. HPLC solvent A and B constitute the mobile phase of RP-HPLC, whereas the column represents the stationary phase. All HPLC solvents should be continuously degassed with helium. TFA is used as an ion-pairing agent, neutralizing the charge of the solute molecules by binding to them, thereby increasing their hydrophobicity.
5. A literature search should be performed in order to use the most efficient protein extraction protocol for your particular cell or tissue type. For many tissues or cell types, reproducible procedures have been described. Further optimization might be necessary to obtain optimal results. Try to keep the extraction protocol as simple as possible to prevent protein loss or the introduction of nonbiological sources of variation.
6. Perform blank runs after initial column equilibration and between two successive separation runs, to check for residual protein on the column (e.g., because of overloading etc.) that could elute from the column in the next separation run. This could result in contamination of your protein fractions from one sample with proteins from the previous sample, thereby introducing nonbiological variation in protein expression levels. Change guard column if necessary.
7. Flow rate is dependent on the type and diameter of column used. Refer to column instructions for optimal flow rate to be used. The elution times referred to in this procedure were optimized for the Poros C_4 column of 4.6 mm internal diameter with a flow rate of 3 ml/min. Separation times should be longer with lower flow rates. Perform a test run to determine the optimal run conditions to obtain the best separation with shortest run times.
8. The amount of protein that can be loaded onto the column is in proportion to the diameter of the column used. Try not to overload the column, as this will result in carry-over of proteins between different separation runs.
9. With this elution gradient, four fractions are collected. However, it is equally possible to increase the number of elution steps, or alternatively run a gradient elution with a continuously and linearly increasing HPLC solvent B concentration, analyzing each collected 1 min step separately. Although requiring a larger number of 2-DE gels, this will result in an even less complex protein composition for each fraction, thus enabling the visualization of a proportionately larger number of spots.
10. A procedure for 2-DE of the obtained protein fractions is not described, as this falls outside the scope of this chapter. Many excellent and detailed protocols can be found, however, in a number of recent papers on this technology *(16–18)*.

When using the classical protein stains, we usually load 100–200 µg protein per gel for silver staining or up to 1 mg for preparative, Coomassie blue stained gels. In most cases, the amount of protein collected after one prefractionation run is sufficient for performing only one or at most two 2-DE experiments. This mandates the use of wide pH gradients, for a full overview of the translated gene products. If enough material is obtained in each fraction to run multiple gels, it could be beneficial to perform IEF separation on several narrow-range pH gradients, thereby further increasing the number of spots visualized in the experiment *(19)*. The number of 2-DE gels to run is thereby evidently multiplied again.

11. This RP-HPLC fractionation can also be used in combination with subsequent 2-D DIGE analysis *(13)*, whereby three samples (one internal standard and two samples to be analyzed) are labeled with three different fluorescent dyes, that do not interfere with the electrophoretic separation but have differing fluorescent characteristics. These samples are then mixed and applied on a single gel. To this end, about 50 µg protein of each sample is labeled with 200–400 pmol dye. In this way, the experimental variation of running different gels for multiple samples is drastically reduced. Labeling the samples can be performed at two steps in the prefractionation procedure (*see* Fig. 1). In our experiments we have always successfully labeled our proteins after RP-HPLC separation and collection of the fractions (for example *see* Fig. 3). But in order to avoid the introduction of experimental variation between the samples because of the HPLC procedure, it could be beneficial to fluorescently label the samples prior to RP-HPLC, mix those samples to be run on the same gel and then perform the prefractionation procedure. In combination with an internal standard, this would further decrease the variability within the experiment. With this approach, both labeled and unlabeled proteins have differences in hydrophobicity, and could potentially elute in different fractions, especially for those proteins eluting at the borders of two fractions. This should not represent a problem for DIGE visualization, but could compromise protein identification, as the larger amount of unlabeled protein could be present on a different gel. Furthermore, to achieve reasonable sensitivity, a larger amount of protein should be labeled with an equally increased amount of dye.

There is one disadvantage of using RP-HPLC in combination with 2-D DIGE, however. When analyzing the most hydrophobic eluted fraction, we observe only very few spots, with a very low sensitivity. This is probably because of the fact that labeling these hydrophobic proteins with CyDyes makes them even more hydrophobic, reducing their solubility and separation ability on 2-DE even further.

Acknowledgments

Gert Van den Bergh is a postdoctoral fellow of the Fund for Scientific Research Flanders (FWO-Vlaanderen), Belgium. We thank Lieve Geenen for critically reading the manuscript.

References

1. Gygi, S. P., Rist, B., Gerber, S. A., Turecek, F., Gelb, M. H., and Aebersold, R. (1999) Quantitative analysis of complex protein mixtures using isotope-coded affinity tags. *Nat. Biotechnol.* **17**, 994–999.
2. Washburn, M. P., Wolters, D., and Yates, J. R., III (2001) Large-scale analysis of the yeast proteome by multidimensional protein identification technology. *Nat. Biotechnol.* **19**, 242–247.
3. Ünlü, M., Morgan, M. E., and Minden, J. S. (1997) Difference gel electrophoresis: a single gel method for detecting changes in protein extracts. *Electrophoresis* **18**, 2071–2077.
4. Tonge, R., Shaw, J., Middleton, B., Rowlinson, R., Rayner, S., Young, J., Pognan, F., Hawkins, E., Currie, I., and Davison, M. (2001) Validation and development of fluorescence two-dimensional differential gel electrophoresis proteomics technology. *Proteomics.* **1**, 377–396.
5. Alban, A., David, S. O., Bjorkesten, L., Andersson, C., Sloge, E., Lewis, S., and Currie, I. (2003) A novel experimental design for comparative two-dimensional gel analysis: Two-dimensional difference gel electrophoresis incorporating a pooled internal standard. *Proteomics* **3**, 36–44.
6. Knowles, M. R., Cervino, S., Skynner, H. A., Hunt, S. P., de Felipe, C., Salim, K., Meneses-Lorente, G., McAllister, G., and Guest, P. C. (2003) Multiplex proteomic analysis by two-dimensional differential in-gel electrophoresis. *Proteomics* **3**, 1162–1171.
7. Van den Bergh, G. and Arckens, L. (2004) Fluorescent two-dimensional difference gel electrophoresis unveils the potential of gel-based proteomics. *Curr. Opin. Biotechnol.* **15**, 38–43.
8. Rabilloud, T., Kieffer, S., Procaccio, V., Louwagie, M., Courchesne, P. L., Patterson, S. D., Martinez, P., Garin, J., and Lunardi, J. (1998) Two-dimensional electrophoresis of human placental mitochondria and protein identification by mass spectrometry: towards a human mitochondrial proteome. *Electrophoresis* **19**, 1006–1014.
9. Murayama, K., Fujimura, T., Morita, M., and Shindo, N. (2001) One-step subcellular fractionation of rat liver tissue using a Nycodenz density gradient prepared by freezing-thawing and two-dimensional sodium dodecyl sulfate electrophoresis profiles of the main fraction of organelles. *Electrophoresis* **22**, 2872–2880.
10. Gronborg, M., Kristiansen, T. Z., Stensballe, A., Andersen, J. S., Ohara, O., Mann, M., Jensen, O. N., and Pandey, A. (2002) A mass spectrometry-based proteomic approach for identification of serine/threonine-phosphorylated proteins by enrichment with phospho- specific antibodies: identification of a novel protein, Frigg, as a protein kinase A substrate. *Mol. Cell Proteomics.* **1**, 517–527.
11. Molloy, M. P., Herbert, B. R., Walsh, B. J., Tyler, M. I., Traini, M., Sanchez, J. C., Hochstrasser, D. F., Williams, K. L., and Gooley, A. A. (1998) Extraction of membrane proteins by differential solubilization for separation using two-dimensional gel electrophoresis. *Electrophoresis* **19**, 837–844.

12. Badock, V., Steinhusen, U., Bommert, K., and Otto, A. (2001) Prefractionation of protein samples for proteome analysis using reversed-phase high-performance liquid chromatography. *Electrophoresis* **22,** 2856–2864.
13. Van den Bergh, G., Clerens, S., Vandesande, F., and Arckens, L. (2003) Reversed-phase high performance liquid chromatography pre-fractionation prior to 2D difference gel electrophoresis and mass spectrometry identifies new differentially expressed proteins between striate cortex of kitten and adult cat. *Electrophoresis* **24,** 1471–1481.
14. Van den Bergh, G., Clerens, S., Cnops, L., Vandesande, F., and Arckens, L. (2003) Fluorescent two-dimensional difference gel electrophoresis and mass spectrometry identify age-related protein expression differences for the primary visual cortex of kitten and adult cat. *J. Neurochem.* **85,** 193–205.
15. Qu, Y., Moons, L., and Vandesande, F. (1997) Determination of serotonin, catecholamines and their metabolites by direct injection of supernatants from chicken brain tissue homogenate using liquid chromatography with electrochemical detection. *J. Chromatogr. B Biomed. Sci. Appl.* **704,** 351–358.
16. Klose, J. (1999) Large-gel 2-D electrophoresis. *Methods Mol. Biol.* **112,** 147–172.
17. Görg, A., Obermaier, C., Boguth, G., Harder, A., Scheibe, B., Wildgruber, R., and Weiss, W. (2000) The current state of two-dimensional electrophoresis with immobilized pH gradients. *Electrophoresis* **21,** 1037–1053.
18. Görg, A., Weiss, W., and Dunn, M. J. (2004) Current two-dimensional electrophoresis technology for proteomics. *Proteomics* **4,** 3665–3685.
19. Stasyk, T. and Huber, L. A. (2004) Zooming in: fractionation strategies in proteomics. *Proteomics* **4,** 3704–3716.

14

Enriching Basic and Acidic Rat Brain Proteins with Ion Exchange Mini Spin Columns Before Two-Dimensional Gel Electrophoresis

Ning Liu and Aran Paulus

Summary

Proteome analysis by two-dimensional gel electrophoresis (2-DGE) faces significant challenges because of the complexity of biological samples. However, the complexity of a protein sample can be reduced prior to 2-DGE by applying protein fractionation. Protein fractionation allows analysis of one protein subset at a time, thereby, increasing the load of proteins of interest, enriching low-abundance proteins, and increasing the resolution of protein spots on a 2-D gel.

Here we describe an ion exchange chromatography based method—the use of anion or cation exchange (AEX or CEX) mini spin columns—for sample fractionation. Using rat brain tissues, we demonstrate that these mini spin columns provide an easy, convenient, and reproducible way of fractionating brain proteins to enrich basic or acidic proteins before 2-DGE.

Key Words: Acidic proteins; basic proteins; 2-D gel electrophoresis; ion exchange chromatography; mini spin column; rat brain; sample fractionation.

1. Introduction

Ion exchange chromatography can be used to concentrate proteins based on their net ionic charge or isoelectric point (pI) at a given pH *(1)*. At the isoelectric point of a protein, its net charge is zero. At a pH higher than the pI of the protein, it is negatively charged and will bind to an AEX resin. At a pH lower than the pI, the protein will be positively charged and will bind to a CEX resin. Binding of the proteins is reversible and absorbed proteins are commonly eluted with salt. Bio-Rad Aurum™ AEX and CEX mini spin columns contain 200 µl of UNOsphere^{TM} Q or S ion exchange media, respectively *(2)*. The sulpho groups ($-SO_3^-$) on the UNOsphere S resin bind proteins with positive net charges, whereas the quaternary ammonium groups ($-N^+(CH_3)_3$) on the UNOsphere Q resin bind proteins with negative net charges *(2)*. The morphology and column behavior of the resins have been characterized *(3–5)*.

Here we describe the procedures for using the Aurum^{TM} ion exchange mini spin columns to enrich acidic or basic rat brain proteins before 2-DGE. The mini spin columns provide a good tool for protein sample fractionation prior to 2-DGE for the following reasons:

1. Compatibility: the Aurum^{TM} ion exchange mini spin columns are compatible with buffers or solutions containing high concentrations of urea and/or thiourea, two compounds widely used in buffers for 2-DGE.
2. Convenience: the operating time with the prepacked mini spin columns is only 15–20 min, making this an easy and convenient method for protein fractionation.
3. High binding capacity: the binding capacities are 12 mg and 36 mg for CEX and AEX mini spin columns, respectively, based on measurement with bovine serum albumin *(6)*.

2. Materials
2.1. Protein Extraction from Tissues

1. Whole rat brain tissues are obtained from Sigma-Aldrich (St. Louis, MO) and stored at –20°C.
2. Dounce tissue grinder, 7 ml (VWR International, Inc., West Chester, PA).
3. Protein lysis buffer: 7 M urea, 2 M thiourea, 2% (3-[3-Cholamidopropyl)dimethylammonio]-1-propane sulfonate (CHAPS), 40 mM Tris-HCl, pH 8.5.
4. Branson Sonifier 450 (VWR).

2.2. Protein Fractionation by CEX Mini Spin Column

1. Aurum™ CEX mini spin columns (Bio-Rad Laboratories, Inc., Hercules, CA).
2. CEX column binding buffer: 7 M urea, 2 M thiourea, 2% CHAPS, 20 mM sodium phosphate, pH 7.0 (*see* Note 1).
3. CEX column elution solution: 7 M urea, 2 M thiourea, 2% CHAPS, 1 M NaCl.

2.3. Protein Fractionation by AEX Mini Spin Column

1. Aurum™ AEX mini spin columns (Bio-Rad).
2. AEX column binding buffer: 7 M urea, 2 M thiourea, 2% CHAPS, 20 mM pyridine, pH 5.0 (*see* Note 1).

3. AEX column elution solution: 7 M urea, 2 M thiourea, 2% CHAPS, 1 M NaCl.

2.4. Protein Sample Reduction and Cleanup before 2-DGE

1. Protein dilution solution: 7 M urea, 2 M thiourea, 2% CHAPS.
2. ReadyPrep reduction-alkylation kit (Bio-Rad).
3. ReadyPrep 2-D cleanup kit (Bio-Rad).
4. *RC DC* protein assay kit II (Bio-Rad).

2.5. 2-D Gel Electrophoresis

1. 11 cm, ReadyStrip IPG strips, pH 3–10 (Bio-Rad).
2. IPG strip rehydration buffer: 7 M urea, 2 M thiourea, 2% CHAPS, 3 mM TBP, 0.2% carrier ampholytes pH 3–10 (Bio-Rad).
3. Equilibration buffer I: 6 M urea, 2% SDS, 0.375 M Tris-HCl, pH 8.8, 20% glycerol, and 2% DTT.
4. Equilibration buffer II: 6 M urea, 2% SDS, 0.375 M Tris-HCl, pH 8.8, 20% glycerol, and 135 mM iodoacetamide.
5. Criterion precast SDS-gels, 8–16% Tris-HCl (Bio-Rad).
6. 1X Tris/glycine/SDS running buffer (Bio-Rad): 25 mM Tris, pH 8.3, 192 mM glycine, 0.1% (w/v) SDS.
7. Agarose solution: 1% agarose, 0.003% bromophenolblue in 1X Tris/glycine/SDS running buffer.
8. Flamingo fluorescent gel stain (Bio-Rad).

3. Methods

3.1. Total Protein Extraction from Rat Brain Tissue

1. Weigh out 250–500 mg of rat brain tissue. The tissue should remain frozen on dry ice before and after weighing.
2. Transfer brain tissue to the Dounce tissue grinder prechilled on ice. Add 5–7 ml of protein lysis buffer and homogenize the brain tissue on ice for 10 min.
3. Transfer the homogenate into a culture tube and sonicate on ice with Branson Sonifier 450 (Branson Ultrasonics Corp., Danbury, CT) for 4 × 30 sec at constant power 10 Watt output at 1-min intervals.
4. Transfer the tissue homogenate (5–7 ml) to a 50 ml Oak Ridge polycarbonate centrifuge tube (VWR) and centrifuge at 16,000g for 30 min at 4°C.
5. Collect the supernatant from the tube, discarding the top lipid layers and the bottom pellets.
6. Determine protein concentration of the sample using *RC DC* protein assay kit (Bio-Rad). The expected concentrations of the protein extract ranges from 10–20 mg/ml.
7. Aliquot the supernatant, freeze on dry ice and store at −70°C.

3.2. Protein Fractionation with CEX Mini Spin Column at pH 7.0

1. Place a CEX column in a 12 × 75 mm test tube and allow the resin to settle for 5 min.
2. Prepare the protein sample: dilute 100 μl (~1–2 mg) of rat brain protein extract into 3.9 ml of CEX binding buffer (*see* Note 2 and Note 3).
3. Remove the column cap, break off the bottom tip, return the column to the test tube allowing the residual buffer to drain by gravity (*see* Note 4).
4. Equilibrate the column with two washes of 1.0-ml each of CEX binding buffer. At the end of the second wash, place the column in an empty 2.0-ml microcentrifuge tube and centrifuge for 10 sec at $1,000g$ to remove excess buffer.
5. Place the column in a clean 12 × 75 mm test tube labeled "unbound." Load 800 μl of rat brain protein sample onto the column and allow the sample to gravity filter through the column (*see* Note 4). Repeat four more times to load all 4 ml of the protein sample (*see* Note 5).
6. Wash the column with 600 μl of CEX binding buffer and collect it into the same unbound test tube.
7. Place the column in a 2.0-ml microcentrifuge tube and add 600 μl of CEX binding buffer. Centrifuge for 20 sec at $1,000g$ and discard the collected wash.
8. Place the column in a new 2.0-ml microcentrifuge tube labeled "bound #1". Add 300 μl of CEX elution solution and centrifuge for 10 sec at $1,000g$. Repeat the elution step and collect the 300 μl of eluant into the same microcentrifuge tube.
9. Place the column in a new 2.0-ml microcentrifuge tube labeled "bound #2" Add 300 μl of CEX elution solution and centrifuge for 10 sec at $1,000g$. Repeat the elution step and collect the 300 μl of eluant into the same microcentrifuge tube.
10. Pool the CEX bound fractions #1 and #2 (total of 1.2 ml).

3.3. Protein Fractionation with AEX Mini Spin Column at pH 5.0

1. Place an AEX column in a 12 × 75 mm test tube and allow the resin to settle for 5 min.
2. Prepare the protein sample: dilute 100 μl (~1–2 mg) of rat brain protein extract into 3.9 ml of AEX binding buffer (*see* Note 2 and Note 3).
3. Remove the column cap, break off the bottom tip and return the column to the test tube allowing the residual buffer to drain by gravity (*see* Note 4).
4. Equilibrate the column with two washes of 1.0-ml each of AEX binding buffer. At the end of the second wash, place the column in an empty 2.0-ml microcentrifuge tube and centrifuge for 10 sec at $1,000g$ to remove excess buffer.
5. Place the column in a clean 12 × 75 mm test tube labeled "unbound." Load 800 μl of protein sample onto the column and allow the sample to gravity filter through the column (*see* Note 4). Repeat four more times to load all 4 ml of protein sample (*see* Note 5).
6. Wash the column with 600 μl of AEX binding buffer and collect it into the same unbound test tube.

7. Place the column in a 2.0-ml microcentrifuge tube and add 600 μl of AEX binding buffer. Centrifuge for 20 sec at 1,000g and discard the collected wash.
8. Place the column in a new 2.0-ml microcentrifuge tube labeled "bound #1." Add 300 μl of AEX elution solution and centrifuge for 10 sec at 1,000g. Repeat the elution step and collect the 300 μl of eluant into the same microcentrifuge tube.
9. Place the column in a new 2.0-ml microcentrifuge tube labeled "bound #2." Add 300 μl of AEX elution solution and centrifuge for 10 sec at 1,000g. Repeat the elution step and collect the 300 μl of elute into the same microcentrifuge tube.
10. Pool the AEX bound fractions #1 and #2 (total of 1.2 ml).

3.4. Protein Sample Reduction and Cleanup before 2-DGE

3.4.1. Preparation of Protein Samples

1. Unfractionated brain proteins: 50 μl (~0.5–1.0 mg) rat brain protein extract with 950 μl of protein dilution solution.
2. CEX bound fraction of rat brain proteins at pH 7.0, 1,000 μl (prepared in Section 14.3.2)
3. AEX bound fraction of rat brain proteins at pH 5.0, 1,000 μl (prepared in Section 14.3.3)

3.4.2. Reduction and Alkylation of the Protein Samples

1. Add 30 μl of alkylation buffer to 1-ml protein sample to adjust the pH to between 8.0 and 9.0.
2. Add 25 μl of 200 mM tributylphosphine (TBP) to 1-ml sample for a final concentration of 5 mM. Vortex and then incubate at room temperature for 30 min.
3. Add 30 μl of 0.5 M iodoacetamide for a final concentration of 15 mM (*see* Note 6). Mix well by vortex. Incubate at room temperature for 1 h.
4. Add 25 μl of 200 mM TBP to quench any unreacted iodoacetamide. Vortex and then incubate at room temperature for 15 min.
5. Centrifuge at 16,000g for 5 min at room temperature to pellet any insoluble material.

3.4.3. Cleanup the Protein Samples with the ReadyPrep 2-D Cleanup Kit

1. Transfer the reduced and alkylated protein sample (~1.1 ml ea.) into a 50-ml Oak Ridge polypropylene centrifuge tube (VWR). Add 3.3 ml of precipitation agent 1 to each tube and mix well by vortexing. Incubate on ice for 15 min.
2. Add 3.3 ml of precipitation agent 2 to each tube. Mix well by vortexing and centrifuge at 16,000g for 5 min at 4°C to form a tight pellet.
3. Remove and discard the supernatant. Centrifuge at 16,000g for 15–30 sec to collect any residual liquid in the tubes and carefully remove them.

4. Add 100 µl of wash reagent 1 on top of the pellet in each tube. Vortex and then centrifuge at 4°C at 16,000g for 5 min. (*see* Note 7).
5. Remove and discard the wash. Add 50 µl of distilled water to each tube and vortex for 10–20 sec (*see* Note 8).
6. Add 1 ml of wash reagent 2 (prechilled at -20°C for at least 1 h) and 5 µl of wash 2 additive to each tube (*see* Note 9). Vortex for 1 min and then incubate the tubes at -20°C for 30 min. Vortex the tubes for 30 sec every 10 min during the incubation period.
7. Centrifuge at 4°C at 16,000g for 5 min to form a tight pellet in each tube. Remove and discard the supernatant. Centrifuge the tubes at 16,000g for 15–30 sec and remove and discard any residual liquid.

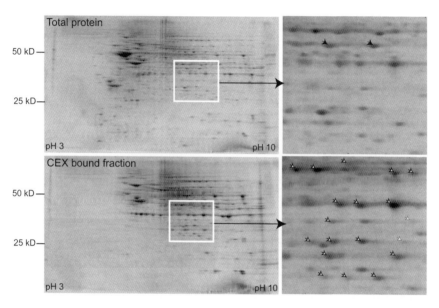

Fig. 1. Enrichment of basic proteins in rat brain after fractionation with the Aurum™ CEX mini spin column at pH 7.0. Compared to the unfractionated rat brain proteins (Total protein, 50 µg), proteins that bound to the CEX mini spin column (CEX bound fraction, 50 µg) focused mostly on the basic side of the 2-D gel. Enlarged views (white squares) reveal that the intensities of many of the proteins increased substantially when compared to the total protein sample (dark empty arrows). Additionally, some proteins that were either very faint or undetectable in the total protein gel were easily detected in the CEX bound fraction gel (light empty arrows), indicating the enrichment of low abundance basic proteins by this method. A few proteins (solid arrows) on the total protein gel were not detected in the CEX bound fraction. Instead, they appeared in the CEX unbound fraction (data not shown). It is possible that the positively charged groups on these proteins were not accessible to the CEX media under the experimental conditions.

8. Air-dry the pellets for no more than 5 min. Resuspend the pellet in each tube with 500 µl of IPG strip rehydration buffer.
9. Determine the protein concentration of the samples using RC DC protein assay (Bio-Rad).

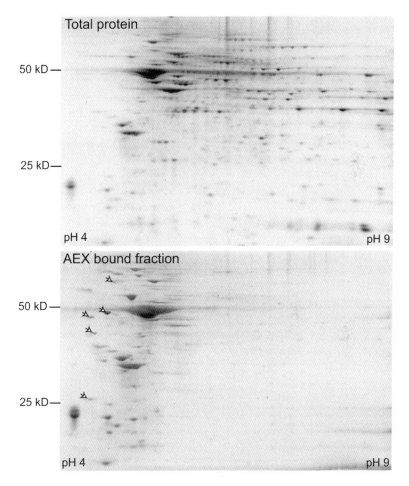

Fig. 2. Enrichment of acidic proteins in rat brain after fractionation with the AEX mini spin column at pH 5.0. Compared to the unfractionated rat brain sample (Total Protein, 50 µg), proteins in the AEX bound fraction (AEX bound fraction, 50 µg) were focused on the acidic side of the 2D gel. Intensities of many protein spots increased after fractionation. Arrows show that several protein spots, which were barely detected in the total protein 2-D gel, were now readily detected in the gel of the AEX bound fraction. Further study demonstrated that the AEX mini spin columns separate the proteins consistently and provide highly reproducible 2-D display of enriched acidic rat brain proteins (see **Note 14**).

3.4.4. 2-D Gel Electrophoresis and Imaging

1. Adjust the protein concentrations of the unfractionated rat brain protein, the CEX bound fraction and the AEX bound fraction to a final concentration of 0.25 mg/ml using IPG strip rehydration buffer.
2. In IPG strip rehydration trays (Bio-Rad), load 200 µl of each sample (50 µg) onto pH 3–10 ReadyStrip IPG strips (3 strips per sample) in. Overlay mineral oil (Bio-Rad) on the IPG strips and allow them to rehydrate overnight.
3. Transfer the rehydrated IPG strips into focusing trays. Set the focusing program on the PROTEAN IEF Cell (Bio-Rad) as following: 250 V, rapid ramp for 125 Vh; 8,000 V rapid ramp for 25,000 Vh; 500 V rapid ramp, infinite (*see* Note 10)
4. After isoelectric focusing, gently remove the mineral oil from the IPG strips. Incubate each IPG strip in 4.0 ml of equilibration buffer I for 20 min at room temperature in rehydration/equilibration trays (Bio-Rad).
5. Discard the equilibration buffer I and incubate each IPG strip in 4.0 ml of equilibration buffer II for 20 min at room temperature.
6. Remove an IPG strip from the equilibration tray, dip it briefly into a graduated cylinder containing 1X Tris/glycine/SDS running buffer, and carefully place the IPG strip in the IPG well of an 8–16% Criterion precast gel. Overlay agarose solution onto the IPG strips (see Note 11).
7. Separate the proteins by SDS-PAGE at 200 V for 1 hr in the Criterion Dodeca Cell (Bio-Rad).
8. After SDS-PAGE, fix the gels in 40% ethanol and 10% acetic acid for 2 h at room temperature (see Note 12).
7. Stain the gels with the Flamingo fluorescent gel stain for 3 h. Cover the gel trays with aluminum foil to limit light exposure.
8. Rinse the gels in distilled water (see Note 13).
9. Use the Molecular Imager PharosFX system (Bio-Rad) to capture 2-D gel images (Figs. 1 and 2).

4. Notes

1. A broad range of buffer systems can be use used with Aurum AEX and CEX mini spin columns *(6)* (Table 1). The buffer system should be chosen based on the pI of the proteins of interest and the purpose of the experiment. The best results of protein fractionation are achieved when buffering ions have the same charge as the functional group on the ion exchanger, e.g., phosphate buffer for CEX columns, or Tris buffer for AEX columns. A buffer concentration of 20 mM is recommended.
2. The AEX and CEX columns are sensitive to salt concentration. The presence of 40 mM Tris in the protein lysis buffer interferes with columns' protein separation performance. Therefore, it is necessary to decrease the Tris concentration in the sample prior to loading onto the columns. In our study, we diluted the Tris concentration to 1 mM. We also determined that the columns perform normally in the presence of 4 mM Tris.

Table 1
Common buffers for ion exchange chromatography (reproduced with permission from Bio-Rad Laboratories).

Cation exchange	Buffering range	Anion exchange	Buffering range
Citric acid	4.2–5.2	Pyridine	4.9–5.6
Acetic acid	4.8–5.2	L-histidine	5.5–6.0
MES	5.5–6.7	Bis-Tris	5.8–7.2
PIPES	6.1–7.5	Imidazole	6.6–7.1
MOPSO	6.5–7.9	Triethanolamine	7.3–8.0
Phosphate	6.7–7.6	Tris	7.5–8.0
TES	7.2–7.8	Bicine	7.6–9.0
HEPES	7.6–8.2	Diethanolamine	8.4–8.8
Tricine	7.8–8.9	Diethylamine	9.5–11.5

3. The AEX and CEX mini spin columns can be applied for protein fractionation for any sample resources including tissues, cell cultures, serum etc. However, crude tissue homogenate may sometimes clog the columns. It is good practice to cleanup the protein sample with the ReadyPrep 2-D cleanup kit to avoid clogging.
4. If the column does not begin to flow after adding protein sample or wash buffer, push the cap back on the column and then remove it again to start the flow.
5. More proteins can be loaded if necessary. The binding capacities of the columns (12.5 mg for CEX and 36 mg for AEX) are much higher than the amount of protein ($\sim$2 mg) applied in our studies.
6. Iodoacetamide is unstable in solution, therefore, requires preparation immediately before use.
7. The volume of wash reagent 1 used should be about 3–4 times of the pellet volume.
8. The amount of water used should be enough to just cover the protein pellet.
9. The volume of the wash reagent 2 must be at least 10 times the volume of water used in the previous step. Add 5 μl of wash additive regardless of the volume of sample.
10. This step of the isoelectric focusing program is optional. It is added to prevent the protein diffusion in the IPG strip after focusing is completed.
11. Melt 1% agarose solution using a microwave oven and apply 1–2 ml liquid agarose solution onto the IPG strip in the well of the Criterion gel. Before the agarose is solidified, gently tap the IPG strip with a pair of forceps to remove any air bubbles trapped between the IPG strip and the gel.
12. The gels may be left in fix solution for up to 24 h. Shortened fix time (less than 2 h) or insufficient fix solution may reduce sensitivity *(7)*.
13. The dye in Flamingo fluorescent gel stain has very low fluorescence when not protein-bound. A destain step is normally not required. However, rinsing the

gels in 0.1% Tween 20 for 10 min will slightly lower background staining and may allow some proteins to be more sensitively detected *(7)*.

14. We have examined the reproducibility of this protein fractionation method. We separated 5 mg of rat brain proteins on three AEX mini spin columns at pH 5.0. 75 µg protein sample from each bound fractions were separated on 2-D gels (triplicate gels for each column). The 2-D gel images were subjected to match analysis by PDQuest software (Bio-Rad). Correlation coefficients of the match-set between the gel groups ranged between 0.97 and 0.98, indicating high reproducibility among the columns.

Acknowledgement

The authors thank Dr. Tim Wehr, Dr. Anton Posch, Julie Hey, MaryGrace Brubacher and Katrina Academia for their critical reviews of this work.

References

1. Walton, H.F. (1968) Ion Exchange Chromatography. *Anal. Chem.* 40, 51R–62R.
2. Liao, J. et al., US Patent No: US 6,423,666 B1
3. Hunter, A.K., Carta, G. (2000) Protein adsorption on novel acrylamido-based polymeric ion-exchangers I. Morphology and equilibrium adsorption. *J Chromatogr. A* 897, 65–80.
4. Hunter, A.K., Carta, G. (2000) Protein adsorption on novel acrylamido-based polymeric ion-exchangers II. Adsorption rates and column behavior. *J Chromatogr. A* 897, 81–97.
5. Hunter, A.K., Carta, G. (2002) Protein adsorption on novel acrylamido-based polymeric ion-exchangers IV. Effects of protein size on adsorption capacity and rate. *J Chromatogr. A* 971, 105–116.
6. Aurum Ion Exchange Mini Kits and Columns Instruction Manual, Bio-Rad bulletin 4110137
7. Flamingo™ Fluorescent Gel Stain Instruction Manuel, Bio-Rad bulletin 10003321

15

Reducing the Complexity of the *Escherichia coli* Proteome by Chromatography on Reactive Dye Columns

Phillip Cash and Ian R. Booth

Summary

High-resolution 2-dimensional gel electrophoresis (2DGE) is a key technology in the analysis of cellular proteomes particularly in the field of microbiology. However, the restricted resolution of 2DGE and the limited dynamic range of established staining methods limit its usefulness for characterising low abundance proteins. Consequently, methods have been developed to either enrich for low abundance proteins directly or to deplete the highly abundant proteins present in complex samples. We present a protocol for affinity chromatography on reactive dye resins for the analysis of the *Escherichia coli* proteome. Using a range of commercially available reactive dye resins in a traditional chromatography system we were able to enrich low abundance proteins to levels suitable for their reliable detection and, most importantly, their identification using standard peptide mass mapping and MALDI-TOF MS methods. Under the chromatography conditions employed up to 4.42% of the proteins present in the total nonfractionated *E. coli* cell lysates bound to the reactive dye column and were subsequently eluted by 1.5 M NaCl. Of the bound proteins approximately 50% were considered to be enriched compared to the nonfractionated cell lysate. The ability to detect low abundance proteins was due to a combination of the specific enrichment of the proteins themselves as well as the depletion of highly abundant cellular proteins, which otherwise obscured the low abundance proteins. There was evidence of some selectivity between the different reactive dye resins for particular proteins. However, the selection of suitable dye resins to selectively enrich for particular classes of proteins remains largely empirical at this time.

Key Words: Chromatography; protein enrichment; proteome; reactive dye columns.

1. Introduction

A key approach for the analysis of prokaryotic proteomes is the separation of the proteins using high-resolution 2-dimensional gel electrophoresis (2DGE). There are, however, well-documented limitations of 2DGE for the analysis of complex proteomes, specifically the separation of proteins with extreme physical properties (for example highly basic or hydrophobic proteins) as well as the detection of low abundance proteins. In the latter case the issue is to be able to recover sufficient material for the identification of the minor relevant proteins. A number of approaches have been used to analyse the low abundance proteins expressed in a cell. These include the application of high protein loads to narrow range immobilized pH gradient gels in the first dimension of 2DGE *(1,2)* and the prefractionation of complex mixtures prior to 2DGE *(3)*. Chromatographic prefractionation has been used to improve the coverage of the cellular proteome when characterized by 2DGE. Fractionation on heparin-actigel columns as well as by chromatofocusing has been used to increase the coverage of the *Haemophilus influenzae* bacterial proteome *(4,5)*. Neither, of these chromatographic methods enriched solely for the low copy number proteins and no single class of protein was selected by either method.

The current chapter describes the application of affinity chromatography on reactive dye resins to prefractionate complex protein mixtures before 2DGE. Although the methodology is applicable to a variety of sample sources, the data presented are drawn from our work on the proteome of *Escherichia coli*. There is a huge dynamic range in protein expression levels within the *E. coli* cell; for example, ribosomal proteins may exist at ~70,000 copies per cell, whereas LacI can be present in as few as 20 copies per cell. Typically, only 50% of the expressed proteins for either *E. coli* or *Sacharomyces cerevisiae* and as low as 20% for mammalian cells are detectable by the standard methods of 2DGE *(6,7)*. The reactive dye resins were used to enrich for the under-represented bacterial proteins that are not detected when analysing total bacterial cell lysates by 2DGE. Reactive dyes have the capacity to assume the polarity and surface geometry for a range of competitive biomolecules and so bind to a variety of proteins *(8)*. The dye compounds bind proteins with high affinity, although the binding capacity and specificity differs between the dye compounds *(8)*. The range of proteins that are capable of binding to the different reactive dyes is largely unknown and most studies have concentrated on their use for purification of specific protein classes. In this context, we have previously demonstrated that Reactive Red 120 can be used to remove abundant proteins from *E. coli* to facilitate the identification of methylglyoxal synthase, an enzyme that is less than 0.1% of the total *E. coli* cell protein complement *(9)*. Early studies described the use of Cibacron-Blue F3-GA to fractionate plasma proteins *(10,11,12)* particularly the depletion of serum albumin thus revealing low abundance proteins by 2DGE. Cibacron-Blue F3-GA also binds to dehydrogenases and kinases *(13,14)*. The other reactive dye compounds are less well characterized. Reactive Blue-2 dye binds to over 60 different proteins *(15)* and the Reactive Brown 10 and Reactive Red 120 dyes bind to tyrosine-tRNA ligases and NADP dependant dehydrogenases respectively *(16,17)*. Jungblut and Klose *(18)* have described a generalized approach to the enrichment of mouse brain proteins by reactive dye compounds. Using a series of five dye compounds they observed that the dyes bound different protein species suggesting that they bind proteins through differing mechanisms. The most selective of the dye columns tested was Cibachron Blue 3GA which bound the smallest portion of the bound sample.

2. Materials

All chemicals are of high purity and obtained from either Sigma or VWR or Fisher unless stated otherwise. Buffers and gel solutions are prepared using MilliQ quality water.

2.1. Preparation of E. coli Proteins

1. French Press, Model FA-073 (SLM Instruments Inc.)
2. Culture medium: Minimal medium based on McIlvaine's buffer at the required pH supplemented with 0.01 M $MgSO_4.7H_2O$, 1 µg/ml thiamine, 1 mg/ml $(NH_4)_2SO_4$, 6 µM $(NH_4)_2SO_4.FeSO_4.6H_2O$ in 1 M HCl and 0.2% glucose. McIlvaine's buffer consists of mixtures of 0.2 M Na_2HPO_4 and 0.1 M citric acid to give solutions of varying pH. 0.2 M K_2HPO_4 at a final concentration of 5 mM was substituted for the equivalent concentration of 0.2 M Na_2HPO_4 to supply essential potassium. This modification provided a means to vary the pH without changing the overall constituents of the media.

2.2. Protein Assay

1. Reagent A: 2% (w/v) Na_2CO_3, 1 M NaOH, 2% (w/v) Na^+/K^+ tartarate. Store at 4°C once prepared.
2. Reagent B: 1% (w/v) $CuSO_4.5H_2O$. Store at 4°C once prepared.
3. Phenol Reagent: Mix 1 ml Folin-Ciocalteau phenol reagent with 14 ml dH_2O. Prepare the phenol reagent on the day of use.

2.3. Affinity Chromatography

1. Purchase reactive dye resins from Sigma UK (Poole, UK) (Table 1) as prepacked columns and use in this format under gravity feed.
2. Fraction collector: LKB 2111 Multirac fraction collector (LKB Instruments, Bromma, Sweden).

3. UV spectrophotometer: Helios g, Unicam.
4. Vivaspin6 tubes (Vivascience, UK) used for concentrating column fractions prior to analysis by either 1-dimensional or 2-dimensional electrophoresis.
5. Column binding buffer: 0.01 M Tris-HCl, pH 7.5, containing 5 mM $MgCl_2$.
6. Column elution buffer: 1.5 M NaCl.
7. Protein sample buffer: 8 M Urea, 2% CHAPS, 40 mM Tris base.

2.4. 1- and 2-Dimensional Gel Electrophoresis

1. Perform one-dimensional gel electrophoresis on commercially available precast 4–12% acrylamide gradient Bis-Tris SDS PAGE gels (Invitrogen, Paisley, UK) using the sample preparation buffer and electrophoresis running buffers from the same supplier.
2. Focus the first dimension IPG gels on a Multiphor II electrophoresis unit cooled to 20°C using a MultiTemp water bath (GE Healthcare, Little Chalfont, UK).
3. Immobilized pH Gradient (IPG) gel strips (7 cm, pH 4-7L) were obtained from GE Healthcare.

Table 1
Properties of the reactive dye compounds and protein recoveries

Column	Dye-ligand	Binding capacity[1]	% Eluted[2]	Percentage enriched[6]
RG-19	Reactive Green 19	3–6 mg HSA[4]/ml	4.42	46
RG-5	Reactive Green 5	3–5 mg HSA[4]/ml	3.50	N.D.[3]
RBr-10	Reactive Brown 10	3–5 mg HSA[4]/ml	1.71	56
RR-120	Reactive Red 120 Type 3000-CL	5–12 mg BSA[5]/ml	2.97	35
CB-3GA	Cibacron Blue 3GA Type 3000-CL	1,000–2,000 U lactic dehydrogenase/ml	3.79	N.D.[3]
RBl-4	Reactive Blue 4	5–7 mg HSA[4]/ml	1.48	58
RBl-72	Reactive Blue 72	3–5 mg HSA[4]/ml	1.76	57
RY-3	Reactive Yellow 3	500–1,000 U citrate synthase/ml	0.62	62
RY-86	Reactive Yellow 86	1 mg HSA[4]/ml	0.38	N.D.[3]

[1] Manufacturer's data
[2] The percentage of the proteins eluted by 1.5 M NaCl for the *E. coli* cell lysates.
[3] N.D.: No Data
[4] HSA: Human serum albumin
[5] BSA: Bovine serum albumin
[6] Percentage of proteins in the bound fraction showing induced levels of >5-fold or unique compared to the nonfractionated samples.

4. IPG reswell buffer: 7 M urea, 2 M thiourea, 4% (w/v) 3-[(3-Cholamidopropyl) dimethylammorio]-1-propanesulfonate hydrate (CHAPS), 0.3% (w/v) DTT, 1% (w/v) IPG Buffer (GE Healthcare).
5. IPG equilibration buffer: 50 mM Tris-HCl, pH 8.8, 6 M urea, 30% (v/v) glycerol, 2% (w/v) SDS.
6. Perform second dimension electrophoresis on SE250 slab gel units (GE Healthcare) cooled with running tap water. Prepare the slab gels as batches using the multi-gel casting unit (GE Healthcare).
7. Second dimension gel solutions. 10% Acrylamide: 16 ml acrylamide monomer solution (30% (w/v) Acrylamide, 0.8% (w/v) Bis-acrylamide), 9 ml 2 M Tris-HCL, pH 8.8, 0.5 ml 10% (w/v) SDS, 21.4 ml MilliQ water, 1.4 ml 75% (v/v) glycerol. The gel solution was polymerized with 100 µl 10% (w/v) Ammonium persulphate and 10 µl TEMED. 15% Acrylamide: 24 ml of acrylamide monomer solution (30% (w/v) Acrylamide, 0.8% (w/v) Bis-acrylamide), 9 ml 2M Tris-HCL, pH 8.8, 0.5 ml 10% (w/v) SDS, 14.5 ml 75% (v/v) glycerol. The gel solution was polymerized with 30 µl 10% (w/v) Ammonium persulphate and 10 µl TEMED. The polymerisation catalysts (ammonium persulphate and TEMED) were added immediately before pouring gels assembled in the casting unit.
8. Second dimension electrophoresis running buffer: 0.4% (w/v) Glycine, 0.05 M Trizma base, 1% (w/v) SDS.

2.5. Peptide Mass Mapping

1. In-gel digestion of selected protein spots, peptide extraction and the preparation of MALDI-TOF MS targets were carried out on ProGEST and ProMS robots (Genomic Solutions, Huntingdon, UK).

3. Methods

3.1. Preparation of Bacterial Cell Extracts

1. A commensal strain of E. coli (strain J1) was used in the following study. Grow the bacteria on a minimal osmotically balanced McIlvaine's medium (203 mosM). (*see* Note 1)
2. Prepare a glucose-limited (0.04% w/v) overnight culture of E. coli in McIlvaine's medium at pH 7.0. Increase the glucose concentration to 0.2% after 16 h of growth and then incubate the cultures further, with shaking, until the OD_{650} doubles (OD_{650} 0.7).
3. Dilute the bacterial cultures 20-fold into fresh, warm McIlvaine's medium, containing 0.2% glucose and grow into midexponential phase (OD_{650} 0.35). The total volume of culture required to prepare the bacterial proteins is 100–250 ml.
4. Add chloramphenicol (12.5 µg/ml final concentration) to prevent further protein synthesis.
5. Collect the bacterial cells by centrifugation at 20,000g for 10 min at 4°C (Sorvall RC-5B, DuPont UK) and suspend the cell pellets in 0.05 volumes of 0.01 M Tris-HCl, pH 7.5, 5 mM $MgCl_2$.

6. Disrupt the bacterial cells disrupted using a French Pressure Cell Press at 18,000 psi.
7. Remove the cell debris by centrifugation at 12,000g for 15 min at 4°C (Sorvall RC-5B, DuPont UK). Clarify the supernatant further by centrifugation at 50,000 rpm for 1 h at 4°C (Beckman TL-100, Beckman-Coulter USA).
8. The supernatant at this stage is taken to represent the "nonfractionated soluble extract" in the following discussion.

3.2. Determination of Protein Concentration

Use the Folin-Ciocalteau method of protein estimation as detailed below to measure protein concentration in the nonfractionated cell extracts using bovine serum albumin as the standard.

1. Prepare three dilutions of the protein samples (1/5 to 1/200, depending upon the cell fraction and estimated concentration of sample) in 0.1 M NaOH, and then pipet 50 μl of each dilution into a flat-bottomed 96-well microtitre plate (Greiner UK Ltd, Gloucestershire) in triplicate. Pipet duplicate, 50 μl samples of BSA protein standards (0, 12.5, 25, 50, 100, 200, 250 μg/ml in 0.1 M NaOH) into wells.
2. Prepare the copper reagent by mixing 50 volumes of reagent A with 1 volume of reagent B, and add 100 μl of this mix into each well. Cover the plate and leave to incubate at room temperature for 10 min. Add 100 μl of Phenol reagent into each well and incubate for a further 15–30 min at room temperature.
3. The optical densities of the solutions were measured in a Labsystems iEMS Reader MF running GENESIS software (Life Sciences International UK Ltd., Basingstoke, Hampshire). Prepare a standard curve for each batch of assays and calculate the protein concentration by reference to this curve.

3.3. Chromatographic Fractionation of Bacterial Proteins

The protocol described uses a large-scale chromatography system for the fractionation of the bacterial protein preparations (see Note 2).

1. Prepacked columns with the individual reactive dye resins were purchased from Sigma as suspensions in 0.5 M NaCl containing 0.02% Thimersol (Table 1). Store the columns (2.5 ml bed volume) +4°C and bring to room temperature before use. Provided the columns are washed correctly after use the same column can be used multiple times. Regenerate the columns by washing the column, from which all protein have been eluted, with 10 column volumes each of: 0.1 M borate, pH 9.8, containing 1 M NaCl, 0.1 M borate pH 9.8, deionized or distilled water, and 2 M NaCl. Between the chromatographic fractionation runs store the columns at 4°C in the presence of 2 M NaCl.
2. Bring the column to room temperature and wash through with eight column volumes of binding buffer to remove unbound dye or storage chemicals.

3. After equilibrating the column with the binding buffer load the cytoplasmic protein suspension onto the column. Typically, load the sample in a volume of 10 ml containing 42–45 mg of protein.
4. Wash unbound proteins through with 12 column volumes of binding buffer at a flow rate of 0.5 ml per min and collect 1.0-ml fractions using an automatic fraction collector.
5. Elute the bound proteins from the column with 1.5 M NaCl (see Note 3). Assay all of the fractions for protein using UV-spectroscopy at OD_{280}. A representative elution profile for proteins fractionated on a Reactive Brown-10 column is shown in Fig. 1.
6. Concentrate and buffer exchange the eluted proteins to protein sample buffer using Vivaspin6 tubes. This process desalts the protein samples as a requirement for subsequent analysis by 2DGE (see Note 4).

3.4. 1- and 2-Dimensional Gel Electrophoresis

3.4.1. 1-Dimensional Electrophoresis in SDS-PAGE

1. Use one-dimensional electrophoresis by SDS-PAGE to monitor the selectivity of the columns as well as to provide a loading guide for the later analysis of the samples by 2DGE. Treat the total cytoplasmic proteins and column fractions with the sample preparation buffer provided by the manufacturer of the electrophoresis gels following their recommendations and load approximately 180 µg of protein for each sample to the gels. The proteins are separated at 200 V for 35 min. The gels are stained as described below for the 2D gels. (see Note 5 and Fig. 2)

3.5. First Dimension of 2DGE

1. Protein samples representing the nonfractionated bacterial cell lysate as well as proteins recovered from the chromatography column fractions are adjusted to a

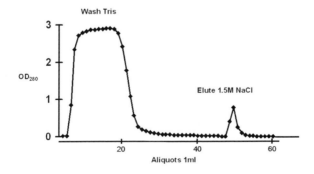

Fig. 1. Elution profile of cellular proteins prepared from *E. coli* J1 from a Reactive Brown-10 column.
Source: Modified from (**19**) with permission

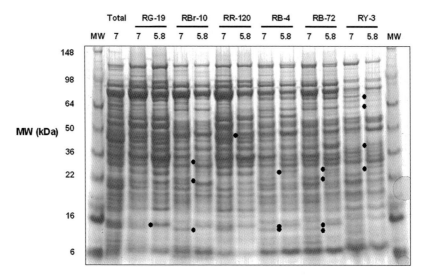

Fig. 2. Analysis of proteins eluted from reactive dye columns by 1-Dimensional gel electrophoresis. Proteins prepared from *E. coli* grown at either pH 7.0 or pH 5.8 were fractionated on the reactive dye columns indicated. The bound proteins were eluted with 1.5 M NaCl and after concentration and desalting analysed by 1DE. Protein differences observed between the profiles of the bacteria grown at either pH 7.0 or pH 5.8 are indicated for each column. ((FIGSRC)Reproduced from (*19*) with permission)

final volume of 140 μl with IPG reswell buffer. Clarify the samples at 11,000*g* for 5 min and use 125 μl of the supernatant to rehydrate the IPG strip for 18 h at room temperature. Typically, approximately 200 μg of protein load to the first dimension IPG gels.

2. Carry out the isoelectric focussing using the Multiphor II flatbed electrophoresis system in three stages. Increase the voltage from 0 to 200 as a linear ramp over 1 min then increase to 3500 V with a linear ramp over 90 min, finally hold the voltage at 3,500 V for 90 min (all stages are at 2 mA and 5 W). Cool the ceramic support plate of the Multiphor II apparatus to 20°C using a MultiTemp water bath.

3. Equilibrate the IPG gel strips in 7-ml IPG equilibration buffer containing 1% (w/v) DTT for 30 min and then in 7-ml equilibration buffer containing 2.5% (w/v) iodoacetamide for another 30 min. The gels are continuously agitated throughout the equilibration process. (*see* Note 6)

3.5.1. Second Dimension of 2DGE

The following protocol for the 2nd-dimension electrophoresis uses the SE250 slab gel apparatus (GE Healthcare).

1. Prepare 10–15% linear gradient acrylamide slab gels for the 2nd-dimension separation as batches of 9 gels as follows. Assemble gel sandwiches of 1 mm thickness into the multigel casting unit as described by the manufacturer.
2. Prepare the acrylamide gradient using a standard linear gradient maker by filling the casting unit from the bottom using a peristaltic pump at a flow rate of 2.5 ml/min. Fill the space below the gel sandwiches with MilliQ water. Immediately after the water has filled the space below the gels, and before air entered the connection tubing, add the gel solutions to the gradient maker. Use 42 ml of each solution with the low density 10% acrylamide gel solution in the mixing chamber of the gradient maker. Pump all of the gel solutions from the gradient maker and, again before air enters the connection tubing, add 75% glycerol to the gradient maker and pump this into the casting unit to displace the acrylamide solution from the space below the gel sandwiches.
3. Allow the slab gels to polymerize overnight and then separate the individual gel sandwiches. Place the equilibrated IPG gels on the surface of the gels.
4. Assemble the slab gels into the individual electrophoresis units. Fill the upper and lower buffer chambers with the 2nd-dimension electrophoresis running buffer. Separate the proteins at 75 V (constant voltage) for 60 min followed by 150 V (constant voltage) for 160 min. Cool the gels with running tap water during electrophoresis.
5. Detect the resolved proteins using Colloidal Coomassie Blue G250 staining *(20)*.
6. Scan the stained gels using a Molecular Dynamics Personal Densitometer at 50 µM resolution to generate 8-bit images and transfer the electronic images to Phoretix 2D Analytical software, version 6.01, (Nonlinear Dynamics, Newcastle, UK) for detailed analysis.

3.5.2. Protein identification by Peptide Mass Mapping

1. Excise the required protein spots stained gels using a clean scalpel and subject to in-gel trypsin digestion. Use the ProGEST and ProMS automated robotic suite for all processing steps for peptide mass mapping.
2. Wash the excised protein spots and then reduce and S-alkylate the proteins before digesting with trypsin (sequencing grade modified trypsin; Promega).
3. Pass an aliquot of the peptide extract through a GELoader tip, which contains a small volume of POROS R2 sorbent (PerSeptive BioSystems, USA). Wash the adsorbed peptides extensively and then elute into 0.5 µl of a saturated solution of α-cyanol-4-hydroxycinnamic acid in 50% (v/v) acetonitrile, 5% (v/v) formic acid.
4. Collect the mass spectra using an Applied Biosystems Voyager-DE STR MALDI-TOF mass spectrometer. Operate the instrument in the reflection delayed extraction mode. Calibrate the spectra internally using the trypsin auto-digestion products.
5. Use the tryptic peptide profiles to search the NCBI nucleotide database using MS-FIT (http://prospector.ucsf.edu/). Use the following search parameters: maximum allowed error of peptide mass 250 ppm, cysteine as S-carbamidomethyl-derivative and oxidation of methionine allowed.

6. The bacterial proteins identified from the enriched fractions following chromatography on the reactive dye columns are displayed on the nonfractionated 2D protein profile of *E. coli* in Fig. 3 and their functions summarized in Table 2.

3.6. Enrichment of E. coli Proteins by Reactive Dye Resins

The use of the reactive dye resins improved the detection of low abundance proteins of *E. coli* and many of the proteins that were enriched were either at undetectable levels in the nonfractionated bacterial proteome or at too low a level for their identification by peptide mass mapping. The enrichment of the bacterial proteins was achieved through two routes. Firstly there was selective binding of the protein to the dye resin and its subsequent elution in 1.5 M NaCl leading to an increased level of the protein in the eluted fraction. This was

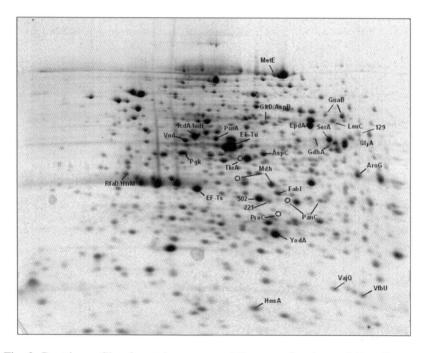

Fig. 3. Protein profile of proteins recovered from nonfractionated *E. coli* grown at pH 7.0. The locations of the proteins identified during the course of this study are indicated. The numbers refer to the spots that were not identified but are referred to in the text. Protein locations indicated by the open circles represent proteins that were detected only under specific enrichment conditions or growth of the bacteria at pH 5.8. These locations were predicted based on the analysis of the protein profiles using the Phoretix 2D™ software.
Source: Reproduced from (*19*) with permission

Table 2
Characteristics of proteins identified by peptide mass mapping and quoted in the text

Gene Name	SWISS-PROT accession number	Protein identification	Function
aroG	P00886	Phospho-2-keto-3-deoxyheptonate aldolase	Aromatic amino acid biosynthesis
aspC	P00509	Aspartate Aminotransferase (transaminase A)	Aspartate biosynthesis
fabI	P29132	Enoyl (Acyl-carrier protein) reductase	Lipid biosynthesis
gdhA	P00370	NADP-specific Glutamate Dehydrogenase	Glutamate biosynthesis
gltD or aspB	P09832	Glutamate Synthase (NADPH) Small Chain, (NADPH-GOGAT)	Nitrogen metabolism
glyA	P00477	SHMT GlyA	One-carbon metabolism
guaB	P06981	Inosine monophosphate dehydrogenase	Nucleoside biosynthesis
hnsA	P08936	DNA binding protein H-NS	DNA binding
icdA or icdE	P08200	Isocitrate dehydrogenase	TCA cycle
leuC	P30127	3-isopropylmalate dehydratase large subunit	leucine biosynthesis
lpdA	P00391	Dihydrolipoamide dehydrogenase	one carbon metabolism
mdh	P06994	Malate dehydrogenase	TCA cycle
metE	P25665	MetE Methionine Synthase B12-indepnt.	methionine biosynthesis
panC	P31663	Pantoate-beta alanine ligase (pantothenate synthetase)	pantothenate biosynthesis
pgk	P11665	phosphoglycerate kinase	glycolysis
proC	P00373	Pyrroline carboxylate reductase	proline biosynthesis
purA	P12283	Adenylosuccinate Synthetase (IMP aspartate ligase)	purine biosynthesis

(Continued)

Table 2
Characteristics of proteins identified by peptide mass mapping and quoted in the text

Gene Name	SWISS-PROT accession number	Protein identification	Function
rfaD or htrM	P17963	ADP-L-Glycero-D-Manno-Heptose-6-Epimerase	lipopolysacharide biosynthesis
serA	P08328	D-3-phosphoglycerate dehydrogenase	one carbon metabolism
EF-TS	P02997	Elongation Factor TS (EF-TS)	protein biosynthesis
EF-Tu	P02990	Elongation factor TU	protein biosynthesis
yajQ	P77482	YajQ	Unknown Function
yfbU	P76482	Protein YFBU	Unknown Function
ynd	P37754	6-phosphogluconate dehydrogenase, decarboxylating	Unknown Function
yodA	P76344	Hypothetical 24.8KDa protein in DCM-SHIA region	Unknown Function

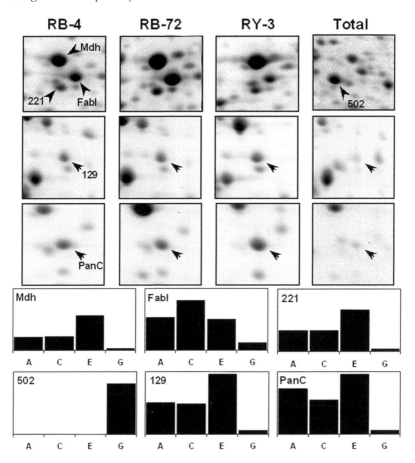

Fig. 4. Chromatographic processing of proteins expressed by *E. coli* grown at either pH 7.0 or pH 5.8. Bacterial proteins were separated on the specified reactive dye columns (RB-4, RB-72 and RY-3) and the proteins eluted by 1.5 M NaCl analysed by 2DGE. The enlarged portions of the protein profiles for three regions of the *E. coli* 2D protein profile are presented for each column bound fraction. The corresponding region of the 2D protein profile for the nonfractionated bacterial cell lysate is also displayed. The histograms show the normalized amounts of each protein present in the bound fractions. The data for PanC were taken from the analysis of *E. coli* grown at pH 5.8 whereas the data for the remaining proteins show the profiles for the bacteria grown at pH7.

observed for PanC (*see* Note 7) as well as an unidentified protein, 129 (Fig. 4). Other proteins showed more modest enrichment levels, which improved the quality of the peptide mass data available for identification (for example Mdh and FabI, Figure 4). Alternatively, there was a depletion of highly abundant proteins, which improved the detection of low abundant proteins, migrating

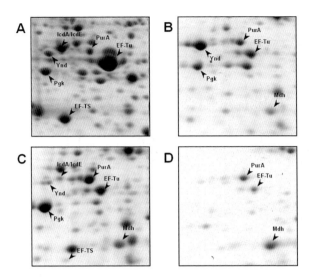

Fig. 5. Chromatographic processing of proteins expressed by *E. coli*. Bacterial proteins were separated on the specified reactive dye columns and the proteins eluted by 1.5 M NaCl analysed by 2DGE. Only enlarged portions of the protein profiles in the region of EF-Tu are presented. The panels show the following protein sources: (A) Nonfractionated bacterial cell lysate, (B) RB-4 (C) RB-72, (D) RY-3.
Source: Reproduced from (**19**) with permission

in the same region of the 2D gel; an example of this was the depletion of EF-Tu from the bound fractions of many of the reactive dye columns (Figs. 5 and 6) (*see* Note 8). In principal these depleted abundant proteins ought to be recovered from the flow-through chromatography fractions, although this was not been formally demonstrated in the current study.

The choice of which of the reactive dye resins was most appropriate for enriching a specific protein or functional class of protein remains empirical at this time. The molecular basis for the interaction of the proteins and individual dyes is poorly understood (*see* Note 9). Considering the enrichment patterns of the identified *E. coli* proteins only the most general conclusions have emerged. The majority of the bacterial proteins were enriched by more than one reactive dye resin (Fig. 6). Three of the bacterial proteins, EF-TS, YodA and EF-Tu, were depleted following chromatography on all of the reactive dye columns screened as part of this study. Both EF-TS and EF-Tu are elongation factors whereas YodA is a hypothetical protein with similarities to *B. subtilis* YrpE, another hypothetical protein. Three proteins AspC, LpdA and Pgk showed similar enrichment profiles with modest or no enrichment by RB-4, RB-72, RR-120, RG-19 and RBr-10 but they were depleted from the bound fraction

Reducing the Complexity of the E. coli Proteome

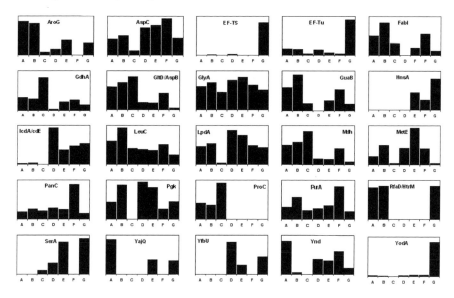

Fig. 6. Expression profiles of identified proteins recovered from reactive dye columns. Whole cell lysates from *E. coli* were analysed on the reactive dye columns and the proteins eluted by 1.5 M NaCl analysed by 2DGE. The 2D protein profiles for the eluted bound fraction were quantified and matched. The histograms show the normalized volumes of the specified proteins in each profile. For those proteins detected as multiple isoforms in the 2D protein profile data for the most abundant form is presented. The reactive dye columns are as follows: (A) RB-4, (B) RB-72, (C) RY-3, (D) RR-120, (E) RG-19, (F) RBr-10, (G) Nonfractionated bacterial cell lysate.
Source: Modified from (*19*) with permission

recovered from RY-3. These three proteins show a diverse range of activities; Pgk is a kinase, LpdA a dehydrogenase and AspC a transaminase. Two identified bacterial proteins, FabI and GuaB, were enriched by RB-4, RB-72 and RBr-10, but there was no enrichment with RY-3 and RG-19 and a complete absence of binding by RR-120. Both of these proteins are involved in reduction pathways utilising NAD^+. Thus, only a limited pattern emerges from the characterization of the proteins and this may be a distinct advantage for a generic technology for sample fractionation. Based on the data available at this time we propose an empirical screen of the reactive dye resins for their abilities to enrich or deplete for specific proteins. There are data available for the binding characteristics of some of reactive dyes that might help direct this screening process (*see* Note 9). Both Stellwagen *(8)* and Scopes *(20)* have previously proposed an empirical approach to identify the appropriate reactive dye compound for a specific purpose. This is a similar strategy to that proposed by Fountoulakis and Takacs *(21)* for cataloguing proteins enriched by specific affinity chromatography systems other than reactive dye resins. As more data become available on the identities of those proteins enriched and depleted by specific reactive dye resins one would expect that a set of guidelines will be developed to provide a more directed strategy for the use of these dyes and that that these guidelines will be applicable to samples of different origins.

4. Notes

1. The bacterial growth conditions presented were specific for the biological system under study rather than being essential for the subsequent chromatographic analyses. The principal aim of the sample preparation method was to achieve a protein suspension with a minimum of particulate material, which could interfere with the chromatography fractionation. Some of the data presented were produced following growth of the bacteria under acid conditions (pH 5.8). This was achieved by modifying the growth media used and had no effect on the later analysis of the bacterial proteins on the reactive dye columns.
2. The current protocol uses a full-scale chromatography system for the fractionation of the bacterial proteins. The possibility of using small scale "spin columns" in microfuge tubes is a potential area of development, which ought to reduce the need for large protein samples as the starting material.
3. Using the separation conditions described, between 0.38% and 4.42% of the proteins present in the nonfractionated cell lysates bound to the column and were subsequently eluted by 1.5 M NaCl (Table 1). These values were believed to be reliable indicators of the amounts of protein bound to the column since no further proteins were detected following elution with 2 M NaCl.
4. The presence of high concentrations of salt in the samples is a potential problem for subsequent analysis by 2DGE. Although IPG gels are capable of processing samples with moderate salt concentrations, steps must be taken where possible to reduce salt levels. This can be achieved using the Vivaspin6 tubes. In addition, an extended low voltage electrophoresis step (200 V for 2 h with the 7-cm IPG strips) can be included at the start of the first dimension electrofocusing to minimize artefacts due to salt in the sample.
5. Even by 1D SDS-PAGE characteristic protein profiles for the eluted proteins were obtained for the reactive dye columns (Fig. 2). The protein profiles obtained for the eluted proteins were clearly distinct from the protein profiles of the nonfractionated extract indicating a selective enrichment of specific proteins species. However there were also some similarities observed between protein profiles obtained from the reactive dye columns. Similar protein profiles were obtained for the two reactive green columns (RG-19 and RG-5), the two reactive yellow columns (RY-3 and RY-86) as well as the two of the reactive blue columns (RB-4 and RB-72) (Fig. 2). These data suggested that related columns had similar binding properties for the bacterial proteins, although the subtle differences present in the protein profiles between these column pairs suggested that further refinement

of the chromatography protocols might enhance the ability of the columns to selectively bind proteins.
6. If required, immediately after the first dimension electrophoresis the IPG gels can be stored frozen at -20°C in clean plastic 15-ml conical tubes for up to 7 d before equilibrating and processing for the second dimension gel electrophoresis. Under these conditions the IPG gel strips are transferred directly to the equilibration buffer when removed from the freezer.
7. The functions of the identified proteins that are described in this chapter are summarized in Table 2.
8. EF-Tu accounted for approximately 10% of the detected proteins in the nonfractionated bacterial proteome whereas after chromatography on all of the reactive dye columns investigated the protein was significantly reduced and in the case of the dye RY-3 EF-Tu represented <0.6% of the bound proteins. This was not due to sample dilution or poor recovery of the proteins, since there was specific enrichment for a number of bacterial proteins on the same columns, for example Ynd and Pgk (Fig. 5).
9. It has been suggested that the dye compounds act as competitive inhibitors for the substrate or coenzyme for a variety of proteins; the reactive dye compound frequently have a higher affinity for the target protein than the competitive molecule *(8,22)*. For some of the reactive dyes specific classes of proteins have been found to bind to the dye compound *(10,12,14,15,23)* and these data provide some guidance on the selection of a specific dye column to form part of an enrichment scheme.

Acknowledgments

We wish to thank Drs Ros Birch and Conor O'Byrne for their helpful discussions of this work. The expert technical assistance of Evelyn Argo and Liz Stewart (Aberdeen Proteome Facility) in analysing proteins by 2DGE and identifying the proteins by peptide mass mapping is greatly appreciated. This work was supported by a grant from the BBSRC. Work in the Aberdeen Proteome Facility is supported in part by grants from Scottish Higher Education Funding Council, BBSRC and Aberdeen University.

References

1. Bjellqvist B, Sanchez JC, Pasquali C, Ravier F, Paquet N, Frutiger S, and Hochstrasser D. (1993) Micropreparative two-dimensional electrophoresis allowing the separation of samples containing milligram amounts of proteins. Electrophoresis. 14, 1375–1378.
2. Westbrook JA, Yan JX, Wait R, Welson SY, and Dunn MJ. (2001) Zooming-in on the proteome: very narrow-range immobilized pH gradients reveal more protein species and isoforms. Electrophoresis. 22, 2865–2871.

3. Cho MJ, Jeon BS, Park JW, Jung TS, Song JY, Lee WK, Choi YJ, Choi SH, Park SG, Park JU, Choe MY, Jung SA, Byun EY, Baik SC, Youn HS, Ko GH, Lim D, and Rhee KH. (2002) Identifying the major proteome components of Helicobacter pylori strain 26695. Electrophoresis. 23, 1161–1173.
4. Fountoulakis M, Langen H, Evers S, Gray C, and Takacs B. (1997) Two-dimensional map of *Haemophilus influenzae* following protein enrichment by heparin chromatography. Electrophoresis. 18, 1193–1202.
5. Fountoulakis M, Langen H, Gray C, and Takacs B. (1998) Enrichment and purification of proteins of Haemophilus influenzae by chromatofocusing. J. Chromatogr. A. 806, 279–291.
6. Herbert, B.R., Sanchez, J.C., and Bini, L. (1997) Two-dimensional electrophoresis: The state of the art and future directions. in Proteome Research: New Frontiers in Functional Genomics. (Wilkins, M.R. Williams, K.L. Appel R.D. and Hochstrasser D.F. Eds.) Springer-Verlag, Berlin. pp 13–33.
7. Celis JE, Rasmussen HH, Madsen P, Leffers H, Honore B, Dejgaard K, Gesser B, Olsen E, Gromov P, Hoffmann HJ, and et al. (1992) The human keratinocyte two-dimensional gel protein database (update 1992): towards an integrated approach to the study of cell proliferation, differentiation and skin diseases. Electrophoresis. 13, 893–959.
8. Stellwagen E. (1990) Chromatography on immobilized reactive dyes. Methods Enzymol. 182, 343–357.
9. Totemeyer S, Booth NA, Nichols WW, Dunbar B, and Booth IR. (1998) From famine to feast: the role of methylglyoxal production in Escherichia coli. Mol. Microbiol. 27, 553–562.
10. Travis J, Bowen J, Tewksbury D, Johnson D, and Pannell R. (1976) Isolation of albumin from whole human plasma and fractionation of albumin-depleted plasma. Biochemical Journal. 157, 301–306.
11. Gianazza E, and Arnaud P. (1982) A general method for fractionation of plasma proteins. Dye-ligand affinity chromatography on immobilized Cibacron blue F3-GA. Biochemical Journal. 201, 129–136.
12. Gianazza E, and Arnaud P. (1982) Chromatography of plasma proteins on immobilized Cibacron Blue F3-GA. Mechanism of the molecular interaction. Biochemical Journal. 203, 637–641.
13. Thompson ST, Cass KH, and Stellwagen E. (1975) Blue dextran-sepharose: an affinity column for the dinucleotide fold in proteins. Proc. Natl. Acad. Sci. U. S. A. 72, 669–672.
14. Lamkin GE, and King EE. (1976) Blue Sepharose: a reusable affinity chromatography medium for purification of alcohol dehydrogenase. Biochem. Biophys. Res. Commun. 72, 560–565.
15. Scopes RK. (1986) Strategies for enzyme isolation using dye-ligand and related adsorbents. J. Chromatogr. A. 376, 131–140.
16. Watson DH, Harvey MJ, and Dean PD. (1978) The selective retardation of NADP+-dependent dehydrogenases by immobilized procion red HE-3B. Biochemical Journal. 173, —596.

17. McArdell JE, Bruton CJ, and Atkinson T. (1987) The isolation of a peptide from the catalytic domain of Bacillus stearothermophilus tryptophyl-tRNA synthetase. The interaction of Brown MX-5BR with tyrosyl-tRNA synthetase. Biochemical Journal. 243, 701–707.
18. Jungblut P, and Klose J. (1989) Dye ligand chromatography and two-dimensional electrophoresis of complex protein extracts from mouse tissue. J. Chromatogr. A. 482, 125–132.
19. Birch RM, O'Byrne C, Booth IR, and Cash P. (2003) Enrichment of Escherichia coli proteins by column chromatography on reactive dye columns. Proteomics 3, 764–776.
20. Scopes RK. (1987) Dye-ligands and multifunctional adsorbents: an empirical approach to affinity chromatography. Anal. Biochem. 165, 235–246.
21. Fountoulakis M, and Takacs B. (1998) Design of protein purification pathways: application to the proteome of Haemophilus influenzae using heparin chromatography. Protein Expr. Purif. 14, 113–119.
22. Denizli A, and Piskin E. (2001) Dye-ligand affinity systems. J. Biochem. Biophys. Methods. 49, 391–416.
23. Kobayashi R, and Fang VS. (1976) Studies on cyclic GMP-dependent protein kinase properties by blue dextran-sepharose chromatography. Biochem. Biophys. Res. Commun. 69, 1080–1087.

16

Reducing Sample Complexity in Proteomics by Chromatofocusing with Simple Buffer Mixtures

Hong Shen, Xiang Li, Charles J. Bieberich, and Douglas D. Frey

Summary

Chromatofocusing has many potential applications in the field of proteomics, such as for the isolation and removal of major sample components to facilitate the analysis of low-abundance components, and for sample prefractionation prior to a subsequent separation using SDS-PAGE, narrow-pI-range 2D-PAGE, or additional chromatography steps. However, the chromatofocusing techniques that are most commonly used employ propriety polyampholyte elution buffers and highly specialized column packings, both of which limit the use of chromatofocusing in practice. To expand the range of application for this technique, this chapter considers chromatofocusing methods which employ common ion-exchange column packings and elution buffers which are simple mixtures of readily available buffering species. Of particular interest is the use of chromatofocusing with a multistep pH gradient for the fractionation of protein mixtures into narrow-pI-range fractions. The cross-contamination characteristics of these fractions using SDS-PAGE are also assessed.

Key Words: Chromatofocusing; multistep pH gradient; prefractionation; proteomics; simple elution buffer; reducing sample complexity.

1. Introduction

In the 1970s, Sluyterman and his colleagues *(1–4)* developed a chromatographic version of isoelectric focusing, termed chromatofocusing, which combines the most favorable attributes of both chromatography and isoelectric focusing. In this method, a retained pH gradient is formed dynamically inside the column (i.e., with no external mixing) to separate amphoteric substances, e.g., proteins, in the order of their isoelectric points (pIs). In the HPLC version of the technique as traditionally practiced, a Mono P column—essentially a weak-base, anion-exchange column manufactured specifically for chromatofocusing by Amersham Biosciences—is first equilibrated with a starting buffer at a high pH value to form the upper limit of the pH gradient. Subsequently, a polyampholyte elution buffer at a low pH value is introduced into the column as a step change at the column entrance, and at the same time the protein sample is injected into the column. A retained linear pH gradient which spans the pH range between the elution and starting buffers is then self-generated inside the column. During chromatofocusing, if a protein is located at a certain position in the column where the pH in the fluid phase is not equivalent to the isoelectric point of the protein, then because of its amphoteric properties, the protein acquires a charge and interacts with the column packing so that it migrates to a single location on the pH gradient. As a result of this unique focusing effect which is not present in standard gradient elution chromatography methods, proteins are eluted as narrow bands in the order of their isoelectric points at a high separation resolution.

Although the theory underlying chromatofocusing is complicated, in practice the method is very simple to perform because the gradients employed are produced inside the column with no external mixing, as opposed to the externally produced gradients generally used in other types of gradient elution chromatography. Because of the ease of operation and the high resolution achieved, chromatofocusing is a powerful tool not only for the purification of a target protein, which is perhaps its most common application, but also for sample prefractionation in proteomics, which is the main application considered here. For example, because of the fact that chromatofocusing separates proteins in order of their pIs, the fractions collected from chromatofocusing should be well suited as a feed material for narrow-pI-range, two-dimensional polyacrylamide gel electrophoresis (2D-PAGE) so that better resolution can be achieved with these types of gels, as compared to the case without prefractionation, because without prefractionation the out-of-range proteins in a narrow-pI-range gel lead to a reduced resolution. In addition, the pH at which the protein elutes during chromatofocusing, although referred to as the isoelectric point above, usually differs somewhat from the true pI, so that it is more properly termed the apparent pI *(1,5,6)*. Consequently, proteins are slightly charged as they migrate down the column during chromatofocusing, which is likely to be advantageous in the current context because this charge tends to enhance protein solubility, as does the low ionic strength, which is also a characteristic of the method. At the same time, the true and apparent pIs are still sufficiently similar so that chromatofocusing is likely to yield fractions highly suitable for various types of subsequent processing, such as the narrow-pI-range gel-based fractionation described above. Chromatofocusing also provides other unique characteristics that can be exploited for various purposes, such as the ability to suppress peak tailing under mass-overloaded conditions, which in turn is likely to reduce fraction cross-contamination when chromatofocusing is used to produce discrete fractions for subsequent analysis. It should be noted that chromatofocusing has been applied in a previous study as a prefractionation method in which a polyampholyte elution buffer was used to form a linear pH gradient on a column packing manufactured specifically for chromatofocusing. Among other results, this previous study demonstrated the ability of chromatofocusing to enrich low-abundance proteins in a bacterial lysate *(7)*.

The usefulness of chromatofocusing for various applications becomes even more apparent if two major constraints characteristic of traditional chromatofocusing are eliminated. The first is the choice of the elution buffer, which in the case of traditional chromatofocusing consists of a polyampholyte mixture. However, polyampholytes tend to form association complexes with proteins, which may complicate subsequent procedures, such as mass spectrometry or the staining method used in gel electrophoresis. The second constraint is the choice of the column packing. There is a common misconception among practitioners of chromatofocusing that a special column packing (such as the Mono P column packing described earlier, or the analogous low-pressure column packings PBE 94 and PBE 118, also manufactured by Amersham Biosciences) must be employed to achieve a retained pH gradient. In actual fact, however, with appropriately designed starting and elution buffers, chromatofocusing can in principle be performed on any type of ion-exchange column packing *(8–10)*.

Chromatofocusing can be performed using a variety of retained pH gradient shapes and with different types of ion-exchange columns. This provides the opportunity to choose a suitable prefractionation strategy for each biological system of interest. However, either computer-aided design software, or considerable experience and knowledge, is required to design a simple elution buffer mixture to achieve a desired pH gradient, the descriptions of which are outside the scope of this chapter *(8,9,11–16)*. To broaden the usefulness of chromatofocusing for the general chromatographic and proteomics communities, and to avoid the complexities associated with understanding the details of the methodology, several buffer "recipes" pertinent to both gradual and multistep pH gradients formed on different types of ion-exchange columns will be described, and the resulting pH gradient achieved will be illustrated. Furthermore, although a gradual pH gradient is useful for resolving very similar proteins, a multistep pH gradient is often more useful for reproducibly fractionating a protein mixture into discrete fractions which have the property that there is minimal cross-contamination so that each protein in the original mixture is largely contained in a single fraction. Because it is this last property which tends to be the most useful characteristic of protein prefractionation methods used for proteomics, chromatofocusing using a multistep pH gradient will be applied to both a bacterial lysate and a model protein mixture, and the extent of fraction cross-contamination will be assessed by sodium dodecyl sulfate-polyacrylamide gel electrophoresis (SDS-PAGE).

2. Materials

2.1. Escherichia coli Lysate Sample Preparation (see Note 1)

1. Wash buffer: 50 mM Tris-HCl, pH 8.0, 50 mM NaCl.
2. Lysis buffer: 150 mM NaCl, 50 mM Tris-HCl, pH 7.5, 0.25% (v/v) Nonidet P 40 (NP40) containing 1 mM protease inhibitor cocktail (Roche, Indianapolis, IN).
3. Deoxyribonucleate 5'-oligonucleotido-hydrolase (DNase I) and Ribonuclease A (RNase A) stock solution (5×): 1 mg/ml DNase I, 0.25 mg/ml RNase A, 50 mM $MgCl_2$, and 500 mM Tris base titrated with HCl to pH 8.0. Store at −20 °C.
4. 0.2-µm polysulfone filter (Millipore, Billerica, MA).
5. 3000 MW cut-off dialysis membrane (Pierce Biotechnology, Rockford, IL).

2.2. Columns for Chromatofocusing

1. ProPac WAX-G, 4 × 50 mm (Dionex, Sunnyvale, CA).
2. ProPac SAX-10, 4 × 250 mm (Dionex, Sunnyvale, CA.)
3. ProPac WAX-10, 4 × 250 mm (Dionex, Sunnyvale, CA).
4. ProPac WCX-10, 4 × 250 mm (Dionex, Sunnyvale, CA).
5. TSK-GEL Q-5PW, 7.5 × 75 mm (Tosoh Bioscience, Montgomeryville, PA).
6. TSK-GEL SP-5PW, 7.5 × 75 mm (Tosoh Bioscience, Montgomeryville, PA).
7. CM-5PW, 5 × 50 mm (Tosoh Bioscience, Montgomeryville, PA).
8. PolyCAT, 4.6 × 100 mm (PolyLC, Columbia, MD).
9. Mono P, 5 × 50 mm (Amersham Biosciences, Piscataway, NJ).

2.3. Buffers for Example 1 (see Figs. 1–2)

1. Starting buffer: 20 mM NaOH titrated with 3-(cyclohexylamino)-1-propane-sulfonic acid (CAPS) to pH 11.
2. Elution buffer: 20 mM NaOH, 60 mM glycine, 50 mM [(2-hydroxy-1,1-bis (hydroxymethyl)ethyl)amino]-1-propanesulfonic acid (TAPS), 30 mM 3-(N-morpholino)propanesulfonic acid (MOPS), 20 mM 2-(N-morpholino) ethanesulfonic acid hydrate (MES), 10 mM acetic acid, and 5 mM lactic acid titrated with HCl to pH 3.0.

2.4. Buffers for Example 2 (see Figs. 3–4)

1. Starting buffer: 20 mM NaOH, 25 mM glycine, and 0.02% n-octyl-β-D-glucopyranoside (*see* Note 2) at pH 10.2.

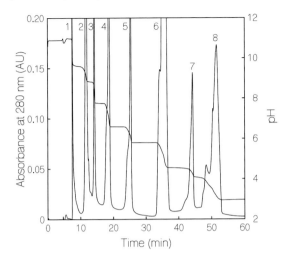

Fig. 1. Elution of 2.6 mg of a protein mixture using an eight-step pH gradient formed on a ProPac WAX-G column 4 × 50 mm, (*see* Note 3), and a TSK-GEL Q-5PW column, 7.5 × 75 mm, in series. The buffer compositions are described in example 1. The protein mixture contained cytochrome *c* (pI = 10.8), α-chymotrypsinogen A (pI = 9.6), myoglobin (pI = 7.3), hemoglobins A, F, S, and C (pIs = 7.1–7.5), rabbit immunoglobin G (pIs = 6.6–6.9), human serum albumin (HSA) (pI = 5.8), β-gluosidase (pI = 3.6), and fetuin (pI = 3.5). The flow rate was 0.5 ml/min, and the absorbance was monitored at 280 nm. The numbers shown indicate the fractions collected, and 10 μl of each fraction was further loaded to a 10% SDS-PAGE gel as shown in Fig. 2.

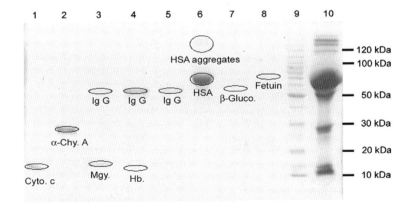

Fig. 2. The fractions separated from a protein mixture using a strong-base anion-exchange column as shown in Fig. 1 examined using a 10% SDS-PAGE gel with Coomassie Blue staining. Lanes 1–8 correspond to fractions 1–8 denoted in Fig. 1, lane 9 shows the protein standards (Invitrogen, Carlsbad, CA), and lane 10 shows the feed sample.

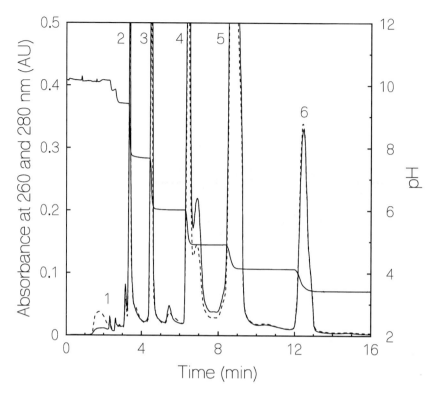

Fig. 3. Elution of 2 mg of an *E. coli* lysate using a six-step pH gradient formed on a ProPac SAX-10 column, 4 × 250 mm. The buffer compositions are as described in example 2. The flow rate was 1.0 ml/min, and the absorbance was monitored at 260 (dotted curve) and 280 (solid curve) nm. The numbers shown indicate the fractions collected, and 15 µl of each concentrated fraction was further loaded onto a 5–20% SDS-PAGE gel as shown in Fig. 4.

2. Elution buffer: 20 mM NaOH, 30 mM tricine, 20 mM MES, 10 mM acetic acid, 8.8 mM formic acid, and 0.02% n-octyl-β-D-glucopyranoside titrated with HCl to pH 3.5.

2.5. Buffers for Example 3 (see Figs. 5–6)

1. Starting buffer: 8 M urea, 20 mM NaOH, 25 mM glycine, and 0.02% n-octyl-β-D-glucopyranoside at pH 10.2.
2. Elution buffer: 8 M urea, 20 mM NaOH, 30 mM tricine, 20 mM MES, 10 mM acetic acid, 8.8 mM formic acid, and 0.02% n-octyl-β-D-glucopyranoside titrated with HCl to pH 3.5.

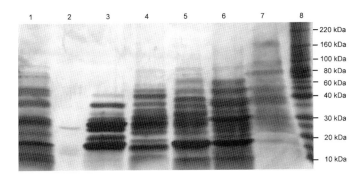

Fig. 4. 5–20% SDS-PAGE gel of the *E. coli* lysate fractions separated on a strong-base anion-exchange column as shown in Fig. 3. Lane 1 shows the feed sample (2 µl), lanes 2–7 correspond to fractions 1–6 denoted in Fig. 3, and lane 8 shows the protein standards (Invitrogen, Carlsbad, CA). The gel was stained with silver stain.

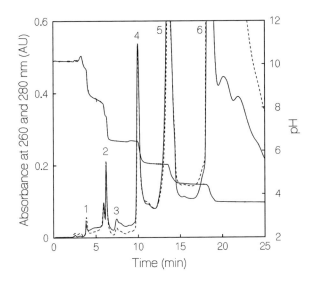

Fig. 5. Elution of 5 mg of an *E. coli* lysate using a six-step pH gradient performed on a ProPac SAX-10 column, 4 × 250 mm (*see* Note 4). The buffer compositions are as described in example 3. The flow rate was 0.5 ml/min, and the absorbance was monitored at 260 (dotted curve) and 280 (solid curve) nm. The numbers shown indicate the fractions collected, and 15 µl of each concentrated fraction was further loaded onto a 12% SDS-PAGE gel as shown in Fig. 6.

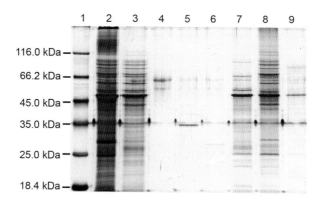

Fig. 6. 12% SDS-PAGE gel of the *E. coli* lysate fractions separated on a strong-base anion-exchange column as shown in Fig. 5. Lane 1 shows the protein standards (Fermentas Life Sciences, Hanover, MD), lane 2 shows the feed sample, lane 3 shows the unbounded proteins during loading, and lanes 4–9 correspond to fractions 1–6 denoted in Fig. 5. The gel was stained with silver stain.

2.6. Retained pH Gradients Formed on Various Columns

2.6.1. Recipes 1–4

See Fig. 7 and Table 1.

2.6.2. Recipes 5-8

See Fig. 8 and Table 2.

2.6.3. Recipes 9–12

See Fig. 9 and Table 3.

3. Methods
3.1. E. coli Lysate Sample Preparation (see Note 5)

1. *E. coli* cells are grown in Luria-Bertani (LB) media at 37°C.
2. The cells are harvested and washed with phosphate buffered saline (PBS) by repeated centrifugation and suspension.
3. The cells are resuspended in lysis buffer and forced through a French Press (Aminco, Silver Spring, MD) at 1,000 psi to break the cells.
4. The lysate is centrifuged (26,000g, 4°C, 30 min) to remove the cell debris.
5. The supernatant is filtered through 0.22-µm PVDF filters (Millipore, Billerica, MA) to remove lipids.

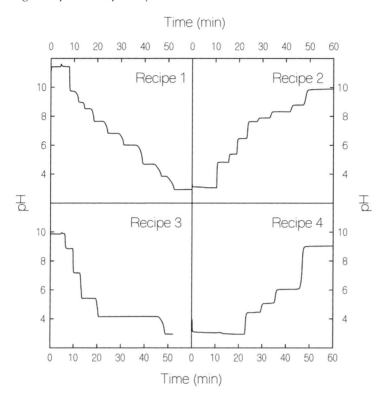

Fig. 7. Top left panel: The pH gradient formed on a TSK-GEL Q-5PW column using recipe 1. The flow rate was 0.5 ml/min. Top right panel: The pH gradient formed on a TSK-GEL SP-5PW column using recipe 2. The flow rate was 0.2 ml/min. Bottom left panel: The pH gradient formed on a ProPac WAX-G and a TSK-GEL Q-5PW column used in series using recipe 3. The flow rate was 0.5 ml/min. Bottom right panel: The pH gradient formed on a TSK-GEL SP-5PW column using recipe 4. The flow rate was 0.2 ml/min. The buffer compositions for recipes 1–4 are described in Table 1.

6. DNase I and RNase A stock solution is diluted with lysate solution to reach the DNase I and RNase A concentrations of 0.2 and 0.05 mg/ml, respectively, and the lysate is incubated at 4°C with this solution for 2 h with agitation.
7. Proteins in the lysate are precipitated with ammonia sulfate at 85% saturation (*see* Note 6), and the protein pellet is recovered by centrifugation (70,000g, 4°C, 30 min).
8. The recovered proteins in the lysate are dialyzed using a 3,000 MW cutoff membrane (Pierce Biotechnology, Rockford, IL) against 1 L of dialysis buffer in a cold room (*see* Note 7) for three times of length 2 h, 2 h, and 12 h.
9. Bicinchoninic acid (BCA) assay (Pierce Biotechnology, Rockford, IL) is performed to measure the protein concentration after dialysis.

Table 1
Recipes for the pH gradients shown in Fig. 7

Recipe	Column	Buffers
1 (see Note 8)	TSK-GEL Q-5PW, 7.5 × 75 mm	Starting buffer: 20 mM NaOH titrated with CAPS to pH 11.4. Elution buffer: 20 mM NaOH, 40 mM glycine, 18 mM tricine, 23 mM N-(1,1-dimethyl-2-hydroxyethyl)-3-amino-2-hydroxypropanesulfonic acid (AMPSO), 13 mM MOPS, 9 mM MES, 6 mM acetic acid, and 4 mM formic acid titrated with HCl to pH 3.0.
2	TSK-GEL SP-5PW, 7.5 × 75 mm	Starting buffer: 20 mM N, O-dimethylhydroxylamine hydrochloride, pH 3.1. Elution buffer: 20 mM N, N-dimethylhydroxylamine hydrochloride, 13 mM bis(2-hydroxyethyl)amino-tris(hydroxymethyl)methane (BIS-TRIS), 10 mM triethanolamine, 7 mM Tris base, 5 mM diethanolamine, and 3 mM ethanolamine titrated with NaOH to pH 10.
3 (see Note 9)	ProPac WAX-0 +TSK-GEL Q-5PW, 7.5 × 75 mm	Starting buffer: 20 mM NaOH and 30 mM glycine, pH 9.9. Elution buffer: 20 mM NaOH, 60 mM tricine, 44 mM MES, and 38 mM acetic acid titrated with trifluoroacetic acid (TFA) to pH 3.0.
4	TSK-GEL SP-5PW, 7.5 × 75 mm	Starting buffer: 20 mM N, O-dimethylhydroxylamine hydrochloride, pH 3.1. Elution buffer: 20 mM N, N-dimethylhydroxylamine hydrochloride, 17.5 mM BIS-TRIS, and 12.1 mM formic acid titrated with ethanolamine to pH 9.

10. Before chromatofocusing, the lysate is dialyzed against the starting buffer for 1 h (see Note 10), and finally filtered through a 0.2-μm polysulfone filter (Millipore, Billerica, MA).

3.2. Procedures for Chromatofocusing (see Note 11)

1. Chromatofocusing can be performed using a standard HPLC or FPLC system and, in its simplest implementation, it can be performed using a single isocratic

Reducing Sample Complexity in Proteomics

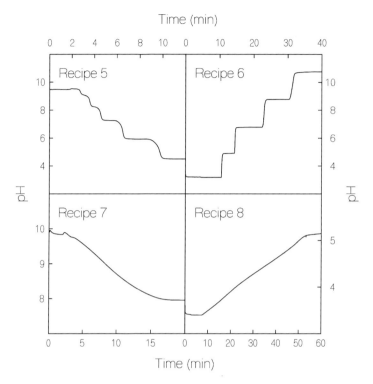

Fig. 8. Top left panel: The pH gradient formed on a Mono P column using recipe 5. The flow rate was 0.5 ml/min. Top right panel: The pH gradient formed on a TSK-GEL SP-5PW column using recipe 6. The flow rate was 0.2 ml/min. Bottom left panel: The pH gradient formed on a Mono P column using recipe 7. The flow rate was 0.5 ml/min. Bottom right panel: The pH gradient formed on a PolyCAT column using recipe 8. The flow rate was 0.5 ml/min. The buffer compositions for recipes 5–8 are described in Table 2.

   pump as shown in Fig. 10. Method development and monitoring is facilitated by the use of a pH measurement flow cell as also shown in Fig. 10 (*see* Note 12).
2. All buffer solutions are prepared using distilled water and are degassed by vacuum filtering using a 47-mm diameter nylon membrane filter with 0.2-μm pores (Whatman, Clifton, NJ).
3. During column equilibration, the injection and bypass valves are set at the 1–2 and 1–6 positions, respectively. The starting buffer, which is connected to the solvent selection valve, is then directed to the column (*see* Note 13) and flow is continued until the pH of the effluent reaches the pH of the starting buffer.
4. The lysate is injected into the column with the starting buffer (*see* Note 14).
5. Both the injection and bypass valves are set at the 1–2 positions. The elution buffer connected to the solvent selection valve is used to purge the column inlet, with about 4 ml of elution buffer used for this purpose (*see* Note 15).

Table 2
Recipes for the pH gradient shown in Fig. 8

Recipe	Column	Buffers
5 (see Note 16)	Mono P, 5 × 50 mm	Starting buffer: 20 mM NaOH and 33 mM 2-(cyclohexylamino)ethanesulfonic acid (CHES), pH 9.5. Elution buffer: 20 mM NaOH, 20 mM MES, 20 mM TAPS, and 20 mM 2-[(2-hydroxy-1,1-bis(hydroxymethyl)ethyl)amino]ethanesulfonic acid (TES) titrated with acetic acid to pH 4.5.
6	TSK-GEL SP-5PW, 7.5 × 750 mm	Starting buffer: 20 mM N, O-dimethylhydroxylamine hydrochloride, pH 3.1. Elution buffer: 20 mM HCl, 7 mM CHES, 29 mM BIS-TRIS, and 11 mM diethanolamine titrated with NaOH to pH 11.
7 (see Note 17)	Mono P, 5 × 50 mm	Starting buffer: 1.8 mM ethanolamine, 2.3 mM diethanolamine, 0.5 mM N-methyldiethanolamine, and 3.5 mM Tris base, pH 9.9. Elution buffer: 1.8 mM ethanolamine, 2.3 mM diethanolamine, 0.5 mM N-methyldiethanolamine, and 3.5 mM Tris base titrated with HCl to pH 8.0.
8	PolyCAT, 4.6 × 100 mm	Starting buffer: 0.1 mM pyroglutamic acid, 8 mM acetic acid, 20 mM MES, and 4.7 mM MOPS, pH 3.5. Elution buffer: 0.1 mM pyroglutamic acid, 8 mM acetic acid, 20 mM MES, and 4.7 mM MOPS titrated with NaOH to pH 5.2.

6. The bypass valve is switched to the 1–6 position, and pH gradient elution is started.
7. The fraction eluted from each pH front is directed to an appropriate collection vial either manually, or by using an automated fraction collector.
8. A small portion of each fraction (2–10 µl) is loaded onto a SDS-PAGE gel to verify the resolution of the separation obtained from chromatofocusing (see Note 18). Three examples are shown in Figs. 1 to 6.

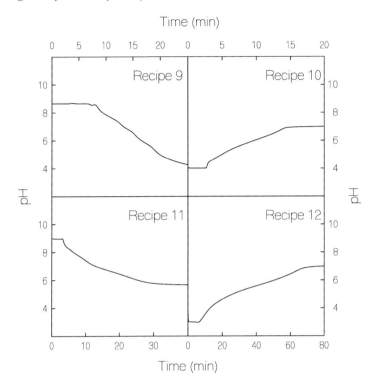

Fig. 9. Top left panel: The pH gradient formed on a Mono P column using recipe 9. The flow rate was 0.5 ml/min. Top right panel: The pH gradient formed on a ProPac WCX-10 column using recipe 10. The flow rate was 0.5 ml/min. Bottom left panel: The pH gradient formed on a ProPac WAX-10 column using recipe 11. The flow rate was 0.5 ml/min. Bottom right panel: The pH gradient formed on a CM-5PW column using recipe 12. The flow rate was 0.2 ml/min. The buffer compositions for recipes 9–12 are described in Table 3.

Table 3
Recipes for the pH gradient shown in Fig. 9

Recipe	Column	Buffers
9	Mono P, 5 × 50 mm	2.5 mM triethylenetetramine, 1.1 mM trimethylamine, 26 mM triethanolamine, 4 mM 1,4-dimethylpiperazine, 3.5 mM piperazine, and 2.8 mM BIS-TRIS, titrated with HCl to pH 8.6 and to pH 3.5 for starting and elution buffers, respectively.

(Continued)

Table 3 (Continued)

Recipe	Column	Buffers
10	ProPac WCX-10, 4 × 250 mm	2 mM pyroglutamic acid, 4 mM MES, 2 mM MOPS, and 6 mM glutaric acid titrated with NaOH to pH 4 and to pH 7 for the starting and elution buffers, respectively.
11	ProPac WAX-10, 4 × 250 mm	3.8 mM triethylenetetramine, 2.9 mM trimethylamine, 3.3 mM triethanolamine, 2.3 mM 1, 4-dimethylpiperazine, 1 mM piperazine, 5.2 mM BIS-TRIS, 15.6 mM N-methyldiethanolamine, 3.8 mM imidazole, 2.1 mM 1-methylpiperazine, and 1.9 mM 1,3-Bis[tris(hydroxymethyl)methylamino]propane (BIS-TRIS propane) titrated with HCl to pH 9.0 and to pH 5.6 for the starting and elution buffers, respectively.
12	CM-5PW, 5 × 50 mm	2 mM pyroglutamic acid, 4 mM MES, 2 mM MOPS, and 6 mM glutaric acid titrated with NaOH to pH 3 and to pH 7 for the starting and elution buffers, respectively.

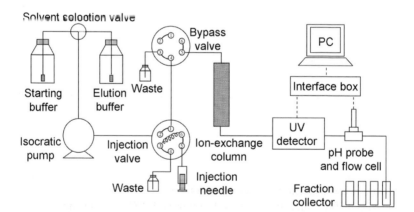

Fig. 10. Equipment for performing chromatofocusing.

4. Notes

1. Unless specified, all the chemicals were purchased from Sigma-Aldrich (St. Louis, MO)
2. The nonionic detergent n-octyl-β-D-glucopyranoside is reported to effectively solubilize proteins *(17)*.

3. The purpose for using a guard column (ProPac WAX-G) in front of the main separation column (TSK-GEL Q-5PW) is to protect the latter column from fouling. The guard column normally does not affect the pH gradient because of its small size.
4. The distortion of the early portion of the pH gradient is likely because of the presence of large amounts of nucleic acids in the feed sample, which results from the omission of the ammonia sulfate precipitation procedure described in step 6 of Section 3.1. The observed ratio of UV absorbances at 260 and 280 nm of 1.2–2.0 indicates that these nucleic acids exit the column primarily in the final peak.
5. This sample preparation protocol is the one used for the *E. coli* cell lysate presented in example 2. Different sample preparation procedures can be used depending on the properties of the sample, as long as two basic rules are followed. First, the additives used for sample preparation should be compatible with the column packing. For example, anionic or cationic detergents should be avoided when anion- or cation-exchange columns are used, respectively. Second, the sample should be in a buffer that is as similar as possible to the starting buffer, which can be accomplished by subjecting the sample to dialysis, to ultrafiltration followed by dilution, or in some cases to simple dilution with the starting buffer.
6. This step may result in a reduced protein recovery, and can be eliminated if DNA and RNA in the cell lysate do not interfere with the protein separation during chromatofocusing
7. The choice of the dialysis buffer depends on the composition of the starting buffer used. Generally, the starting buffer adjusted to a neutral pH can be used as the dialysis buffer because this tends to minimize the effect of the sample on the pH gradient while also limiting sample exposure to extremes in pH.
8. This nine-step pH gradient can also be performed on other types of strong-base anion- exchange columns, e.g., ProPac SAX-10.
9. This pH gradient is specially designed for the prefractionation of blood serum samples. The fourth front followed by the long plateau is used to focus and completely isolate serum albumin which normally constitutes 60% of a serum sample.
10. This step is performed to help ensure that the proteins in the sample are bound to the column packings as strongly as possible. An alternative method to accomplish this is to use ultrafiltration with a 3,000 MW cutoff membrane in an ultrafilter microcentrifuge tube operated at 14,000g and 4°C for 100 min. The retentate can then be diluted to the desired protein concentration with the starting buffer.
11. This procedure can be applied to any type of chromatofocusing method.
12. Before performing chromatofocusing, the pH meter and probe should be calibrated.
13. In principal, any type of ion-exchange column can be used. However, for the specific case of multistep pH gradient chromatofocusing, strong-acid or strong-base columns are preferred over weak-acid or weak-base columns because the pH fronts are shaper for the first two types of columns.

14. The pH of the effluent may fluctuate during sample injection. It is recommended to maintain starting buffer flowing through the column after sample injection until the pH signal is stable. The amount of sample loading depends on the loading capacity of the column. For example, in the case of ProPac SAX-10 column the sample loading amount was varied from 0.01 to 1 mg for the case of a microbore column with dimensions of 25 × 0.1 cm, and from 0.1 to 16 mg for the case of a standard analytical-scale column with dimensions of 25 × 0.4 cm.
15. The purpose of this and the next step is to ensure that the step change in the buffer composition occurs immediately before the column inlet so that the pH fronts constituting the multistep pH gradient are as sharp as possible throughout the chromatofocusing process. However, this step and the use of bypass valve can be omitted if a sufficiently long column is used because the self-sharpening nature of the pH fronts that form will cause these fronts to eventually become sharp inside the column despite the presence of upstream extra-column band broadening.
16. In general, the individual pH fronts formed on a weak-base anion-exchange column are not as sharp as those formed on a strong-base column. However, weak-base ion-exchange columns may impart additional control over the shape of the pH gradient which may be desirable in some situations. This multistep pH gradient is applicable to this latter type of column as well.
17. This pH gradient is designed for the prefractionation of alkaline proteins because 2D-PAGE tends to exhibit poor resolution of proteins with extremely alkaline pI values.
18. After this prefractionation method, proteins in each fraction can be precipitated with trichloroacetic acid (TCA) at 16% (w/v) final concentration, and then recovered by centrifugation (15,000g, 4°C, 10 min). The recovered proteins can further be separated using 2D-PAGE.

Acknowledgements

This work has been supported by grant CTS-0442072 from the National Science Foundation. The authors also thank Remco van Soest and Chris Pohl (Dionex/LC Packings, Sunnyvalle, CA) for providing technical support and for donating several of the columns used in this study.

References

1. Sluyterman, L. A. A., and Elgersma, O. (1978) Chromatofocusing: Isoelectric focusing on ion-exchange columns. I. General principles. *J. Chromatogr.* **150**, 17–30.
2. Sluyterman, L. A. A., and Wijdenes, J. (1978) Chromatofocusing: isoelectric focusing on ion-exchange columns. II. Experimental verification. *J. Chromatogr.* **150**, 31–44.

3. Sluyterman, L. A. A., and Kooistra, C. (1989) Ten years of chromatofocusing: a discussion. *J. Chromatogr.* **470**, 317–326.
4. Sluyterman, L. A. A., and Kooistra, C. (1990) Change of counter ion concentration and of resolving power in a chromatofocusing run. *J. Chromatogr.* **519**, 217–220.
5. Frey, D. D., Narahari, C. R., and Bates, R. C. (2001) Chromatofocusing. In: *Encyclopedia of Life Sciences*. John Wiley and Sons, http://www.els.net/ [doi:10.1038/npg.els.0002681].
6. Shen, H., and Frey, D. D. (2004) Charge regulation in protein ion-exchange chromatography: Development and experimental evaluation of a theory based on hydrogen ion Donnan equilibrium. *J. Chromatogr. A* **1034**, 55–68.
7. Fountoulakis, M., Langen, H., Gray, C., and Takács, B. (1998) Enrichment and purification of proteins of *Haemophilus influenzae* by chromatofocusing. *J. Chromatography A*. **806**, 279–291.
8. Frey, D. D., Narahari, C. R., and Butler, C. D. (2002) General local-equilibrium chromatographic theory for eluents containing adsorbing buffers. *AIChE J.* **48**, 561–571.
9. Bates, R. C., Kang, X., and Frey, D. D. (2000) High-performance chromatofocusing using linear and concave pH gradients formed with simple buffer mixtures I. Effect of buffer composition on the gradient shape. *J. Chromatogr. A* **890**, 25–36.
10. Kang, X., and Frey, D. D. (2003) High-performance cation-exchange chromatofocusing of proteins. *J. Chromatogr. A* **991**, 117–128.
11. Frey, D. D., Barnes, A., and Strong, J. (1995) Numerical studies of multicomponent chromatography using pH gradients. *AIChE J.* **41**, 1171–1183.
12. Frey, D. D. (1996) Local-Equilibrium behavior of retained pH and ionic strength gradients in preparative chromatography. *Biotechnol. Prog.* **12**, 65–72.
13. Strong, J. C., and Frey, D. D. (1997) Experimental and numerical studies of the chromatofocusing of dilute proteins using retained pH gradients formed on a strong-base anion-exchange column. *J. Chromatogr. A* **769**, 129–143.
14. Bates, R. C., and Frey, D. D. (1998) Quasi-linear pH gradients for chromatofocusing using simple buffer mixtures: local equilibrium theory and experimental verification. *J. Chromatogr. A* **814**, 43–54.
15. Narahari, C. R., Strong, J. C., and Frey, D. D. (1998) Displacement chromatography of proteins using a self-sharpening pH front formed by adsorbed buffering species as the displacer. *J. Chromatogr. A* **825**, 115–126.
16. Kang, X., and Frey, D. D. (2002) Chromatofocusing using micropellicular column packings with computer-aided design of the elution buffer composition. *Anal. Chem.* **74**, 1038–1045.
17. Chong, B. E., Yan, F., Lubman, D. M., and Miller, F. R. (2001) Chromatofocusing nonporous reversed-phase high-performance liquid chromatography/electrospray ionization time-of-flight mass spectrometry of proteins from human breast cancer whole cell lysates: a novel two-dimensional liquid chromatography/mass spectrometry method. *Rapid Commun. Mass Spectrom.* **15**, 291–296.

17

Fractionation of Proteins by Immobilized Metal Affinity Chromatography

Xuesong Sun, Jen-Fu Chiu, and Qing-Yu He

Summary

It is widely known that the progress of proteomics mostly depends on the development of more advanced and sensitive protein separation technologies. Immobilized metal affinity chromatography (IMAC) is a useful protein fractionation method used to enrich metal associated proteins. IMAC represents an affinity separation approach based on the interaction between proteins and metal ions immobilized on a solid support. By changing various metal ions and other experimental conditions such as pH and elution composition, IMAC can selectively isolate metal-binding protein fractions for further specific proteomic analysis. The combination of IMAC with other protein analytical technologies has been successfully applied in characterizing metalloproteome and post-translational modifications. In the future, newly developed IMAC techniques integrated with other proteomic methods will significantly contribute to the further development of proteomics.

Key Words: Affinity chromatography; chelator competition; electrophoresis; IDA; IMAC; imidazole competition; prefractionation; proteomics; SDS-PAGE.

1. Introduction

Immobilized metal affinity chromatography (IMAC) is a powerful protein fractionation method used to enrich metal-associated proteins and peptides *(1)*. It has great merits including easy regeneration, longevity and stability from proteolytic degradation. This technology is based on the interaction between peptides/proteins and metal ions immobilized on matrix. Beginning from the 1990s, IMAC has been increasingly employed as a prefractionation tool in the field of proteomic research *(2–4)*.

To date, the most popular metal ions used in IMAC for proteomic research are Cu^{2+}, Ni^{2+}, Zn^{2+}, Co^{2+}, Fe^{3+} and Ga^{3+} *(5,6)*. These transition metal ions can chemically interact with resin-immobilized chelating ligands which are commercially available, such as iminodiacetic acid (IDA), nitrilotriacetic acid (NTA) and dipicolylamine (DPA). Generally speaking, proteins/peptides with electron-donating side chains, especially those containing surface-exposed atoms of N, S and O, can be bound to resin-immobilized metal ions under mild conditions. Usually, it is the best to perform IMAC in the presence of high ionic strength (0.5-1.0 M NaCl) binding buffer to minimize nonspecific electrostatic interactions. Based on the nature of the interactions between metal ions and proteins, the target proteins can be selectively eluted from IMAC by an elution buffer with a suitable pH (protein protonation affecting the binding capability) or with an appropriate concentration of imidazole (competitive binding). Some elution protocols may apply strong chelating compounds, such as EDTA, to elute proteins with a strong affinity for IMAC.

2. Materials

2.1. Cell Culture and Lysis

1. LB medium: 10 g of Tryptone, 5 g of Yeast extracts, and 5 g of NaCl, dissolved in 950 ml of deionized H_2O, adjust the pH to 7.0 with 5 N NaOH (approx 0.2 ml) and final volume to 1 L with deionized H_2O. Sterilize by autoclaving for 20 minutes at 15 psi (1.05 kg/cm^2) on liquid cycles.
2. Lysis buffer: identical with the respective binding buffers for the specific kinds of IMAC discussed below (*see* Note 1).
3. Liquid nitrogen.
4. Triton X-100.
5. Water-bath set at 37°C.
6. Sonicator.
7. Centrifuge with a speed over 12,000 *g*.

2.2. Materials for Ni^{2+}, Cu^{2+}, Zn^{2+} and Co^{2+}-IMAC

1. Stationary phase: agarose immobilized IDA (See Fig. 1). Store at 4°C.
2. Metal ions: 50 mM solutions of $NiSO_4$, $CuSO_4$, $ZnSO_4$, and $CoSO_4$ (*see* Note 2).
3. Binding buffer: 20 mM sodium phosphate (7.8 mM NaH_2PO_4, 12.2 mM Na_2HPO_4), 0.5 M sodium chloride, pH 7.0 (*see* Note 3).
4. Low pH elution buffers: 100 mM sodium acetate, 0.5 M sodium chloride, pH 5.8 and 3.8. Store at 4°C.
5. Imidazole competition elution buffer: 20 mM sodium phosphate, 0.5 M sodium chloride, 500 mM imidazole, pH 7.0. Store at 4°C.
6. Chelator competition elution buffer: 20 mM sodium phosphate, 0.5 M sodium chloride, 50 mM EDTA, pH 7.0. Store at 4°C.
7. Regeneration buffer: identical with chelator competition elution buffer.

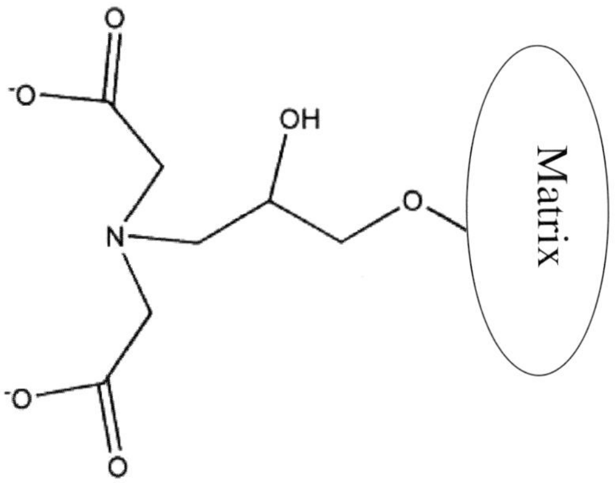

Fig. 1. The structure of iminodiacetic acid (IDA) resin. IDA is immobilized to a matrix, normally dry silica. A metal ion, such as Ni^{2+}, can be bound to IDA, which is a tri-dentate chelator and will occupy three of the binding sites of the metal ion. The remaining coordination sites of the metal ion can be occupied or exchanged with histidine or cysteine residues of the targeted proteins.

2.3. Materials for Fe^{3+}-IMAC

1. Stationary phase: agarose immobilized IDA. Store at 4°C.
2. Metal ion: 50 mM ferric chloride.
3. Binding buffer: 100 mM sodium acetate, pH 3.3. Store at 4°C.
4. Elution buffer: 20 mM sodium phosphate, pH 7.0. Store at 4°C.
5. Regeneration buffer: 50 mM Tris, 0.5 M sodium chloride, 0.1 M EDTA, pH 7.5. Store at 4°C.

2.4. Columns and Equipment

1. 1-cm inner-diameter, 5-cm long columns, for an analytical procedure (Bio-Rad Laboratories, Inc., Hercules, CA).
2. Peristaltic pump.
3. Simple gradient forming device to hold a 20 column volumes (CVs) (if stepwise elution is unsuitable).
4. Ultraviolet (UV) detector (280 nm) and pH monitor.

2.5. SDS-Polycarylamide Gel Electrophoresis (SDS-PAGE)

1. Separating gel: 1.5 M Tris-HCl, pH 8.8, 10% SDS, *N,N,N,N'*-Tetramethyl-ethylenediamine (TEMED) (store at room temperature), 30% acrylamide (store at 4°C), 10% Ammonium per-sulfate (APS) (store at -20°C).
2. Stacking gel: 1.0 M Tris-HCl, pH 6.8, 10% SDS, TEMED (store at room temperature), 10% APS (store at -20 °C), 30% acrylamide (store at 4°C).

3. 1× Tris-glycine electrophoresis buffer: 25 mM Tris, 250 mM glycine (electrophoresis grade) pH 8.3 and 0.1% (w/v) SDS.
4. 2× SDS Gel-loading buffer: 100 mM Tris-HCl (pH 6.8), 4% SDS (electrophoresis grade), 0.2% bromophenol blue, 20% (v/v) glycerol, 200 mM dithiothreitol or β-mercaptoethanol. Add the thiol reagents from 1 M (dithiothreitol) or 14 M (β-mercaptoethanol) stocks just before use.
5. Coomassie blue staining: Dissolve 0.25 g of Coomassie Brilliant Blue R-250 in 90 ml of methanol:H_2O (1:1, v/v) and 10 ml of glacial acetic acid. Filter the solution through a Whatman No.1 filter to remove any particles.
6. Destaining solution: 30 ml methanol, 10 ml acetic acid, 60 ml H_2O.

2.6. Silver Staining

1. Fixing solution: 40 ml ethanol, 10 ml glacial acetic acid, 50 ml H_2O.
2. Incubation solution: 4.1 g sodium acetate, 10 ml ethanol, 0.2 g sodium thiosulfate, adjust volume to 100 ml with H_2O.
3. Silver nitrate solution: 0.1 g silver nitrate, 20 µL formaldehyde, adjust volume to 100 ml with H_2O.
4. Development solution: 2.5 g sodium carbonate, 10 µL formaldehyde, adjust volume to 100 ml with H_2O.
5. Stop solution: 1.46 g $Na_2EDTA \cdot 2H_2O$, adjust volume to 100 ml with H_2O.

3. Methods

3.1. Preparation of Samples

1. Culture *Escherichia coli* wherever applicable in 100 ml of LB medium in a rotating shaker at 37°C for about 16 h (*see* Note 4).
2. Harvest cells by centrifugation at 5,000g for 20 min at 4°C.
3. Wash once the pellet with 10 ml of binding buffer (as listed in 17.2.2.3 or 17.2.3.2 chosen to fit with the columns being used), and then resuspend in 10 ml of the same buffer. Vortex 5 min to ensure complete suspension.
4. Triple freezing-thawing intermittently in liquid nitrogen for 2 min and 37°C water-bath for 15 min (*see* Note 5).
5. Sonicate for 10 min at 5-sec pulse to facilitate the cell rupture and break down the DNA (to reduce the viscosity).
6. Add 0.1% (v/v) triton X-100 and shake at 4°C for 30 min.
7. Centrifuge the lysate at 12,000 g for 30 min at 4°C to remove debris, and collect the supernatant as soluble proteins (*see* Note 6).

3.2. Immobilization of Metal Ions and Column-Packing Procedures.

1. Equilibrate the immobilized IDA at room temperature and completely resuspend the immobilized IDA slurry with 30-sec vortex (*see* Note 7).
2. Suspend the gel and transfer to a suction flask and degas the gel slurry.

Fractionation of Proteins by IMAC

3. Add the gel slurry to column with the column outlet closed. Allow gel to settle in the column for about 30 min. Then, open the outlet to allow flow.
4. When the desired volume of gel has been packed, insert the column adapter. Pump H_2O through the column at twice the desired end flow rate for several minutes. Readjust the column adapter until it just touches the settled gel bed.
5. Add 3 CVs of prepared metal ion solution to the column; wash the column with 5 CVs of H_2O to remove any nonbound metal ions.
6. Wash the column with at least five CVs of 20% ethanol and store at 4°C.

3.3. Binding of Target Proteins from Cell Extracts to the Column.

1. Connect packed column to a peristaltic pump, simple gradient forming device, UV detector and pH monitor.
2. Equilibrate the column with 10 CVs of binding buffer until A_{280} is at baseline.
3. Apply the prepared cell lysate sample to the column.
4. Wash the column with 10 CVs of binding buffer to remove unbound proteins.

3.4. Elution of IMAC-Retained Proteins and Column Regeneration

3.4.1. Elution of proteins from columns immobilized with Ni^{2+}, Cu^{2+}, Zn^{2+} and Co^{2+}.

1. Low pH elution: Wash the column with 20 CVs of low pH elution buffer with linear gradient from pH 5.8 to 3.8 and collect the fractions representing target proteins as illustrated by SDS-PAGE (discussed in 17.3.4) by monitoring the UV absorbance at 280 nm; or wash the column with 10 CVs of low pH elution buffer (pH 3.8) to collect all the proteins retained on the columns.
2. Imidazole competition elution: Wash the column with 5 CVs of Imidazole competition elution buffers containing 2 mM imidazole to remove nonspecifically bound proteins. Then, elute bound proteins with imidazole competition elution buffer containing 500 mM imidazole. For a highly resolving elution, apply with 20 CVs of a linear elution gradient from 2 to 500 mM imidazole (*see* Note 8).
3. Chelator competition elution: Wash the column with 5 CVs of chelator competition elution buffer. (*see* Note 9).

3.4.2. Elution of Phosphoproteins from Column Immobilized with Fe^{3+}

Apply 8 CVs of elution buffer for column immobilized with Fe^{3+}. (*see* Notes 10 and 11).

3.4.3. Column Regeneration

Wash column with 5 CVs of regeneration buffer, then wash it with 10 CVs of ultra pure H_2O. (*See* Note 12).

3.5. SDS-PAGE

1. Prepare a 1.5 mm thick, 12% separating gel by serially mixing 3.3 ml ultra pure H_2O, 4.0 ml 30% acrylamide, 2.5 ml 1.5 M Tris-HCl (pH 8.8), 0.1 ml 10% SDS, 0.1 ml APS and 4 μL TEMED. Pour the gel, leaving about 1 cm space for stacking gel and fill with water saturated isobutanol. The gel will polymerize in about 30 min. Drain off the isobutanol and wash the top of the gel 3 times with distilled H_2O.
2. Prepare the stacking gel by mixing 1.4 ml of H_2O, 0.33 ml of 30% acrylamide, 0.25 ml of 1.0 M Tris-HCl (pH 6.8), 0.2 ml of 10% SDS, 0.2 ml of APS, and 2 μl of TEMED. Quickly pour stacking gel to the surface of the separating gel and insert the comb. Wait the gel to polymerize for about 30 min.
3. While the stacking gel is polymerizing, resuspend protein samples purified from IMAC column in the same volumes of 2× SDS-loading buffer, and heat the solution at 100°C for 5 min to denature the proteins.
4. Prepare the 1×SDS-running buffer by diluting the 10×SDS-running buffer with ultrapure H_2O.
5. After polymerization of stacking gel is complete, carefully remove the comb and wash the wells with the running buffer.
6. Pour the running buffer to the upper and lower chambers of the gel-running tank. Load protein samples premixed with SDS-loading buffer into the bottom of each well side by side and prestained molecular weight markers into a neighboring well. Load equal volumes of 1× SDS-loading buffer into unused wells.
7. Connect to a power supply and run through the stacking gel with a constant of 15 mA per gel, then adjust power to 30 mA per gel through the separating gel until bromophenol blue reaches the bottom of the separating gel.
8. Remove the glass plates from the electrophoresis apparatus and cut a corner of gel to mark the orientation of the gel.
9. Stain the gel with Coomassie Blue by immersing the gel with 5 volumes of staining solution or fix the gel with fixation solution for silver staining on a slowly rotating platform for 4 h at room temperature.
10. Destain the gel by immersing gel in destaining solution on a slowly rocking platform for 4–8 h, changing the destaining solution three or four times, store the gel in water in a sealed plastic bag.
11. To make a permanent record, either scan the stained the gel or dry the gel.

3.6. Silver Staining

1. Fix the gel for 30 min to 12 h at room temperature with gentle shaking in 5 gel volumes of fixing solution.
2. Discard the fixing solution and add 5 gel volumes of incubation solution. Incubate the gel at room temperature with gentle shaking.
3. Discard the incubation solution and add 10 gel volumes of H_2O. Incubate the gel with H_2O for 10 min at room temperature with gentle shaking. Repeat 4 times.

Fractionation of Proteins by IMAC

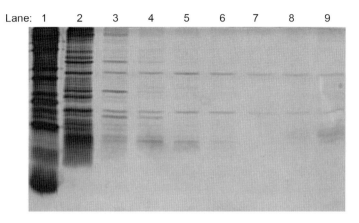

Fig. 2. Separation of nickel binding proteins by Ni^{2+}-IMAC with imidazole from *E.coli* BL21. SDS-PAGE stained with silver staining. Lane 1: marker; Lane 2: whole cell lysate; Lane 3-9: cluents with 60, 100, 150, 250, 300, 400, and 500 mM imidazole, respectively.

4. Discard the H_2O and add 5 gel volumes of silver nitrate solution. Incubate the gel for 30 min at room temperature with gentle shaking.
5. Discard the silver nitrate solution and wash gel with H_2O.
6. Add 5 gel volumes of fresh developing solution. Gently shake the gel at room temperature and carefully observe the gel. The protein profiles should appear within a few minutes depending on the protein amount and component of loading sample.
7. Discard the developing solution and add 5 gel volumes of stop solution to quench the reaction. Gently shake the gel at room temperature for 10 min.
8. Discard the stop solution and wash the gel with H_2O three times.
9. Follow steps that described in 17.3.5.11. An example is shown in Fig. 2.

4. Notes

1. The methods described in this chapter emphasize the specific use of agarose-immobilized IDA metal chelating groups. Normally, these methods are all acceptable for use with a wide variety of different immobilized metal-chelate affinity adsorbents.
2. All solutions should be prepared with ultra pure 18.2 MΩ-cm water, unless otherwise stated.
3. All chemicals used are the highest grade reagents commercially available.
4. This protocol can be used for other cell cultures.
5. Other lysis methods can be chosen for this protocol such as osmotic lysis, enzymatic lysis and French pressure cell. If cell lysate is dissolved in other buffers, desalt it to binding buffer appropriate for the used column by HiTrap

Desalting column or PD-10 column (Amersham Biosciences) before loading to IMAC column.

6. It is recommended to filter all the buffers and sample by passing them through a 0.45-μm filter before applied to the column to prevent column jam, which decreases the binding efficiency and leads to increased background pressure.
7. If prepacked IMAC column is used, the experiments can be performed at room temperature using a FPLC system (ÄKTAexplorer, Amersham Biosciences). The system consisted of dual syringe pumps, gradient mixer, UV monitor (280 nm), fraction collector, UNICORN data collection, and archive software.
8. If A_{280} is monitored, note that imidazole absorbs at this wavelength and the baseline should only be relative to the absorbance of the gradient buffer to elute the target proteins.
9. The chelator competition elution scheme will strip the metal ions from the column. The column has to be reequilibrated with metal ions before used again.
10. To facilitate the elution of some proteins, additives such as urea, ethylene glycol, detergents and alcohols, can be included in the column binding and elution buffer.
11. The ferric ions are present in the sample collected from IMAC-Fe^{3+} and will interfere with both the Coomassie blue staining and BCA protein assay. The ferric ions should be removed by dialysis or desalting.
12. For longer storage of column, metal ions should be stripped off from the column

Acknowledgement

This investigation was partially supported by grants from Hong Kong University funding (No. 200511159099 to Q.Y.H.) and the Area of Excellence Scheme of the Hong Kong University Grants Committee.

References

1. Sun,X., Chiu,J.F., and He,Q.Y. (2005) Application of immobilized metal affinity chromatography in proteomics. *Expert Rev.Proteomics.*, **2**, 649–657.
2. Posewitz,M.C. and Tempst,P. (1999) Immobilized gallium(III) affinity chromatography of phosphopeptides. *Analytical Chemistry*, **71**, 2883–2892.
3. Scanff,P., Yvon,M., and Pelissier,J.P. (1991) Immobilized Fe^{3+} Affinity Chromatographic Isolation of Phosphopeptides. *Journal of Chromatography*, **539**, 425–432.
4. Neville,D.C.A., Rozanas,C.R., Price,E.M., Gruis,D.B., Verkman,A.S., and Townsend,R.R. (1997) Evidence for phosphorylation of serine 753 in CFTR using a novel metal-ion affinity resin and matrix-assisted laser desorption mass spectrometry. *Protein Science*, **6**, 2436–2445.
5. Ueda,E.K., Gout,P.W., and Morganti,L. (2003) Current and prospective applications of metal ion-protein binding. *J Chromatogr.A*, **988**, 1–23.
6. Gaberc-Porekar,V. and Menart,V. (2001) Perspectives of immobilized-metal affinity chromatography. *J.Biochem.Biophys.Methods*, **49**, 335–360.

18

Fractionation of Proteins by Heparin Chromatography

Sheng Xiong, Ling Zhang, and Qing-Yu He

Summary

Heparins are negatively charged polydispersed linear polysaccharides which have the ability to bind a wide range of biomolecules including enzymes, serine protease inhibitors, growth factors, extracellular matrix proteins, DNA modification enzymes and hormone receptors. In this chromatography, heparin is not only an affinity ligand but also an ion exchanger with high charge density and distribution. Heparin chromatography is an adsorption chromatography in which biomolecules can be specifically and reversibly adsorbed by heparins immobilized on an insoluble support. An advantage of this chromatography is that heparin-binding proteins can be conveniently enriched using its concentration effect. This is especially important for separating low abundance proteins for the analysis in two-dimensional electrophoresis (2DE) or other proteomics approaches. Heparin chromatography is a powerful sample-pretreatment technology that has been widely used to fractionate proteins from extracts of prokaryotic organism or eukaryotic cells. As an example, the fractionation of fibroblast growth factors (FGFs) from the extract of mouse brain microvascular endothelial cells (MVEC) is now introduced to demonstrate the procedure of heparin chromatography.

Key Words: Affinity chromatography; heparin; ion exchange; protein fractionation; proteomics.

1. Introduction

Detection of low abundance proteins is a challenge to proteomics analysis *(1)*. Addressing this problem requires either the specific depletion of high abundance proteins or optimized protein fractionations based on affinity, charge, size or hydrophobicity *(2,3,4)*. As a classic separation technology, column chromatography has been widely used to enrich or fractionate target proteins *(5,6)*.

From: *Methods in Molecular Biology, vol. 424: 2D PAGE: Sample Preparation and Fractionation, Volume 1*
Edited by: A. Posch © Humana Press, Totowa, NJ

Heparin chromatography is an adsorption chromatography in which molecules to be purified are specifically and reversibly adsorbed by heparin immobilized on an insoluble support *(7)*. In general, heparin chromatography is a method of affinity chromatography used to purify or fractionate biological substances that can interact with heparin. This method has been extensively developed following the progress of chromatographic media and technologies. Because heparin chromatography has concentration effect, heparin-binding proteins can be conveniently enriched from the cell extract and large volume of samples can be easily processed. This is especially important when low abundance proteins need to be analyzed by 2DE or other proteomics technologies *(8)*. The high selectivity of this chromatography is derived from the natural specificities of the interaction between heparin and target molecules.

Heparins are negatively charged polydispersed linear polysaccharides. They are composed of 1–4 linked disaccharide repeat units containing an uronic acid and an amino sugar. Heparin molecule is a glycsoaminoglycan consisting of alternating hexuronic acids and D-glucosamine residues. The hexuronic acids are D-gluconic acid and its C-5 epimer L-iduronic acid. Heparin is heavily sulfonated, carrying sulfamino groups on the C2 of the glucosamine (*N*-sulfation) and ester sulfates (*O*-sulfation) on the C1 of the iduronic acid and the C6 of the glucosamine residue. With these acidic sulphuric groups, heparins are strongly negatively charged polyanions. The constituent monosaccharides of heparin are shown in Fig. 1.

Many studies have demonstrated that heparin has the ability to bind a wide range of biomolecules, including enzymes (mass cell proteases, lipoprotein lipase, coagulation enzymes, and superoxide dismutase), serine protease inhibitors (antithrombin and protease nexins), growth factors (fibroblast growth factor, Schwann cell growth factor, and endothelial cell growth factor), extracellular matrix proteins (fibronectin, vitronectin, laminin, thrombospondin, and collagens), nucleic acid-binding protein 9 initiation factors, elongation factors, restriction endonucleases, DNA ligase, DNA and RNA polymerases; hormone receptors (oestrogen and androgen receptors), and lipoproteins *(9)*.

Heparin is not only an affinity ligand with biological specificity, but also an ion exchanger with high-charge density and distribution *(10)*. In this kind of chromatography, heparin covalently immobilized on a porous bead matrix has two modes of interaction with proteins in sample solutions: (i) acting as a specific affinity ligand, and (ii) serving as a cation exchanger through its high content of anionic sulphate groups. By controlling elution conditions, the bound biomolecules can be selectively desorbed from heparin in bioactive form. Heparin chromatography is a powerful technology for separating various target proteins *(9,11)* and has been widely used to fractionate proteins from extract of prokaryotic organism or eukaryotic cells *(8,12,13)*. Molecules of the

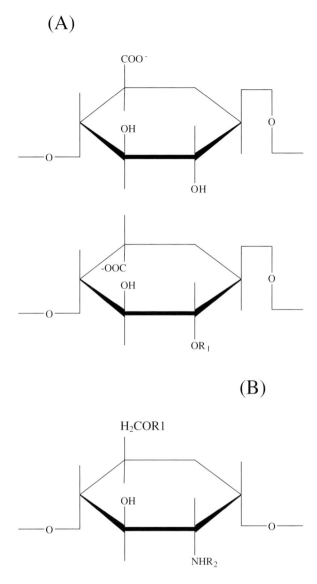

Fig. 1. Structure of a heparin polysaccharide, consisting of alternating hexuronic acid (**A**) and D-glucosamine resides (**B**). The hexuronic acid can either be D-glucuronic acid (top) or its C-5 epimer, L-iduronic acid. R_1=-H or $-SO_3^-$; R_2= $-SO_3^-$ or $-COCH_3$.

fibroblast growth factor (FGF) family are representative of heparin-binding proteins (14,15). As an example, the fractionation of FGFs from the extract of mouse brain microvessel endothelial cells (MVEC) is now introduced to demonstrate the procedure of heparin chromatography (16).

2. Materials
2.1. Cell Culture and Protein Extraction (see Note 1)
1. Cell line: microvessels endothelial cells (MVECs), established from the grey matter of the cerebral cortex of mouse brain.
2. Medium: minimal essential medium (MEM), pH 7.4 (Gibco/BRL, Bethesda, MD).
3. F-12 nutrient mixture 50% (w/v) (Invitrogen, Carlsbad, CA).
4. Plasma derived equine serum 11% (v/v) (Invitrogen, Carlsbad, CA).
5. Heparin 50 g/L (Sigma, St. Louis, MO).
6. Antibiotics: 100 U/ml penicillin, and 100µg/ml streptomycin (Life Technologies Bethesda Research Laboratories, Grand Island, NY).
7. Angiogenesis inhibitor: 100 mM homocysteine (Sigma, St. Louis, MO) (see Note 2).
8. Extraction buffer: 10 mM Tris-HCl, pH 7.6, 10 mM KCl, 5 mM MgCl.
9. Nuclei isolation buffer: 10 mM Tris-HCl, pH 7.6, 10 mM KCl, 5 mM MgCl, 0.35 M sucrose.

2.2. Chromatography Media

As shown in Table 1, several types of media can be used for heparin chromatography *(17)*. With differences in their rigidity, coupling arm, binding capacity and pH stability, these media can be used to separate different proteins. For example, Heparin Sepharose 6 Fast Flow can be applied to fractionate FGFs and other heparin-binding proteins from MVECs, and has been used to purify recombinant human FGF-2 from *Escherichia coli* in our previous study.

2.3. Chromatrography Buffers
1. Starting buffer: PBS, 0.01 M NaHPO, 0.15 M NaCl, pH 7.0 (*see* Note 3).
2. Elution buffer: PBS, 0.01 M NaHPO, 2.0 M NaCl, pH 7.0 (*see* Note 4).
3. Sanitization buffer: 0.1 M NaOH and 20% ethanol (12 h) or 70% ethanol (1 h).
4. Cleaning-in-place buffer: 0.1 M NaOH.
5. Storage buffer: 0.05 M sodium acetate in 20% ethanol.

3. Methods
3.1. Cell Culture and Total Protein Extraction
1. Isolate mouse brain endothelial cells from the grey matter of the cerebral cortex, directly plate the cells in the culture medium containing 50% (v/v) MEM, 50% (w/v) F-12 nutrient mixture (Ham), 11% (v/v) plasma derived equine serum, 50mg/ml heparin plus antibiotics/antimycotic *(18)*.
2. Incubate the cells at 37°C with 5% (v/v) CO_2 until confluent monolayers (10–14 days) are formed *(19)*.

Table 1
Media for heparin chromatography

Media	Matrix	Binding capacity	Company
HeparinSepharose 6 Fast Flow	Highlycross-linked agarose, 6%	Min. 2 mg AT III/ml medium	Amersham Bioscience Inc.
Heparin Sepharose CL-6B	Highlycross-linked agarose, 6%	2 mg AT III /ml medium	Amersham Bioscience Inc.
PROSEP-Heparin Chromatography Media	PROSEP	3 mg AT III/ml medium	Millipore Inc.
Affi-Gel Heparin Gel	agarose	> 1.2 mg AT III/ml medium	Bio-Rad Laboratories

[a] AT: antithrombin

3. When MVECs reach about 90% confluence, wash the attached cells with PBS (pH 7.4) and culture in serum free media for 24 h at 37°C, and then incubate with homocysteine at concentrations 6, 12, 20, 40 mM for 24h *(16,19)*.
4. Harvest the cells, lyse them with extraction buffer plus 0.3% (w/v) Triton X-100 (about 1×10^7 cells per ml buffer), and then homogenize the cell extract in an ice-cold tissue grinder. 30-50 passes with the grinder are recommended *(4,6,20,21)*.
5. Add 0.5 volume of nuclei isolation buffer to remove the nuclei by centrifugation at 4°C, 700g for 10 min. The supernatant contains the cytoplasmic protein fraction.
6. Exchange the buffer system of the cytoplasmic protein extraction by dialysis against starting buffer at 4°C for 6–8 h (*see* Note 5).

3.2. Starting/Wash Buffer Preparation

1. Use salts with high purity (e.g., analytical grade) to prepare chromatography buffers according the recipe (*see* Note 4).
2. Degas and filter (0.22 μm or 0.45 μm membrane) all buffers before use.

3.3. Column Preparation

1. Weigh out about 3 g Heparin Sepharose 6 Fast Flow gel. Swell the freeze-dried powder in distilled water or starting buffer. After the gel is swelled, wash it and suspend it in water or starting buffer, degas and pour it into the column (10/16 mm) as slurry. In addition to this, prepacked columns are also commercially available.

2. Connect the column with the purification apparatus and check the system.
3. Equilibrate the column with starting buffer at least 3 bed volumes. A 10/16 column packed with 5 ml Heparin Sepharose 6 Fast Flow is equilibrated with about 20 ml starting buffer at flow rate of 150 cm/h.

3.4. Sample Preparation and Loading

1. Filter the MVECs extract (about 8 ml) by using 0.22 μm or 0.45 μm membranes before injection to avoid column plugging.
2. Apply the total protein extraction to the column with starting buffer at flow rate of 100 cm/h (*see* Notes 6 and 7).

3.5. Wash/Elution

1. Wash the column with 15–25 ml starting buffer until the O.D. 280 nm reaches the baseline to remove impurities or unbound materials from the column (*see* Note 8).

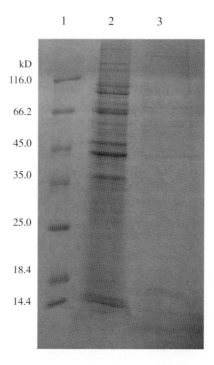

Fig. 2. SDS-PAGE analysis the total proteins of MVECs. Lane 1, protein molecular weight marker. Lane 2, total proteins of MVECs. Lane 3, heparin-binding proteins of MVECs.

Fractionation of Proteins by Heparin Chromatography 219

2. Elute the heparin-binding proteins with elution buffer (starting buffer plus 2 M NaCl) until the O.D. 280 nm reaches the baseline (*see* Note 9).
3. Collect fraction (about 1–2 ml) of the heparin-binding proteins (*see* Note 10).
4. The high salt eluted proteins should be desalted before applied on 1-DE or 2-DE. Fig. 2 displays the 1-DE result of MVECs total proteins before and after the heparin chromatography.

3.6. Column regeneration and storage

1. High salt buffer containing 2.0 M NaCl with the volume of 3–5 times of column should be used to remove strongly adsorbed proteins from the column. Here, 20–30 ml elution buffer is used to regenerate the column (*see* Note 11).
2. Re-equilibrate the column with starting buffer for next purification.
3. For short-term storage, keep the column in starting buffer at 4°C.
4. For long-term storage, store the column at 4°C with 0.02% sodium azide. The media of Sepharose 6 Fast Flow can be long-term stored with 0.05 M sodium acetate containing 20% ethanol.

4. Notes

1. In general, soluble extraction of homogenized tissue (*4,20*), cytoplasmic (membrane or nuclear) proteins of eukaryotic cells (*6*), total proteins or fraction of bacteria (*8*) can be used as the start sample to run heparin chromatography. For insoluble proteins, such as many membrane proteins or inclusion bodies in the cytoplasm of recombinant bacteria, solublization in 8 M urea or 6 M guanidine hydrochloride is difficult to bind heparin because the binding depends on the three-dimensional (3D) structure of proteins.
2. Besides homocysteine, other angiogenesis inhibitor, endostatin or ginseng extract Rg3 can also be used. Because the angiogenesis inhibitors down-regulate the expression of FGFs, pretreatment of the cell extract to enrich FGFs is certainly necessary to perform a comparative proteomics analysis.
3. Heparin is negatively charged and acts as a cation exchanger; salt (0.1–0.2 M NaCl or KCl) should be added to prevent nonspecific adsorption. To avoid nonspecific interactions, the binding buffer should be at least with ionic strength of 0.15 M. However, if the proteins of interest bind to heparin mainly by cation exchange, starting buffer with lower ionic strength may be feasible. In most cases, 10–50 mM phosphate or Tris can be used as buffer system. Here, 10 mM phosphate plus 0.15 M NaCl is applied.
4. In most cases, a wash / elution buffer is prepared by adding higher concentration of salts into the starting buffer. NaCl or KCl (up to 1.5–2 M) is the most commonly used salts for elution, although other salts can alternatively be used. Here, 2.0 M of NaCl is used to elute the bound proteins. In some cases, lower concentration of NaCl can also be used. This depends on the property of the sample and the resolution of 2-DE.

5. Dissolve sample in the starting buffer, or exchange sample buffer system against the starting buffer in the case of a large volume of sample to be load (>25% of the column volume). Otherwise, binding efficiency may decrease heavily.
6. Determine the maximum load based on the property of media. In most cases, this data is listed in the manual of media you chose. Lower loading achieves better resolution. For most proteins, experiments determining loading capacity can be performed in necessary.
7. Apply sample to column in a flow rate slower than that of equilibration.
8. In general, 3–5 bed volumes of starting buffer should be used to wash out the contaminants. Samples with high concentration of contaminants may require longer washing time.
9. After removing impurities and unbound materials, specific elution may be achieved by using heparin as a competing agent in the starting buffer when heparin functions as an affinity ligand. However, in many cases where heparin functions as a cation exchanger, elution is more usually performed by increasing the ionic strength of the eluting buffer in continuous or stepwise gradient elution.
10. The loading of columns and the eluting condition depend upon the total mass of samples applied, not on the concentration of samples. Very dilute samples can be concentrated during the process of loading.
11. If necessary, 8 M Urea, 6 M guandine or 0.1 M NaOH can be used to remove precipitates or denatured proteins, whereas 0.1–0.5% nonionic detergent can be used to remove hydrophobically bound proteins.

References

1. Rocken, C., Ebert, M.P., Roessner, A. (2004) Proteomics in pathology, research and practice. *Pathol. Res. Pract.* **200**, 69–82.
2. Betgovargez, E., Knudson, V., Simonian, M.H. (2005) Characterization of proteins in the human serum proteome. *J. Biomol. Tech.* **16**, 306–310.
3. Yuan, X., Desiderio, D.M. (2005) Proteomics analysis of prefractionated human lumbar cerebrospinal fluid. *Proteomics* **5**, 541–550.
4. Jarrold, B., DeMuth, J., Greis, K., Burt, T., Wang, F. (2005) An effective skeletal muscle prefractionation method to remove abundant structural proteins for optimized two-dimensional gel electrophoresis. *Electrophoresis* **26**, 2269–2278.
5. Linke, T., Ross, A.C., Harrison, E.H. (2006) Proteomic analysis of rat plasma by two-dimensional liquid chromatography and matrix-assisted laser desorption ionization time-of-flight mass spectrometry. *J. Chromatogr. A* [Epub ahead of print].
6. Guerrera, I.C., Predic-Atkinson, J., Kleiner, O., Soskic, V., Godovac-Zimmermann, J. (2005) Enrichment of phosphoproteins for proteomic analysis using immobilized Fe(III)-affinity adsorption chromatography. *J. Proteome. Res.* **4**, 1545–1553.
7. Farooqui, A.A. (1980) Purification of enzymes by heparin-sepharose affinity chromatography. *J. Chromatogr.* **184**, 335–345.

8. Fountoulakis, M., Takacs, B., Langen, H. (1998) Two-dimensional map of basic proteins of Haemophilus influenzae. *Electrophoresis* **19**, 761–766.
9. Karlsson, G., Winge, S. (2004) Separation of latent, prelatent, and native forms of human antithrombin by heparin affinity high-performance liquid chromatography. *Protein Expr. Purif.* **33**, 339–345.
10. Staby, A., Sand, M.B., Hansen, R.G., Jacobsen, J.H., Andersen, L.A., Gerstenberg, M., et al. (2005) Comparison of chromatographic ion-exchange resins IV. Strong and weak cation-exchange resins and heparin resins. *J. Chromatogr. A* **1069**, 65–77.
11. Srivastava, P.N., Farooqui, A.A. (1979) Heparin-sepharose affinity chromatography for purification of bull seminal-plasma hyaluronidase. *Biochem. J.* **183**, 531–537.
12. Iida, T., Kamo, M., Uozumi, N., Inui, T., Imai, K. (2005) Further application of a two-step heparin affinity chromatography method using divalent cations as eluents: purification and identification of membrane-bound heparin binding proteins from the mitochondrial fraction of HL-60 cells. *J Chromatogr. B Analyt. Technol. Biomed. Life Sci.* **823**, 209–212.
13. Marques, M.A., Espinosa, B.J., Xavier da Silveira, E.K., Pessolani, M.C., Chapeaurouge, A., Perales, J., et al. (2004) Continued proteomic analysis of Mycobacterium leprae subcellular fractions. *Proteomics* **4**, 2942–2953.
14. Faham, S., Hileman, R.E., Fromm, J.R., Linhardt, R.J., Rees, D.C. (1996) Heparin structure and interactions with basic fibroblast growth factor. *Science* **271**, 1116–1120.
15. Katoh, Y., Katoh, M. (2005) Comparative genomics on FGF7, FGF10, FGF22 orthologs, and identification of fgf25. *Int. J. Mol. Med.* **16**, 767–770.
16. Shastry, S., Tyagi, N., Hayden, M.R., Tyagi, S.C. (2004) Proteomic analysis of homocysteine inhibition of microvascular endothelial cell angiogenesis. *Cell. Mol. Biol. (Noisy-le-grand)* **50**, 931–937.
17. Amersham Bioscience Inc. Introduction of heparin sepharose 6 fast flow.
18. Audus, K.L., Borchardt, R.T. (1987) Bovine brain microvessel endothelial cell monolayers as a model system for the blood-brain barrier. *Ann. N. Y. Acad. Sci.* **507**, 9–18.
19. Shastry, S., Tyagi, S.C. (2004) Homocysteine induces metalloproteinase and shedding of beta-1 integrin in microvessel endothelial cells. *J. Cell. Biochem.* **93**, 207–213.
20. Shen, H., Cheng, G., Fan, H., Zhang, J., Zhang, X., Lu, H., et al. (2006) Expressed proteome analysis of human hepatocellular carcinoma in nude mice (LCI-D20) with high metastasis potential. *Proteomics* **6**, 528–537.
21. Hauck, S.M., Schoeffmann, S., Deeg, C.A., Gloeckner, C.J., Swiatek-de Lange, M., Ueffing, M. (2005) Proteomic analysis of the porcine interphotoreceptor matrix. *Proteomics* **5**, 3623–3636.

IV

FRACTIONATION OF PROTEINS BY ELECTROPHORESIS METHODS

19

Fractionation of Complex Protein Mixtures by Liquid-Phase Isoelectric Focusing

Julie Hey, Anton Posch, Andrew Cohen, Ning Liu, and Adrianna Harbers

Summary

Protein fractionation is essential to uncovering low-abundance proteins in complex protein mixtures. Many common methods and techniques are used to fractionate proteins, including chromatography (size exclusion, affinity, ion exchange, etc.), electrophoresis, and solution chemistry. Regardless of the method employed, the ultimate goal of protein fractionation is to enable more protein analysis by today's current proteomics technologies, such as one- (1-DGE) or two-dimensional gel electrophoresis (2-DGE) and liquid-chromatography and tandem mass spectrometry (LC-MS/MS).

The MicroRotofor™ isoelectric focusing (IEF) cell fractionates proteins in free solution according to their isoelectric point (pI). We demonstrate the ability of the MicroRotofor to enrich low-abundance proteins in mouse brain tissue, thus enabling further identification of potential biomarker candidates.

Key Words: Carrier ampholytes; 1-D electrophoresis; 2-D electrophoresis; fractionation; liquid phase isoelectric focusing; low-abundance proteins; mass spectrometry; protein purification; proteomics; Rotofor.

1. Introduction

Fractionation by liquid-phase isoelectric focusing (IEF) is particularly beneficial for those proteins that are insoluble or otherwise do not separate well in other, gel-based IEF media. IEF in solution enables fractionation of proteins in their native state by isoelectric point (pI). Because ampholytes are used to generate the pH gradient used for separation, a continuous and, if desired, customized pH gradient may be formed, permitting true screening of a protein sample by pI.

The technology behind the design of the MicroRotofor cell was pioneered by Bier et al. *(1,2)* and was originally commercialized as the Rotofor® cell, which performs liquid IEF-based fractionation of 60 ml samples. The Micro-Rotofor cell accommodates 2.5 ml samples, enabling application of the powerful Rotofor technology to smaller volumes of precious samples, the development of separation protocols, and the refractionation of fractions obtained from runs on the Rotofor or Mini Rotofor cells. The cell uses a cylindrical focusing chamber divided into ten compartments (each containing a single fraction) by nine parallel, monofilament polyester screens. Temperature regulation, which is critical for reproducibility and maintenance of protein integrity during IEF, is provided by a Peltier-driven cooling block. Oscillation, or rocking, of the focusing chamber within the cooling block and around the focusing axis stabilizes the sample against convective and gravitational disturbances. After focusing, the solution in each compartment is rapidly collected without mixing using the vacuum-assisted harvesting apparatus that is supplied with the cell.

Use of the MicroRotofor cell can enrich low-abundance proteins *(3)* to enhance the results obtained from downstream applications as varied as one- and two-dimensional gel electrophoresis, chromatography, and mass spectrometry *(4–13)*.

2. Materials

2.1. Preparation of Complex Samples under Denaturing IEF Conditions

1. Lysis buffer: 7 M urea, 2 M thiourea, 2–4% (v/v) CHAPS, 1% dithiothreitol (DTT), 0.5% (v/v) carrier ampholytes, pH 3–10 (*see* Note 1).
2. Carrier ampholytes (Bio-Lyte), pH 3–10 (Bio-Rad).

2.2. Protein Extraction from Rat Brain Tissues

1. Whole rat brain tissues (Sigma-Aldrich, St. Louis, MO) stored at -20°C.
2. Dounce tissue grinder, 7 ml (VWR International, Inc., West Chester, PA).
3. Protein lysis buffer: 7 M urea, 2 M thiourea, 2% CHAPS, 40 mM Tris-HCl, pH 8.5.
4. Branson Sonifier 450 (Branson Ultrasonics Corp., Danbury, CT).
5. 50-ml Oak Ridge polycarbonate centrifuge tube (VWR International, Inc., West Chester, PA).
6. *RC DC* protein assay kit (Bio-Rad Laboratories, Hercules, CA).
7. Rehydration buffer: 7 M urea, 2 M thiourea, 2% CHAPS.
8. Carrier ampholytes (Bio-Lyte), pH 3–10 (Bio-Rad).

2.3. MicroRotofor Cell Set-Up

1. MicroRotofor cell (Bio-Rad), illustrated in Fig. 1.
2. High voltage power supply (*see* Note 2).
3. Vacuum source (*see* Note 3) and tubing.
4. MicroRotofor focusing chamber.
5. Anodic and cathodic ion exchange membranes (each one piece).
6. Anodic electrolyte solution: 0.1 M H_3PO_4 (*see* Note 4).
7. Cathodic electrolyte solution: 0.1 M NaOH (*see* Note 4).

2.4. Postfractionation Quality Control and Analysis

1. pH-meter with micro tip.
2. Standard 1-D SDS-PAGE equipment and buffers.
3. Standard 2-D electrophoresis equipment and buffers.
4. ReadyPrep 2-D cleanup kit (Bio-Rad).
5. *RC DC* protein assay kit from (Bio-Rad).

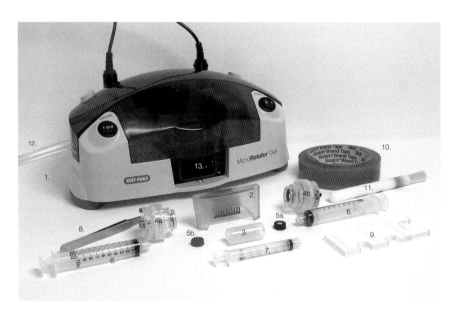

Fig. 1. MicroRotofor components and accessories: MicroRotofor chassis and lid (*1*), harvesting tray (*2*), focusing chamber (*3*), cathode assembly (4a), anode assembly (4b), cathode membrane, (5a), anode membrane, (5b), 10-ml syringes (*6*), 3-ml syringe (*7*), forceps (*8*), assembly tool (*9*), sealing tape (*10*), cleaning brush (*11*), vacuum hose (*12*), vacuum chamber (*13*).

3. Methods

3.1. Preparation of Complex Samples under Denaturing IEF Conditions

Denaturing running conditions are usually applied if the MicroRotofor is used to fractionate complex protein samples to enrich low-abundance proteins in proteomics research.

1. Prepare complex protein sample (e.g., whole cell lysate or grinded tissue) in denaturing lysis buffer (as used for 2-D electrophoresis) at a concentration of about 1–2 mg/ml. The loading volume for the separation chamber is about 2.5 ml. For optimal results the samples should be salt-free or limited to 10 mM (*see* Note 5).
2. Add carrier ampholytes, pH 3–10, to the protein sample solution. The final carrier ampholyte concentration should be around 2% (v/v) (*see* Note 6).
3. Vortex thoroughly and spin the sample for 15 min at 15,000g (20°C).
4. Remove the supernatant for fractionation in the MicroRotofor cell.

3.2. Total Protein Extraction from Rat Brain Tissue

1. Weigh out 250–500 mg of rat brain tissue. The tissue should remain frozen on dry ice before and after weighing.
2. Transfer the brain tissue to the Dounce tissue grinder prechilled on ice. Add 5–7 ml of protein lysis buffer and homogenize the brain tissue on ice for 10 min.
3. Transfer the homogenate into a culture tube and sonicate on ice with the Branson Sonifier 450 for 4 × 30 sec at constant power, 10 Watt output at 1 min intervals.
4. Transfer the tissue homogenate (5–7 ml) to a 50 ml Oak Ridge polycarbonate centrifuge tube and centrifuge at 16,000g for 30 min at 4°C.
5. Collect the supernatant from the tube, discarding the top lipid layers and the bottom pellets.
6. Determine the protein concentration of the sample using the *RC DC* protein assay kit (Bio-Rad). The expected concentrations of the protein extract ranges from 10–20 mg/ml.
7. Aliquot the supernatant, freeze on dry ice and store at −70°C.
8. When ready for MicroRotofor fractionation, take an aliquot and dilute it to a concentration of 1 mg/ml with rehydration buffer. The dilution will reduce the salt concentration for optimal focusing conditions.
9. Add carrier ampholytes, pH 3–10, to a final ampholyte concentration of 2% (v/v)

3.3. MicroRotofor Cell Set-Up

1. The ion exchange membranes are shipped dry and must be equilibrated overnight in the appropriate anodic and cathodic electrolyte solutions before first use (*see* Note 7). For rat brain sample, equilibrate the anode membrane (red) in 0.1 M H_3PO_4 and the cathode membrane (black) in 0.1 M NaOH (*see* Note 8).

2. Rinse the equilibrated ion exchange membranes with deionized water.
3. Place the anode membrane with the red casing on a flat surface and push one end of the focusing chamber onto it. Repeat with the cathode membrane with the black casing in the other end.
4. Assemble both electrode assemblies by placing the electrode and electrolyte components together loosely with the captive, threaded sleeve. Do not over tighten because some adjustment may be needed later.
5. Attach the anode assembly (red) to the end of the focusing chamber containing the anode membrane (red casing). Repeat with the cathode assembly (black casing) (*see* Note 9).
6. Use the assembly tool to ensure that the electrode assemblies are securely attached to the focusing chamber.
7. Align one row of ports on the focusing chamber with the vents on both electrode assemblies. Rotate the outer electrode component of each electrode assembly within the threaded sleeve until the vent and the ports are aligned. These aligned ports are now the sample loading ports.
8. Holding both the focusing chamber and the electrode assemblies in place, tighten the threaded sleeve around the assemblies, making sure to maintain the alignment of the loading ports and the electrolyte vents. Be careful not to over tighten

3.4. MicroRotofor Sample Loading

The focusing chamber features two rows of ports. The row that is aligned with the vents on the electrode chambers will be used for sample loading. The row on the opposite side of the focusing chamber will be used for harvesting the fractions. The harvesting ports have to be sealed with sealing tape before the sample is loaded. After the sample is loaded, the sample loading ports will also be sealed with a piece of sealing tape.

1. Use the assembly tool to cut two pieces of sealing tape by positioning the tape across the assembly tool covering all three cutting grooves. Cut the tape with a cutting blade at all three grooves to generate two strips of tape.
2. Seal the lower row of ports, the harvesting ports, in the focusing chamber with a strip of sealing tape. Make sure to cover all of the ports, and not to extend the tape beyond the focusing chamber (*see* Note 10).
3. Fill the 3-ml syringe with sample. Make sure there are no air bubbles inside the syringe. Slowly load the sample through the centermost sample loading port of the focusing chamber. As the center channel fills, the sample will slowly spread to adjacent channels. All the channels can be filled in this manner (*see* Note 11).
4. When the sample is loaded, make sure that all the channels are filled and that no bubbles remain. Air bubbles will disrupt the electric field, which can lead to poor separation.

5. Dry the outside surface of the focusing chamber and seal the row of sample loading openings with the second piece of sealing tape. Again, make sure to cover all the ports, not to extend the tape beyond the focusing chamber, and not to overlap with the harvesting port sealing tape.

3.5. MicroRotofor Focusing Run

It is recommended to bring the cooling block up to the correct temperature before beginning the focusing run. Turn the power and cooling on at least 15 min before beginning the focusing run.

1. Open the cooling cradle block by unscrewing the block screw.
2. Making sure the power is off, place the focusing assembly, with the vents and colored plugs on the electrode chambers facing up, into the running station of the chassis. Make sure the anode side (red) is to the left and the cathode side (black) is to the right (when facing the MicroRotofor).
3. Gently push the anode side (red) of the focusing assembly into the anode connection on the chassis until it is completely retracted.
4. Lower the focusing assembly into the cooling block and slide the cathode end of the assembly into the notch on the cathode end of the chassis. If necessary, rotate the focusing assembly until the slots on the cathode chamber align with the notch on the chassis. Alternatively, turn power on to the oscillating motor and wait until the notch is in a better position to connect the focusing assembly.
5. Using 10-ml syringes, add 6 ml 0.1 M H_3PO_4 through the vent hole of the anode chamber (red) and 6 ml 0.1 M NaOH through the vent hole of the cathode chamber (black).
6. Close the cooling block cover and tighten the screw.
7. Place the lid on the chassis and attach the power cord to the back of the MicroRotofor chassis and connect it to an electrical outlet.
8. Turn the power switch to the ON position to start the oscillating motor.
9. Set the cooling switch to the II (20°C) position. See Table 1 for detailed cooling setting information.
10. Attach the leads from the MicroRotofor lid to the power supply and perform the run at 1 Watt constant. A typical run is usually completed in 2–3 hours (*see* Note 12). To monitor the progress of a run under constant power, observe the voltage increase over time. The run is complete when the voltage stabilizes. At that point, allow the run to continue for 30 min before harvesting. Longer run times do not improve focusing and may result in a collapse of the pH gradient.

3.6. Harvesting the Fractions after a MicroRotofor Run

Once the IEF run is complete, fractions should be harvested as quickly as possible to avoid diffusion of the separated proteins. Throughout the following steps, minimize movement of the focusing chamber to avoid diffusion.

Table 1
MicroRotofor cooling setting

Ambient Temp.	Temperature inside the focusing chamber	
	Cooling setting I	Cooling setting II
4°C	6 ± 2°C	Not applicable
15°C	7 ± 2°C	17 ± 2°C
19°C	10 ± 2°C	20 ± 2°C
22°C	10 ± 2°C	20 ± 2°C
26°C	11 ± 2°C	20 ± 2°C
30°C	15 ± 2°C	21 ± 2°C
35°C	20 ± 2°C	25 ± 2°C
>35°C	We recommend running the MicroRotofor in a cold room to maintain an internal temperature of 6 ± 2°C	We recommend running the MicroRotofor Cell at Cooling setting I to maintain a temperature of 20–25°C

1. Turn the power supply off and disconnect it from the MicroRotofor cell.
2. Connect the MicroRotofor cell to a vacuum source. We recommend installation of a vacuum trap between the cell and the vacuum system (*see* Note 13).
3. Turn off power to the oscillating motor and cooling block on the MicroRotofor cell, and remove the lid from the chassis.
4. Make sure the harvesting tray is in place and flush against the sealing gasket.
5. Open the cooling block cover and, using forceps, remove the sealing tape from the sample loading ports.
6. Apply a vacuum to the chassis.
7. Remove the focusing assembly from the running station. First, lift up gently on the cathode (black) end to dislodge the connector from the notch in the chassis. Then remove the anode end (red).
8. With the row of sample loading ports facing up, position the focusing assembly in the harvesting station. Two sides of the focusing chamber are somewhat flattened to correctly orient the focusing assembly within the harvesting station and to help align the harvesting needle array with the sealed harvesting ports of the focusing chamber. Do not puncture the sealing tape covering the harvesting ports.
9. Taking care to not cover any of the loading ports with your fingers, use both hands to press down evenly and firmly on the focusing assembly so that all the needles penetrate the sealing tape and harvesting ports simultaneously. At the same time, press the harvesting tray against the seal of the vacuum assembly with your thumbs.

10. Continue to press down on the focusing chamber for several seconds to aspirate the ten fractions into the harvesting tray (*see* Note 14).
11. Remove the harvesting tray and turn off the vacuum source.
12. Transfer the fractions to microtubes or other containers with a syringe or standard pipet. For storage in the harvesting tray, seal the tray with the sealing film and store as appropriate.

3.7. Quality Control and Analysis of the Individual Rotofor Fractions

After fractionation and harvesting, the ten fractions are ready for further analysis by one-dimensional SDS-PAGE electrophoresis (1-DGE) or two-dimensional gel electrophoresis (2-DGE).

3.7.1. Analysis of the Individual Micro Rotofor Fractions by 1-D Electrophoresis

For protein purification, each fraction is analyzed to determine which fraction(s) contain(s) the protein of interest. There are many different methods

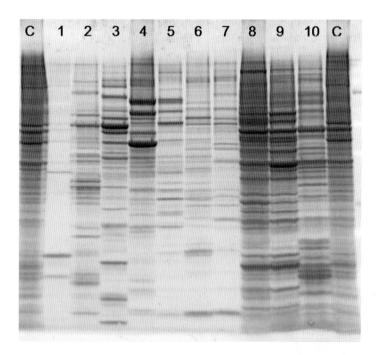

Fig. 2. SDS-PAGE of MicroRotofor fractionated HeLa cells dissolved in 7 M urea, 2 M thiourea, 4% CHAPS, 1% DTT, 2% Bio-Lyte ampholyte, pH 3-10 (total protein concentration: 1 mg/ml). C= unfractionated, crude sample; 1–10= Fraction number.

available and the best method is dependent on the protein being analyzed. 1-D electrophoresis using the Experion™ automated electrophoresis system, SDS-PAGE analysis, or pH 3–10 IEF gel analysis will give an accurate representation of the fractionation.

1. Measure the pH of each fraction and graph the pH profile of the MicroRotofor run as a function of the fraction number (*see* Note 15).
2. Measure total protein for each fraction to determine recovery and yield (*see* Note 16).
3. Run all or selected fractions on an Experion system, SDS-PAGE or IEF. Typical examples of analysis of MicroRotofor fractions with a 1-D SDS-PAGE gel or an Experion system are illustrated in Figs. 2 and 3, respectively.
4. Gel bands can be excised for further analysis by LC-MS/MS *(14)*.

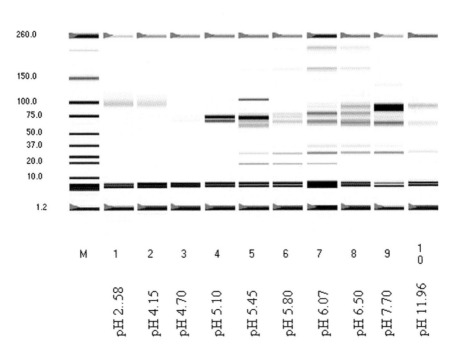

Fig. 3. Experion system analysis of MicroRotofor fractionated human serum in 7 M urea, 2 M thiourea, 4% CHAPS, 2% Bio-Lyte ampholyte, pH 4–7. Each fraction was diluted 1:10 with deionized H_2O. M= Molecular weight marker; 1–10 = Fraction number.

3.7.2. Analysis of the Individual MicroRotofor Fractions by 2-D Electrophoresis

Fractionation of complex samples with the MicroRotofor cell can be used to reduce sample complexity, thus allowing improved resolution and relatively larger sample load for low-abundance proteins for 2-D electrophoresis.

1. Measure the pH of each fraction and graph the pH profile of the MicroRotofor run as a function of the fraction number.
2. Measure total protein for each fraction to determine recovery and yield.
3. Use the ReadyPrep 2-D cleanup kit for protein concentration and buffer exchange (*see* Note 17).
4. Analyze the treated fractions by 2-D electrophoresis. Fig. 4 provides an indication concerning the fractionation and enrichment quality of the MicroRotofor

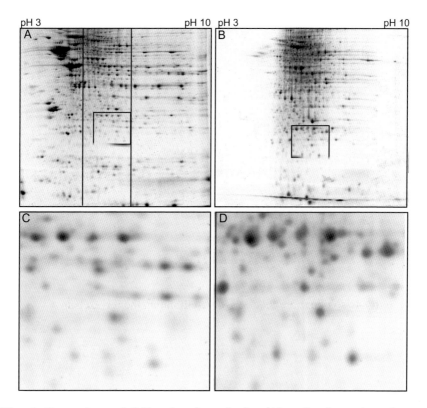

Fig. 4. Comparison of 2-D gels of rat brain. (A) unfractionated total protein (500 μg/gel); (B) MicroRotofor fraction 5 (pH 7.5, 300 μg/gel); (C) enlarged region (square) of unfractionated total protein gel (D) enlarged region (square) of MicroRotofor fraction showing enrichment of many spots.

4. Notes

1. During isoelectric focusing (IEF), proteins become concentrated at their isoelectric point (pI), where they are uncharged. The lack of electrostatic repulsion may cause some proteins to precipitate by a phenomenon known as "pI fallout" that is common to all IEF methods. A number of different reagents may be added to the sample to maintain solubility of proteins during focusing: (i) Chaotropes: addition of up to 8 M urea (or 7 M urea plus 2 M thiourea) is recommended to improve protein solubility in cases where native fractionation is not required; (ii) Detergents: Addition of nonionic detergents, such as CHAPS, CHAPSO, ocytlglucoside (OGS), digitonin, or Triton X-114 is also valuable in maintaining the solubility of focused proteins. The concentration of these detergents is generally held between 0.1 and 4%. Ionic detergents like SDS cannot be used because they interfere with isoelectric focusing; (iii) Reducing agents: Reducing agents, such as beta-mercaptoethanol (BME), tributylphosphate (TBP) and dithiothreitol (DTT) may be added to the protein sample to help break disulfide bonds from forming between and among proteins; (iv) Ampholytes: Solubility may sometimes be maintained by increasing the ampholyte concentration in the sample. Usually, 2–3% ampholyte is used for fractionation, but up to 8% may be used. Ampholytes may be added during protein extraction; alternatively, the protein sample buffer may be exchanged for ampholyte-containing fractionation buffer by dialysis; (v) Glycerol: Addition of glycerol from 5 to 25% (v/v) in the sample is also highly effective for maintaining the solubility and stability of proteins. Glycerol stabilizes water structure and the hydration shell around proteins.
2. The power supply must meet the following requirements: (i) The power supply must be capable of power control at 1 W constant power or if constant power mode is not available, the power supply must have the ability to program multiple step constant voltage methods; (ii) The power supply must be capable of supplying up to 1000V; (iii) The power supply must be capable of operating at low currents 1–15 mA.
3. A house vacuum or vacuum pump is required for optimal fraction collection with the harvesting apparatus. A vacuum trap, installed between the vacuum source and the MicroRotofor cell, is also recommended.
4. Electrolyte solutions are required for equilibrating the ion exchange membranes before use and for filling the electrode chambers. Generally, 6 ml each of 0.1 M H_3PO_4 and 0.1 M NaOH is sufficient for both applications. Alternative electrolyte solutions may also be used and are listed in Table 2.
5. Samples should be desalted (for example, by dialysis or Bio-Gel® P-6 chromatography) before ampholyte addition to ensure that the nominal pH range of the ampholyte will extend over the full length of the focusing chamber and that the maximum voltage can be applied. It is best to limit salt concentrations to 10 mM for optimal fractionation. However, because the maximum salt capacity will vary with the application, optimal running conditions must be determined empirically. Note that during focusing, all salts will migrate to the compartments

Table 2
Alternative electrolytes for the MicroRotofor cell

pH Range of Ampholyte	Anode Electrolyte	Cathode Electrolyte
3–5	0.5 M acetic acid	0.25 M HEPES
4–6	0.5 M acetic acid	0.5 M ethanolamine
5–7	0.1M glutamic acid	0.5 M ethanolamine
6–8	0.1M glutamic acid	0.1 M NaOH
7–9	0.25 M MES	0.1 M NaOH
8–10	0.25 M MES	0.1 M NaOH

adjacent to the anode and cathode, effectively desalting the sample. Additional ampholytes may sometimes be added to maintain solubility of proteins that require solutions of high ionic strength. Buffers should also not be added in concentrations greater than 10 mM as they add to the conductivity of a sample and decrease resolution. Also, buffering solutions may flatten the pH gradient in the region of the pKA of the buffer.

6. The final concentration of ampholyte used in the MicroRotofor cell depends on the protein concentration in a given sample (see Table 3), but may be increased as required to maintain protein solubility. Up to 8% (w/v) ampholyte concentrations have been used for various applications. Low ampholyte concentrations (~1%) permit higher applied voltages and are recommended if refractionation is not required. Higher ampholyte concentrations provide better buffering, help maintain solubility and are required if refractionation will be performed. The proper choice of ampholyte is critical for good results. The pI of the protein(s) of

Table 3
General recommendations for ampholyte concentrations to be used in the MicroRotofor cell

Purpose		Ampholyte concentration, % (v/v)
Fractionation	Protein concentration (mg/ml)	
	> 2.0	2.0
	1.0	1.5
	0.5	1.0
	0.25	0.5
Refractionation		$\geq$ 2.0*
Improve solubility		Increase, up to 8%**

* ($\geq$2%) ampholyte is needed for the initial fractionation.
** If protein precipitation occurs during a run, sample solubility may be maintained with higher ampholyte concentrations.

interest should fall in the middle of the ampholyte range used for fractionation. Ampholytes with narrow pH ranges are most efficient for separating the protein from the bulk of its contaminants. A narrow pH range of ampholytes that span the pI of the protein of interest should be used.
7. Equilibrated membranes, stored in electrolyte solution or deionized water, can be used for 4–5 runs. The membranes can be stored indefinitely when dry. After the membranes are equilibrated, they cannot be allowed to dry out and must be kept moist. The membranes may be stored in electrolyte or in distilled water between runs. If they dry out, membranes may crack and cause electrolyte to leak into the focusing chamber.
8. For pH 3–10 fractionation runs, the recommended electrolyte solutions for the anode and cathode chambers are 0.1 M H_3PO_4 and 0.1 M NaOH, respectively. For narrow range pH fractionation runs, spanning two pH units, alternative electrolytes are listed in Table 2.
9. Ion exchange membrane polarity is essential for the instrument to work properly. To facilitate correct assembly, all anode chamber and ion exchange membrane components are color-coded in red and cathode components, in black.
10. Excess tape, or tape not properly positioned, interferes with the cooling cradle during oscillation which can result in sample leaking from the focusing chamber.
11. If air bubbles are introduced, or are "caught", into the focusing chamber, the sample will not spread to adjacent channels as described. Make sure to dislodge and remove all air bubbles from the chamber. You can dislodge air bubbles by aspirating the sample from a channel and reloading the sample into the focusing chamber.
12. Samples with different protein loads, higher or lower carrier ampholyte concentrations, or other pH ranges may display slightly different current ranges and may require longer run times for optimal resolution.
13. Keep the vacuum valve closed until after the run is complete and the fractions are ready for harvesting.
14. If the sample contains high foaming detergents, such as Triton, we recommend removing the harvesting tray immediately after harvesting.
15. One of the quickest and easiest ways to identify which fractions contain a protein of interest is to measure the pH of the fractions and select those with the pH that is closest to the pI of the protein. The pH profile of the fractions is also one of the best indicators of the efficacy of a fractionation on the MicroRotofor cell; in most cases where the cell and its components have been properly maintained and samples have been properly prepared, a linear pH gradient will be established among the fractions. In addition, the pH range of each fraction can be estimated from the pH range of the ampholyte blend that was selected for the fractionation. For example, if pH 3–10 ampholytes were selected for fractionation, the 7 pH unit range would be separated across 10 fractions, and the estimated pH range for each fraction would be ~0.7 pH units.
16. To determine the recovery or protein yield of the fractions, measure the total protein in each fraction. Because ampholytes interfere with several commonly

used protein quantitation assays, such as the method developed by Bradford, it is recommended to use a method validated for use with ampholyte-containing solutions, such as the RCDC™ protein assay (Bio-Rad). Based on the method developed by Lowry, this assay is compatible with a broader range of reagents and so is used for protein quantitation directly in complex sample solutions. If ampholytes must be removed before downstream applications, it may be best to quantitate proteins after the removal step as well.

17. Ampholytes typically have a molecular weight of about 300–1,200 Daltons and form weak electrostatic complexes with proteins that may interfere with the measurement of protein amounts and activity. Though many applications can tolerate the presence of ampholytes in protein solutions, a number of applications, such as amino acid analysis, RP-HPLC, and mass spectrometry, cannot. Several methods for separating ampholytes from focused proteins can be applied: (i) Precipitation of protein by addition of ammonium sulfate, alcohol, or trichloroacetic acid (TCA) is an effective, inexpensive means of removing ampholytes and other contaminants from protein samples. These methods provide the added advantage of concentrating the protein before further analysis; (ii) The ReadyPrep™ 2-D cleanup kit is a prepackaged kit that provides a TCA-like precipitation of protein from small (<100 μl) sample volumes; (iii) Centrifugal filtration is an easy, effective means of removing ampholytes and other contaminants from protein samples. Centrifugal filtration also results in concentration of the protein sample; (iv) Dialysis is a simple method for ampholyte removal when concentration of proteins is not required or desired. Adjust the sample to 1 M NaCl to effectively strip electrostatically bound ampholytes from proteins by ion exchange. For example, add either 10× phosphate-buffered saline (PBS) or 10 M NaCl to the sample at 1/9 the volume of the sample. The salt will compete with the ampholytes for sites on the protein. Dialyze the sample into the buffer that is appropriate to your downstream application; (v) Chromatographic techniques, such as gel filtration, ion exchange, hydroxyapatite, affinity chromatography, can be used to separate proteins from ampholytes.

References

1. Bier, F. F., Egen, N. B., Allgyer, T. T., Twitty, G. E., and Mosher, R. A. (1979) New developments in isoelectric focusing, in *Structure and biological function* (Gross, E., and Meierhofer, J., Eds.), pp. 79–89, Pierce Chemical Company.
2. Bier, F. F., and Egen, N. B. (1979) Large-scale recycling electrofocusing in *Electrofocus/78* (Haglund, H., and Westerfield, J. G., Eds.), Elsevier, New York.
3. Righetti, P. G., Castagna, A., Herbert, B., and Candiano, G. (2005) How to bring the "unseen" proteome to the limelight via electrophoretic pre-fractionation techniques. *Biosci Rep* 25, 3–17.
4. Davidsson, P., Folkesson, S., Christiansson, M., Lindbjer, M., Dellheden, B., Blennow, K., and Westman-Brinkmalm, A. (2002) Identification of proteins in

human cerebrospinal fluid using liquid-phase isoelectric focusing as a prefractionation step followed by two-dimensional gel electrophoresis and matrix-assisted laser desorption/ionisation mass spectrometry. *Rapid Commun Mass Spectrom* 16, 2083–8.
5. Davidsson, P., Paulson, L., Hesse, C., Blennow, K., and Nilsson, C. L. (2001) Proteome studies of human cerebrospinal fluid and brain tissue using a preparative two-dimensional electrophoresis approach prior to mass spectrometry. *Proteomics* 1, 444–52.
6. Davidsson, P., and Nilsson, C. L. (1999) Peptide mapping of proteins in cerebrospinal fluid utilizing a rapid preparative two-dimensional electrophoretic procedure and matrix-assisted laser desorption/ionization mass spectrometry. *Biochim Biophys Acta* 1473, 391–9.
7. Davidsson, P., Puchades, M., and Blennow, K. (1999) Identification of synaptic vesicle, pre- and postsynaptic proteins in human cerebrospinal fluid using liquid-phase isoelectric focusing. *Electrophoresis* 20, 431–7.
8. Peirce M.J., Wait R., Begum S., Saklatvala J. and Cope A.P. (2004) Expression profiling of lymphocyte plasma membrane proteins. *Mol.Cell. Proteomics* 3: 56–65.
9. Hamler R.L., Zhu K., Buchanan N.S., Kreunin P., Kachman M.T., Miller F.R. and Lubman D.M. (2004) A two-dimensional liquid-phase separation method coupled with mass spectrometry for proteomic studies of breast cancer and biomarker identification. Proteomics 4: 562–577.
10. Wang M.Z., Howard B., Campa M.J., Patz, E.F. and Fitzgerald M.C. (2003) Analysis of human serum proteins by liquid-phase isoelectric focusing and matrix-assisted laser desorption/ionization-mass spectrometry. *Proteomics* 3: 1661–1666.
11. Harper R.G., Workman S.R., Schuetzner S., Timperman A.T. and Sutton J.N. (2004) Low-molecular-weight human serum proteome using ultrafiltration, isoelectric focusing, and mass spectrometry. *Electrophoresis* 25: 128–133.
12. Xiao Z., Conrads T.P., Lucas D.A., Janini G.M., Schaefer C.F., Buetow K.H., Issaq H.J. and Veenstra T.D. (2004) Direct ampholyte-free liquid-phase isoelectric peptide focusing: application to the human serum proteome. *Electrophoresis* 25: 128–133.
13. Janini G.M., Conrads T.P., Veenstra T.D. and Issaq H.J. (2003) Development of a two-dimensional protein-peptide separation protocol for comprehensive proteome measurements. *J Chromatography* B 787:43–51.
14. Bindschedler L.V., Palmblad M. and Cramer R. (2006) First-dimension separation with the MicroRotofor cell prior to SDS-PAGE and LC-MS/MS analysis. Bulletin 5451, Bio-Rad Laboratories, Inc.

20

Microscale Isoelectric Focusing in Solution

A Method for Comprehensive and Quantitative Proteome Analysis Using 1-D and 2-D DIGE Combined with MicroSol IEF Prefractionation

Mee-Jung Han and David W. Speicher

Summary

Current methods for quantitatively comparing complex protein profiles such as two-dimensional gel electrophoresis (2-DE), 2-D differential in-gel electrophoresis (DIGE), and liquid chromatography (LC)-mass spectrometry (MS) have limited resolution and dynamic ranges and therefore detect only a small portion of complex proteomes. To enhance protein profiling of complex samples, including human cell lines, tissue specimens, and plasma samples, complex proteomes can be prefractionated with microscale solution isoelectric focusing (MicroSol IEF). MicroSol IEF is compatible with most downstream proteome analysis methods including narrow range 2-D gels and 1-D gels followed by LC-MS/MS or LC/LC-MS/MS. This chapter describes the use of MicroSol IEF followed by 1-D and 2-D DIGE. The method has the advantage of more extensive proteome coverage compared with conventional 2-D DIGE alone. Furthermore, the use of fluorescent labeling before MicroSol IEF avoids any complications resulting from slight run-to-run variations during MicroSol IEF fractionation or the subsequent 2-D gel separations. The combination of DIGE and MicroSol IEF produces a powerful method for more comprehensive and quantitative comparison of protein profiles of complex proteomes.

Key Words: Comprehensive proteome analysis; 1-D DIGE; 2-D DIGE; ZOOM IEF fractionator; MicroSol IEF; protein profiling; quantitative proteome analysis; sample prefractionation.

1. Introduction

Proteomics has emerged as an indispensable methodology for large-scale protein analyses in functional genomics *(1)*. 2-DE has dominated proteome profile analysis for more than 30 years *(2,3)*, and it is still a core technology for quantitative comparisons of proteins from two or more closely related experimental samples. Unfortunately, the conventional 2-DE method is technically challenging and has a number of shortcomings including gel-to-gel variability, difficulty detecting some types of proteins, and inadequate capacity for the most complex proteomes such as mammalian cells and tissues. For example, most mammalian cells probably contain more than 20,000 unique protein components as a result of mRNA alternative splicing, and posttranslational modifications *(4,5)*, whereas large format 2-D gels typically resolve only about 1,500 to 2,000 spots or less. Furthermore, many pairs of gels are required to establish statistically significant differences in protein expression as each gel contains inherent experimental variations that limit image matching and detection of real biological differences as compared with gel-to-gel variation. In addition, silver staining, which has been widely used for high-sensitivity protein visualization on 2-D gels has a very limited dynamic range of spot detection that greatly complicates quantitative analysis. To overcome these disadvantages, 2-D DIGE was first introduced by Ünlü et al. *(6)*, and had been further developed by GE Healthcare (Chalfont St Giles, Bucks, UK; formerly Amersham Biosciences, Uppsala, Sweden). A variety of applications of the 2-D DIGE technique have been reported involving comparative protein profiling studies including identification of cancer-specific protein markers *(7–9)*. However, similar to a traditional 2-D gel, 2-D DIGE still detects only a small portion of complex proteomes.

One strategy for enhancing the capacities of 2-D gels involves parallel separation of replicate aliquots of unfractionated samples with or without fluorescent/isotope tags on a series of narrow pH range IPG (immobilized pH gradient) gels *(10–12)*. The main benefit of narrow pH range IPG gradients is that the total number of protein spots per pH unit that can be separated should increase because of higher spatial resolution. However, in practice, this approach results in only moderate increases in proteins detected compared to a single broad pH range gel. Narrow pH range IPG gels readily produce variable, unreliable separations when high protein loads of unfractionated complex samples are analyzed because proteins with pIs outside the IPG strip pH range usually cause massive precipitation and aggregation *(13,14)*.

A superior strategy for enhancing the separation capacity of 2-D gels is to prefractionate complex samples to reduce complexity and facilitate detection of low abundance proteins. Unfortunately, most prefractionation methods such as gel filtration, ion exchange, subcellular fractionation and selective solublization are low or moderate resolution techniques that result in substantial cross-contamination of many protein components among two or more fractions and, consequently, complicate quantitative comparisons. In contrast, preparative solution based IEF methods produce high resolution separations. Although some IEF devices produce large sample volumes that necessitate sample concentration before 2-D gel analysis, which can result in low, variable yields *(15,16)*, in contrast, the MicroSol IEF method can fractionate samples into well-defined pH pools on a scale compatible with high sensitivity proteome studies *(13,14,17–20)*. This method separates proteins under denaturing conditions based on charge in a series of small (typically 650 µl) sealed separation chambers without cross-flow or external mixing but with large pore partition membrane disks to ensure that all proteins, including large proteins are effectively separated. These partition membrane disks are available from Invitrogen Life Technologies. Alternatively, if special pH ranges are needed for a specific experimental design, custom partition membrane disks can be made with different acrylamide gel concentrations *(19,20)*. One feature of the MicroSol IEF method is that the total number of separation chambers, their pH ranges, and their volumes can be easily adjusted to fit requirements of different proteomes and research goals. Hence, MicroSol IEF prefractionation enhances protein profiling of complex proteomes such as human cell extracts and biofluids *(13,14,17–21)*. Moreover, MicroSol IEF prefractionation is compatible with most downstream proteome profiling methods, including 1-D gel, narrow pH range 2-DE, 2-D DIGE, and LC-MS/MS methods.

One very powerful downstream separation mode for MicroSol IEF fractions is to directly apply and separate each fraction on an appropriate narrow pH range IPG strip using sample amounts that are more than 10 times higher than for unfractionated samples. The combination of MicroSol IEF prefractionation, high protein loads on narrow range IPG gels, and use of multiple protein stains can dramatically increase the detection dynamic range and the total number of proteins detected. In addition, this method greatly conserves proteome samples compared with direct analyses of unfractionated samples on a series of narrow pH range 2-D gels. However, subtle variations in MicroSol IEF separations will result in artifactual spot differences on subsequent 2-D gels when conventional methods are used. To circumvent these problems, we have combined DIGE technology with MicroSol IEF prefractionation for more reliable comparative protein profiling of complex eukaryotic cells (Fig. 1). This approach has a number of advantages including: (i) high reproducibility because reference samples are included in each MicroSol IEF run; (ii) gel-to-gel variability is eliminated; (iii) more in-depth analysis of complex proteomes compared with 2-D DIGE alone, because MicroSol IEF plus use of multiple slightly overlapping IPG strips resulting in an extremely long effective IEF separation distance; (iv) a much wider linear dynamic range is achieved because much larger total

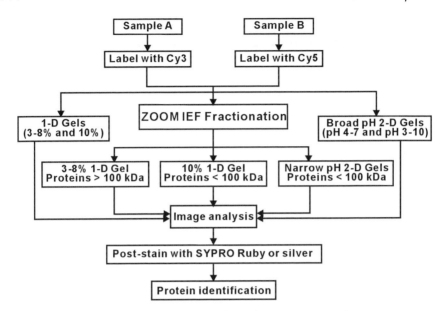

Fig. 1. A strategy combining MicroSol IEF prefractionation with 1-D and 2-D DIGE for more comprehensive and quantitative gel-based analysis of complex proteomes. In this scheme, two different samples are labeled with Cy3 and Cy5 and are then mixed. A portion of the Cy-labeled mixed sample is subjected to fractionation followed by 1-D and 2-D gels, while the remaining Cy-labeled samples are analyzed without fractionation for comparative purposes. After acquiring fluorescent images, gels are stained with a general protein stain, e.g., SYPRO Ruby or silver to facilitate excising spots of interest.

amounts of samples are separated compared with unfractionated samples, and the fluorescent dyes have much wider linear responses than alternative stains and (v) the number of large, high-resolution, narrow-range gels that must be run are greatly reduced. This chapter describes a new method using DIGE labeling combined MicroSol IEF and subsequent analysis on 1-D and narrow range 2-D gels for more comprehensive and quantitative protein profile analysis of complex proteomes.

2. Materials

A major advantage of the ZOOM IEF Fractionator is its flexibility. The pH ranges and the actual number of fractions can be readily altered depending on various proteome samples. The Fractionator device, associated buffers, reagents and a selection of partition membrane disks with pH's of 3.0, 4.6, 5.4, 6.2, 7.0, 9.1, 10.0, and 12.0 are available from Invitrogen Life Technologies.

A photograph of the commercial MicroSol IEF device, the ZOOM IEF Fractionator, produced by Invitrogen Life Technologies, is shown in Fig. 2

2.1. Protein Sample Labeling

1. N-Hydroxy succinimidyl ester-derivatives of the cyanine dyes Cy2, Cy3 and Cy5 (GE Healthcare, UK) and dimethylformamide (DMF, Sigma, St. Louis, MO).
2. Reaction buffer: 8 M urea, 2 M thiourea, 4% (w/v) 3-[(3-cholamidopropryl) dimthylammonio]-1-propanesulfonic acid (CHAPS), 30 mM Tris (adjust to pH 8.5 with HCl if necessary). Store in aliquots at -20°C.
3. 10 mM Lysine (Sigma) for stopping labeling reaction.

2.2. The ZOOM IEF Fractionator

1. ZOOM IEF Fractionator (Invitrogen Corporation, Carlsbad, CA).
2. A series of partition membrane disks (ZOOM disks; Invitrogen Corp.)
3. Anode buffer: 50× Novex IEF Anode buffer (Invitrogen Corp.) diluted to 1× and with addition of 8 M urea and 2 M thiourea (final concentrations). Adjust to pH 3.0 with 50×Novex IEF buffer if necessary.

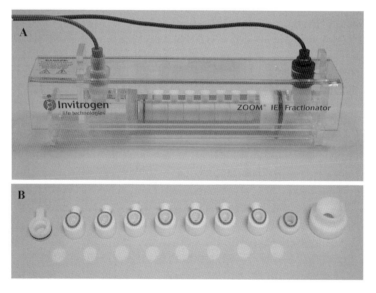

Fig. 2. The ZOOM IEF Fractionator. (**A**) A completely assembled device with seven separation chambers. (**B**) The internal components of the device required for prefractionation: top row (from left to right)–anode end sealer, seven oval shaped separation chambers with caps and O-rings, cathode end sealer, and cathode end screw cap; bottom row–eight porous partition membrane disks. Reproduced from Ref. *(18)* with permission from Elsevier.

4. Cathode buffer-Extended Range for use with membrane disks up to pH 12.0: 10× Novex IEF Cathode buffer pH 9–12 (Invitrogen Corp.) diluted to 1× and with addition of 8 M urea and 2 M thiourea (final concentration). Adjust to pH 12.0 with NaOH if necessary.
5. Sample buffer: 8 M urea, 2 M thiourea, 4% (w/v) CHAPS, 1% (w/v) dithiothreitol (DTT), 1% (v/v) pH 3–7 and 1% (v/v) pH 7–12 of ZOOM Focusing buffers (Invitrogen Corp.).
6. 0.22 μm Ultrafree-MC microfilter unit (Millipore Corporation, Bedford, MA).
7. Power supply capable of operating at low currents (0.1–1 mA) and up to at least 1,200 V (for example, Electrophoresis Power Supply-EPS 3501 XL, GE Healthcare).

3. Methods
3.1. Prelabeling Samples with Cyanine Dyes

Before MicroSol IEF, protein samples are labeled with spectrally resolvable CyDyes (GE Healthcare), that is, Cy2, Cy3 and Cy5 (*see* Chapter 4.1). Typically, 50 μg of protein from each sample (e.g., control or experimental sample) is minimally labeled with 400 pmol of Cy2, Cy3, or Cy5 reagent in a lysis buffer containing 8 M urea, 2 M thiourea, 4% (w/v) CHAPS, and 30 mM Tris (pH 8.5) (*see* Note 1). Labeling reactions are performed on ice in the dark for 30 min and then quenched with a 50-fold molar excess of 10 mM lysine for 10 min on ice in the dark. The labeled samples are stored at −80°C until needed.

3.2. Assembling the ZOOM IEF Fractionator

The ZOOM IEF Fractionator can be assembled in a number of different configurations depending on the pH range. Even though the typical arrangement of chambers, partition membrane disks, and spacers for a ZOOM IEF Fractionator uses five separation chambers (*see* Note 2), the extremely basic proteins could be further separated using pH 9.1 and pH 12.0 partition membrane disks as shown in Fig. 3 to produce a total of six pH range fractionations or up to seven fractions can be obtained if custom partition membrane disks are used. The unit is assembled before loading samples by following the manufacturer's instructions. As an example, the method for the configuration shown in Fig. 3 is briefly described below.

1. Insert the anode end sealer into the chamber assembly tube followed by a spacer.
2. Insert a separation chamber, which contains a port plug and an O-ring. The O-ring should be facing down. The use of a spacer between the electrode chamber and this first potential separation chamber means that in this configuration the first "separation chamber" will actually contain electrode buffer (*see* Fig. 3).

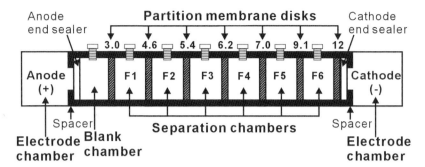

Fig. 3. Schematic illustration of the ZOOM IEF Fractionator configured for six pH ranges. The device consists of seven small Teflon chambers (total volume ~ 750 µl and maximum sample loading volume ~ 650 µl) separated by seven partition membrane disks (pH values shown above the partitions). In this configuration of six fractions, the one blank chamber is coupled to the anode electrode chamber. pHs of the membrane disks (crosshatched vertical rectangles) are shown above each partition.

3. Add a membrane disk pH 3.0 on the anode end sealer using forceps. The membrane disks are labeled with the pH to eliminate the potential that membrane disks will be inserted out of order.
4. Add the next separation chamber, making sure that the port plug is in place and the O-ring is facing down. Push the chamber into the chamber assembly tube until it is flush with the top of the assembly tube.
5. Repeat steps 3 and 4 for the pH 4.6, 5.4, 6.2, 7.0, 9.1, and 12.0 membrane disks and separation chambers.
6. Insert the cathode end sealer next to the last separation chamber.
7. Attach the cathode end screw cap onto the assembly tube and insert the cathode electrode chamber.
8. Load 1×Novex IEF Anode (pH 3.0) and Cathode buffer (pH 12.0) into electrode chambers.

3.3. Loading CyDye Labeled Protein Samples in ZOOM IEF Fractionator

Two or three protein samples labeled with different CyDye reagents are mixed, reduced by adding DTT and sample buffer reagents to have a final concentration of 8 M urea, 2 M thiourea, 4% (w/v) CHAPS, 1% (w/v) DTT, and 1% (v/v) pH 3–7 and 1% (v/v) pH 7–12 of ZOOM Focusing buffers (Invitrogen Corp.) (*see* Note 3). Depending upon the number of separation chambers and the amount and volume of the sample to be used, the sample can be further diluted with additional sample buffer to a final volume equal to the total volumes of the chambers that will be loaded with sample (*see* Note 4). Specifically, samples can be loaded into a single separation chamber, into several separation chambers, or into all separation chambers (*see* Note 5). For example, if a six-chamber separation strategy is used as shown in Fig. 3 and a total of 1.5 mg of Cy-labeled mixed samples are loaded into all six chambers, a final sample volume of 3.9 ml (6 × 650 µl) would be used. Immediately before loading, the sample should be filtered through a 0.22 µm Ultrafree-MC microfilter unit (Millipore Corp.) by centrifugation at 16,000g for 10 min at room temperature to remove any particulate material. After all chambers are filled with either sample in sample buffer or sample buffer only, it is very important that all port plugs be tightly sealed to form an air-tight seal. Do not over-tighten the chamber assembly as over-tightening will distort the fill ports and keep plugs from properly sealing.

3.4. Performing Solution IEF Fractionation

Solution IEF conditions and total time required are very dependent on sample ionic strength, protein concentration, ampholyte concentration, and number of chambers loaded with the samples. Typically, prefractionation of a complex proteome such as mammalian cell extracts requires a power supply with a capacity of at least 1,200 V and that can operate at currents as low as 0.1 mA. If the power supply has current and power limit capacities, set maximum current at 1 mA and maximum power at 1 W. The critical factors are to limit maximum power to 1 W or less and to ultimately reach and hold 1,200 V until a stable low current (<0.5 mA) is reached. For example, 1.5 mg of Cy-labeled cell extracts can be separated with maximum limits of 1 mA and 1 W until a maximum voltage of 1,200 V is reached. The separation is then continued until a stable low current is reached (about 2 h at 1,200 V to reach ~ 0.4 mA). The total focusing time is about 5–6 h.

3.5. Collecting Fractions and Eluting Proteins Trapped in Partition Membrane Disks

After the MicroSol IEF run is complete, fractionated samples are removed through the fill ports using a pipet, transferred to 1.5-ml microcentrifuge tubes, and volumes are measured. Volumes in separation chambers may be reduced slightly because of electroosmosis during the separation. However, reduction of any fraction to less than 50% of its original volume indicates poor sealing of that chamber, which readily compromises the separation; hence in this case, the sample should be discarded and a new MicroSol IEF separation should be performed. After removing fractions from the separation chambers, a small volume of sample buffer (~ 100 µl) is used to rinse the inside walls of the chambers and membrane disk surfaces. This chamber rinse is added to the original fraction to minimize protein losses. Fractions are then adjusted to equal volumes (e.g., 700 µl). A few proteins, primarily those with pIs equal to the pHs of partition membrane disks are retained in the membrane disks. They can be separately extracted using two sequential 30 min extractions with 350 µl sample buffer with shaking at room temperature. The two sequential extracts (~ 700 µl) are then pooled. All samples, i.e., the fractions and partition membrane disk extracts, can be used immediately for further analyses or stored as aliquots at −80°C until needed.

3.6. Analyzing Samples from the ZOOM IEF Fractionator

3.6.1. Initial Evaluation of Fractionation Efficiency on 10% 1-D Gels

After MicroSol IEF prefractionation, each chamber should contain only proteins with pIs between the pHs of the boundary membrane disks of that chamber. An initial rapid assessment of protein recoveries in fractions and membrane disk extracts as well as separation quality and reproducibility can be made by analyzing equal volumes of fractions and membrane disk extracts on 10% Tris-Tricine 1-D gels (Invitrogen Corp.) and medium range 2-D minigels. When DIGE labeling is used, protein patterns can be visualized without additional staining. If desired, total protein amounts in each lane can be estimated and compared using 1-D gel image analysis software such as Discovery Series Quantity One (Bio-Rad Laboratories, Hercules, CA). These 1-D gels are also useful for rapid preliminary comparison of reproducibility between replicate runs. Representative results from duplicate separations of Cy-labeled human melanoma cell extracts using the ZOOM IEF Fractionator (Invitrogen Corp.) configured with 6 separation chambers are shown in Fig. 4. Overall patterns of proteins are quite reproducible, although a few subtle differences could be detected. The majority of proteins were recovered in fractions 2-5 which include pI's between 4.6 and 9.1 (Fig. 4A). As expected, a few proteins were trapped in partition membrane disks (Fig. 4B). After acquiring all fluorescent images for the CyDyes, these gels were stained with SYPRO Ruby (Molecular Probes, Invitrogen Corp.) or silver stain (Invitrogen Corp.) to aid in spot excision for identification using tandem mass spectrometry. To determine whether proteins in each fraction are within the expected pH ranges, a small amount of each fraction should be run on appropriate 3 pH unit wide 2-D minigels (*see* Note 6).

3.6.2. Large-Pore 1-D Gel

To analyze large proteins (>100 kD), the fractionated Cy-labeled samples are also separated on 3–8% gradient gels (Invitrogen Corp.). These large pore 1-D gels effectively separate large proteins up to about 500 kD and therefore

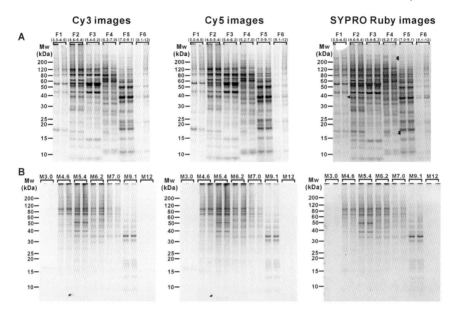

Fig. 4. Initial evaluation of MicroSol IEF fractions on 1-D gels. An initial rapid assessment of MicroSol IEF separation quality, reproducibility, and amount of protein in each fraction can be performed by analyzing equivalent aliquots of each partition membrane disk extract on 10% NuPAGE Bis-Tris 1-D gels. The Cy3 and Cy5 images were acquired from each single gel using mutually exclusive excitation/emission wavelengths of 540/590 nm for Cy3, and 625/680 nm for Cy5 using a ProXPRESS Proteomic Imaging System (PerkinElmer, Boston, MA). After imaging for CyDye components, the gels were stained with SYPRO Ruby (Molecular Probes). The SYPRO Ruby images were acquired on the same imager using 480/680 nm. The replicate separations were relatively reproducible, however, a few differences between replicates could be detected at this level and several of these are highlighted by arrows on the SYPRO Ruby image. F1-F6, MicroSol IEF fractions 1-6, respectively (*see* Fig. 3). All samples had the same final total volume (700 µL) and equal volumes of each pool (7.5 µL) were analyzed. The amount of each fraction on these gels was estimated using the Coomassie Plus protein assay (Pierce Biotechnology, Inc., Rockford, IL): F1 (2.5 µg), F2 (3.5 µg), F3 (3.5 µg), F4 (3.2 µg), F5 (3.8 µg), and F6 (1.6 µg).

complement data from narrow pH range 2-D gels, because 2-D gels do not reliably separate proteins larger than 80–100 kD (*17,18*). As shown in Fig. 5, when unfractionated cell extracts were analyzed, some co-migratory high molecular mass bands could mask quantitative changes in underlying minor proteins. Differences in levels of specific large proteins could be more readily detected after MicroSol IEF fractionation. These comparisons of large proteins show that MicroSol IEF coupled with quantitative image analyses of large

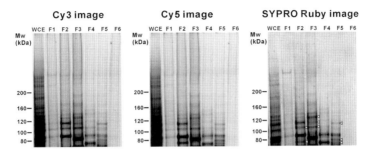

Fig. 5. Analysis of large proteins in human melanoma cells after MicroSol IEF fractionation on 3-8% NuPAGE Tris-Acetate gels. Some specific large proteins showing differential expression are highlighted by white arrows on the SYPRO Ruby image. WCE (whole cell extract), Cy-labeled samples before fractionation (4 μg); F1-F6, MicroSol IEF fractions 1-6, respectively (see Fig. 3). All lanes were loaded with the identical volumes (32 μl). The amount of each fraction loaded on the gel was: F1 (10.6 μg), F2 (15.0 μg), F3 (14.7 μg), F4 (13.4 μg), F5 (16.3 μg), and F6 (6.7 μg). The methods used for image acquisition are described in the legend of Fig. 4.

pore 1-D gels is an effective method for detecting changes of large proteins in protein profile studies.

3.6.3. Narrow pH Range 2-D Gel

Prefractionation of complex proteomes followed by separation of individual fractions on an appropriate series of narrow range IPG strips can dramatically improve detection, resolution, and dynamic range of detected proteins. MicroSol IEF interfaces seamlessly with subsequent narrow pH range IPG gels because similar sample buffers are used and fractions usually do not need to be concentrated before 2-D gels. To maximize spot separation and resolution of individual fractions, the IPG gels should have pH ranges as narrow as possible and be as long as possible. Ideally, these IPG gels should be slightly wider (typically ± 0.1 pH units) than the pH range of each fraction to prevent loss of proteins near the boundaries of the 2-D gels, while maximizing separation distance as much as possible. Usually, the strips can be trimmed after IEF to fit the next smaller size 2-D gels when MicroSol IEF fractions are substantially narrower than available commercial IPG strips. The effective total IEF separation length of fractions from 3.0-10.0 can be as high as about 85 cm when a series of commercial narrow range 24-cm IPG strips are used for IEF and are then trimmed to 18-cm *(17,18)*. This strategy maximizes separation distances as well as maximizing throughput, because running second dimension gels is the major bottleneck in 2-D gels, and smaller second dimension gels require less time and reagents than larger gels. For example, as noted above, 24-cm 1 pH unit strips can be trimmed to 18 cm, or 18-cm 1 pH unit strips can be trimmed to fit Criterion gels (Bio-Rad Laboratories), which are particularly easy to run in large numbers. In addition, these gels require much less sample for a given staining method compared with larger 2-D gels. A representative gel of fraction 4 is shown in Fig. 6. As expected, the proteome profiles of fractionated proteins were greatly enhanced on the narrower pH range gels. The results clearly show that the 2-D separations of fractionated Cy-labeled samples provide greatly improved overall protein resolution, recovery and consistency compared with unfractionated samples. In addition to increased separation distances, this approach allows for higher sample loads, which increases dynamic range and detection of lower-abundance proteins and enables detection of far more total proteins. The images from these 2-D gels can then be combined to produce a composite 2-D gel image with an effective IEF separation distance between pH 3 to 9.1 of about 55 cm using Criterion gels.

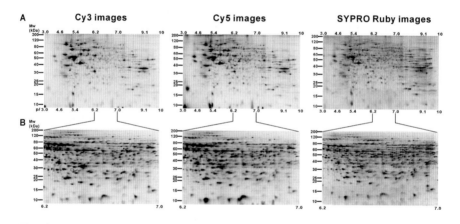

Fig. 6. Illustration of the enhanced protein profiles of human melanoma cell extracts obtained after MicroSol IEF prefractionation. (**A**) 2-D gel of the unfractionated samples of Cy3-labeled 1205LU (high potential metastatic cell line) and Cy5-labeled WM793 melanoma cells (low potential metastatic cell line). The mixed unfractionated Cy-labeled samples were separated on 11-cm IPG strips (pH 3-10) followed by SDS-PAGE using 10% Tris-Tricine gels cast in a Criterion gel cassette (Bio-Rad Laboratories). (**B**) 2-D gel of fraction 4 from MicroSol IEF separation of Cy-labeled samples (pH 6.2-7.0) and separated on a pH 6.2-7.5 (18 cm) IPG strip. After IEF, the strip was trimmed to 12.5 cm to fit into a Criterion gel cassette (the length of the pH 6.2-7.0 region is actually 11 cm). The corresponding pH region (pH 6.2-7.0) of the 2-D gel of the unfractionated sample is indicated for comparison. The unfractionated Cy-labeled 2-D gel contained 20 µg and the fractionated sample was derived from 100 µg of sample. The methods used for image acquisition are described in the legend of Fig. 4.

Image analysis programs can be used to generate volume ratios for each spot which describe the amount of a particular spot contributed by each test sample, and thus enable expression differences to be identified and quantified. This technique is particularly useful for those applications that require direct accurate differential quantitative comparisons.

4. Notes

1. The protein solution for CyDyes labeling should not contain excessive amounts of components that can react with amine specific reagents or high levels of reducing reagents such as DTT because these will compete with the proteins for the CyDyes. The pH of the protein solution should be maintained between pH 8.0–9.0 at 0°C, by including a low concentration of a suitable buffer such as Tris (10–40 mM) or sodium bicarbonate (5 mM). Higher buffer concentrations may interfere with CyDye reactions or negatively affect IEF. Two common buffer compositions for CyDye labeling that are compatible with MicroSol IEF are: 1) 8 M urea, 4% (w/v) CHAPS, and 30 mM Tris (pH 8.5) and 2) 8 M urea, 2 M thiourea, 4% (w/v) CHAPS, 5 mM magnesium acetate, and 10 mM Tris (pH 8.5).
2. A useful feature of the MicroSol IEF method is its flexibility. The total number of separation chambers, their pH ranges, and their volumes can be readily adjusted to meet properties of different proteomes and requirements of different research goals. A typical arrangement of the ZOOM IEF Fractionator uses five separation chambers and six partition membrane disks (e.g., partition membrane disk pH 3.0, 4.6, 5.4, 6.2, 7.0, and 10.0) *(17,20)*. In this case, cathode buffer is prepared by diluting 10× Novex IEF Cathode buffer pH 3–10 (Invitrogen Corp.) to 1× and addition of 7 M or 8 M urea and 2 M thiourea. For this pH range, the pH of the cathode buffer should not be adjusted.
3. The MicroSol IEF prefractionation method is typically conducted under denaturing conditions similar to that used for subsequent 2-DE, e.g., 8 M urea, 2 M thiourea, 4% (w/v) CHAPS, 0.2–1% IPG buffer (carrier ampholytes, typically 0.5%), and 1% (w/v) DTT. In particular, the presence of amphoytes or their equivalent is necessary in both systems, otherwise the current during electrophoresis would be too low to efficiently separate the proteins. Amphoytes can also minimize precipitation/aggregation of proteins during IEF. Typically, 0.2–0.5% (v/v) IPG buffer (pH 3–10) is used for MicroSol IEF with five chambers spanning the pH 3–10 range (*see* Note 2). Because IPG buffers are fractionated along with the proteins during MicroSol IEF, additional IPG buffers must be added, e.g. an additional 0.5–2% (v/v) final concentration before performing 2-D gels. Alternatively, 1% (v/v) pH 3–7 and 1% (v/v) pH 7–12 ZOOM Focusing buffers (Invitrogen Corp.) can be used for the pH range separation of 3–10. When pH's above 10 are used, it is essential that these ZOOM Focusing buffers be used rather than amphoytes. ZOOM Focusing buffers are proprietary low molecular weight buffer formulations that provide improved focusing and resolution of proteins and, most importantly, extend to pH 12.

4. The maximum protein loading capacity for MicroSol IEF prefractionation depends upon the sample type, sample purity (e.g., high salts, lipids, and nucleic acids), loading position, and the number of separation chambers used. Typically, at least 3 mg of complex samples can be effectively separated into five to six pH range fractions. In some cases, separations can be improved by removing excess lipids or nucleic acids. Higher sample loads tend to result in protein precipitation/aggregation on the partition membrane disk surfaces. High sample loads or high initial sample conductivity will also markedly increase focusing times. The pIs of some proteins will exactly match the pH of any partition membrane disk, which may be a critical factor in limiting sample load. At high concentrations, these proteins may precipitate and block the pores with subsequent deposition of a few proteins on the membrane disk surfaces or they may overwhelm the buffering capacity of the immobilines incorporated into the membrane disk. Increasing the partition gel pore size, cross sectional area, or buffering capacity of the membrane disks might further increase protein load capacity, but there are trade-offs with these strategies. Sample load capability is rarely an important factor because the present load capacity matches well with the maximum feasible loads for subsequent narrow pH range 2-D gels *(13,14)*.

5. In general, a sample is loaded into all separation chambers to allow it to be diluted to the largest possible volume. But the most acidic and most basic proteins in sample loaded into the most basic and acidic separation chambers, respectively, must migrate a larger distance than if the entire sample is placed in one or several central chambers only. It may also be advantageous to avoid loading some types of samples into the most acidic and most basic separation chambers to avoid possible precipitation of proteins at these pH extremes during the separation process. In addition, there may be sample-specific benefits to selected sample loading. For example, albumin in serum constitutes more than 50% of total protein and normally severely restricts the amount of serum that can be applied to a 2-D gel *(13,14)*. Therefore, a good experimental design is to isolate albumin in a single separation chamber with a final very narrow pH range to enhance detection of low abundant proteins in other fractions; i.e., mouse serum albumin can be sequestered in a single narrow range pool of pH 5.4–5.8 *(13)*.

6. Fractionated samples and extracts from partition membrane disks can be evaluated on broad or mid pH range 2-D minigels to verify that good separation of proteins based on pI have been achieved; i.e., the vast majority of proteins in a fraction should have pIs within the expected range defined by the partition membrane disks bordering that separation chamber. For example, with the five fraction separation from pH 3.0–10 described in Note 2, fraction 1 is analyzed on a pH 3–6 IPG strip (11 cm; Bio-Rad), fractions 2 and 3 are separated on pH 4–7 IPG strips (11 cm; Bio-Rad or GE Healthcare), fraction 4 is analyzed on a pH 5–8 IPG strip (11 cm; Bio-Rad), and fraction 5 is analyzed on pH 6–11 IPG strips (11 cm; GE Healthcare) followed by 10% SDS gels using Criterion cassettes.

Acknowledgments

This work was supported in part by the National Institutes of Health Grants CA92725 and CA77048 to D.W.S., and institutional grants to the Wistar Institute including an NCI Cancer Core Grant (CA10815), and the Commonwealth Universal Research Enhancement Program, Pennsylvania Department of Health.

1. Wilkins, M. R., Sanchez, J.C., Gooley, A. A., Appel, R. D., Humphery-Smith, I., et al. (1996) Progress with proteome projects: Why all proteins expressed by a genome should be identified and how to do it. *Biotechnol. Genet. Eng. Rev.* **13**, 19–50.
2. Klose, J. (1975) Protein mapping by combined isoelectric focusing and electrophoresis in mouse tissue: A novel approach to testing for induced point mutations in mammals. *Humangenetik* **26**, 231–243.
3. O'Farrell, P. H. (1975) High resolution two-dimensional electrophoresis of proteins. *J. Biol. Chem.* **250**, 4007–4021.
4. Godley, A. A. and Packer N. H. (1997) The importance of protein co- and post-translational modifications in proteome projects, in *Proteome Research: New Frontiers in Functional Genomics* (Wilkins, M. R., Williams, K. L., Appel, R. D., and Hochestrasser, D. F., eds.), Springer, Berlin, pp. 65–91.
5. Miklos, G. L. and Rubin, G. M. (1996) The role of the genome project in determining gene function: Insights from model organisms. *Cell* **86**, 521–529.
6. Ünlü, M., Morgan, M. E., and Minden, J. S. (1997) Difference gel electrophoresis: a single gel method for detecting changes in protein extracts. *Electrophoresis* **18**, 2071–2077.
7. Hanlon, W. A. and Griffin, P. R. (2004) Protein profiling using two-dimensional gel electrophoresis with multiple fluorescent tags, in *Proteome Analysis: Interpreting the Genome* (Speicher, D. W., ed.), Elsevier Science, NY, pp. 75–91.
8. Friedman, D. B., Hill, S., Keller, J. W., Merchant, N. B., Levy, S. E., Coffey, R. J., and Caprioli, R. M. (2004) Proteome analysis of human colon cancer by two-dimensional difference gel electrophoresis and mass spectrometry. *Proteomics* **4**, 793–811.
9. Zhou, G., Li, H., DeCamp, D., Chen, S., Shu, H., Gong, Y., et al. (2002) 2D differential in-gel electrophoresis for the identification of esophageal scans cell cancer-specific protein markers. *Mol. Cell. Proteomics* **1**, 117–124.
10. Wildgruber, R., Harder, A., Obermaier, C., Boguth, G., Weiss, W., Fey, S. J., et al., (2000) Towards higher resolution: two-dimensional electrophoresis of *Saccharomyces cerevisiae* proteins using overlapping narrow immobilized pH gradients. *Electrophoresis* **21**, 2610–2616.
11. Corthals, G. L., Wasinger, V. C., Hochstrasser, D. F., and Sanchez, J. (2000) The dynamic range of protein expression: A Challenge for proteome research. *Electrophoresis* **21**, 1104–1115.
12. Poznanovic, S., Wozny, W., Schwall, G. P., Sastri, C., Hunzinger, C., Stegmann, W., et al. (2005) Differential radioactive proteomic analysis of

microdissected renal cell carcinoma tissue by 54 cm isoelectric focusing in serial immobilized pH gradient gels. *J. Proteome Res.* **4,** 2117–2125.
13. Zuo, X., Echan, L., Hembach, P., Tang, H. Y., Speicher, K. D., Santoli, D. et al. (2001) Towards global analysis of mammalian proteomes using sample prefractionation prior to narrow pH range two-dimensional gels and using one-dimensional gels for insoluble and large proteins. *Electrophoresis* **22,** 1603–1615.
14. Zuo, X. and Speicher, D. W. (2002) Comprehensive analysis of complex proteomes using microscale solution isoelectrofocusing prior to narrow pH range two-dimensional electrophoresis. *Proteomics* **2,** 58–68.
15. Herbert, B. and Righetti, P. G. (2000) A turning point in proteome analysis: Sample prefractionation via multicompartment electrolyzers with isoelectric membranes. *Electrophoresis* **21,** 3639–3648.
16. Righetti, P. G., Castagna, A., and Herbert, B. (2001) Prefractionation techniques in proteome analysis: A new approach identifies more low-abundance proteins. *Anal. Chem.* **73,** 320–326.
17. Speicher, D. W., Lee, K., Tang, H. Y., Echan, L., Ali-Khan, N., Zuo, X., Hembach P. (2004) Current cties, in *Advances in Mass Spectrometry* (Ashcroft, A. E., Brenton, G., and Monaghan, J. J. eds.), Elsevier Science, NY, pp. 37–57.
18. Zuo, X., Lee, K., and Speicher, D. W. (2004) Electrophoretic prefractionation for comprehensive analysis of proteomes, in *Proteome Analysis: Interpreting the Genome* (Speicher, D. W., ed.), Elsevier Science, NY, pp. 93–117.
19. Zuo, X. and Speicher, D. W. (2004) Microscale solution isoelectrofocusing: a sample prefractionation method for comprehensive proteome analysis, in *Methods Molecular Biology, vol. 244: Protein Purification Protocols* (Cutler, P., ed.), Humana, Totowa, NJ, pp. 361–375.
20. Zuo, X., Lee, K., and Speicher, D. W. (2005) Fractionation of complex proteomes by microscale solution isoelectrofocusing using ZOOM IEF fractionators to improve protein profiling, in *The Proteomics Protocols Handbook* (Walker, J. M., ed.), Humana, Totowa, NJ, pp. 97–118.

21

Prefractionation, Enrichment, Desalting and Depleting of Low Volume and Low Abundance Proteins and Peptides Using the MF10

Valerie Wasinger, Linda Ly, Anna Fitzgerald, and Brad Walsh

Summary

The success of proteomics relies heavily in the ability to characterize very diverse species of proteins. This diversity stems from a proteins physicochemical properties, its copy number and abundance and its association with other proteins. Prefractionation and simplification of biological samples prior to downstream MS analysis is showing some virtue in obtaining greater depth of protein analysis. The MicroFlow MF10 is a prefractionation device for low volume, low abundance complex samples that can also enrich for very specific species of proteins based on charge and/or size either in native or denaturing format. It has also been used to desalt and deplete samples of contaminating ions or proteins. Although this instrument is only in its infancy in terms of exploring its capabilities, the technology has been used successfully for the fractionation of plasma proteins Wasinger(1), Omenn GS(2), as well as purification of human growth hormone Catzel D(3) antibodies Cheung(4) and IgY Gee(5) and other complex samples.

Key Words: Abundant protein removal; albumin; depletion; desalting; concentrating; electrophoresis; prefractionation; proteomic.

1. Introduction

One of the greatest contributions of electrophoresis to the life sciences is the increase in knowledge of the dynamic play of protein expression. The ability to resolve the collective variability of proteins in a sample to provide a snapshot of cellular activity and distinguish translational modifications, isoforms, relative abundance, altered molecular interactions, complex formation and breakdown, and presence or absence of proteins has become the hallmark of the proteomics field. This variability in addition to the dynamic range of protein expression and the detection range of technologies is a reason for a lack of completion of a proteome and is fuelling the drive for the development of novel tools and technologies. These technologies simplify the complexity of samples while concentrating species, resulting in a larger number of peptides and proteins being identified *(1–6)*.

The MicroFlow MF10 uses the well-established technique of electrophoresis to move charged molecules in solution and was first described by Horvath and co-workers in 1994 *(7)*. Low volume as well as low concentration samples of peptides and proteins can be fractionated based on pH and mass; and desalted using native, denaturing, or denaturing and reduced conditions. A schematic of the instrument is provided in Fig. 1. Separation is carried out in an electrolytic cell, consisting of between 2 to 6 fractions. The membranes and cartridges are disposable and manufactured to create turbulence within the fractions to limit

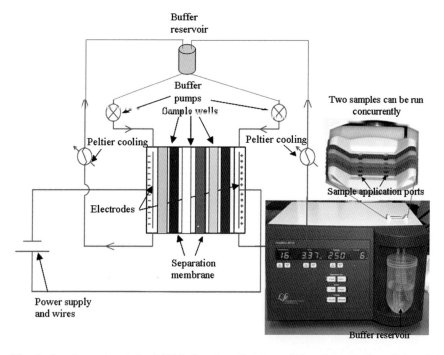

Fig. 1. A schematic of the MF10 showing division of the electrolytic cell into 2 to 6 chambers and circulation of buffer ions. Samples can be run in parallel within the one system. Inset shows front of instrument and the membranes positioned to create the fractionation chambers.

protein build up and membrane fouling. The membranes separate the sample from the other fractions. The fractions and sample are in turn constrained from the running buffer and electrodes by 5kDa restriction membranes. Samples can be run in parallel within the one system. The cathode and anode are housed behind fraction 1 (front) and 6 (back) respectively and preceded by the restriction membrane. The fractions are divided by the membranes, through which buffer ions can pass. Migration of proteins is dependent on the separation membrane pore size and/or pH of buffer. In addition, the size, shape, and charge/mass ratio of the proteins will also affect separation.

As proteins are amphoteric molecules with ionisable acidic and basic groups, the balance of charges dictates how the proteins will migrate during electrophoresis. Migration is influenced by the pH of buffers and a proteins isoelectric point (pI), the pH at which a protein carries no net charge. Proteins in a buffer with a pH greater than their pI will be negatively charged and migrate toward the anode whereas proteins in a buffer less than their pI will be positively charged and migrate toward the cathode during electrophoresis. Selection of buffers compatible with the MF10 are typically of low conductivity (<0.6 mS), weak acids and bases of low Molarity (20–100 mM) and based on their dissociation constants +/– 1pH unit from their pKas. Buffers provide a liquid medium in which proteins are stably maintained, heat can be transferred and current can flow between the two electrodes. A list of common buffers are described in Table 1. The greater the charge on a protein the faster it will migrate during electrophoresis.

The fundamental driving force in electrophoresis is the voltage applied to the electrolytic cell. There are two well known equations that describe the forces contributing to migration in electrophoresis. These are Ohm's Law which relates voltage (V) to current (I) measured in amperes, and resistance (R), measured in ohms ($V = I \times R$) and the second equation relates power (W) to voltage and current and describes the amount of heat generated ($W = V \times I$). By keeping one of these parameters constant, the other parameters can be calculated. Resistance does not remain constant, but changes as buffer ions or salts move through the system. Other factors contributing to increased rate of transfer are: increasing voltage; lower protein concentrations; reduced levels of contaminants (low salt concentration); and low conductivity. Figure 2 depicts this.

1.1. Size Separations

Buffers that impart a like charge on all proteins allow separation based on the pore size of a membrane rather than protein charge. By selecting a buffer with a high pH, all proteins with pI<the buffer pH will have an overall negative charge

Table 1
List of common buffers. pH is achieved by titration

Stable range pH 3.0–4.0	Lactic Acid (MW 90.1) (88mM stock) pK at 25°C = 3.8		GABA (MW 103.12) pK at 25°C = 4.1	
pH	Final mM		Final mM	
3.0	76		12	
3.4	63		25	
3.6	54		34	
3.8	44		44	

Stable range pH 4.0–5.5	Acetic Acid (MW 60.1) (100mM stock) pK at 25°C = 4.7	GABA (MW 103.12) pK at 25°C = 4.1
pH	Final mM	Final mM
4.0	83	17
4.2	76	24
4.4	66	34
4.6	56	44
4.8	44	56
5.0	33	67
5.2	24	76
5.5	14	86

Stable range pH 6.7–7.5	HEPES (MW 238.3) pK at 25°C = 7.5	Imidazole (MW 68.08) (100mM stock) pK at 25°C = 7.0
pH	Final mM	Final mM
6.7	34	66
7.1	56	44
7.3	67	33
7.5	76	24

Stable range pH 7.6–9.0	Boric Acid (MW 61.83) (100mM stock) pK at 25°C = 9.2	Tris (MW 121.14) pK at 25°C = 8.1
pH	Final mM	Final mM
7.6	97.5	2.5
7.8	96	4
8.0	94	6
8.2	91	9
8.4	86	14
8.6	80	20
8.8	71	29
9.0	61	39

Stable range pH5.5–6.5	MES (MW 231.26) pK at 25°C = 6.1	Histidine (MW 155.16) pK at 25°C = 6.2	Stable range pH 9.2–10.2	Tris (MW 121.14) pK at 25°C = 8.1	EACA (MW 131.17) (100mM stock) pK at 25°C = 10.75
pH	Final mM	Final mM	pH	Final mM	Final mM
5.5	80	20	9.24	3	97
5.7	71	29	9.4	4	96
5.9	61	39	9.63	7	93
6.1	50	50	9.8	10	90
6.3	38	62	10.0	15	85
6.5	28	72	10.2	22	78

Stable range pH7.4	Phosphate Buffered Saline (1L)				
Chemical	Amount (g)				
NaCl	8				
Na$_2$HPO$_4$	1.15				
KCl	0.2				
KH$_2$PO$_4$	0.2				

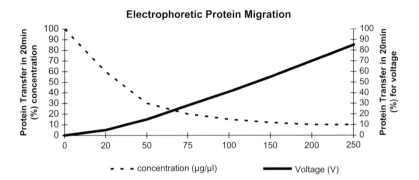

Fig. 2. Electrophoretic protein migration. Protein transfer under an electric field is dependent on the concentration; decreasing as concentration increases (dashed line). Increasing the voltage will increase protein transfer (solid line).

and migrate toward the anode. Proteins are transferred through the membrane which acts like a molecular sieve and are trapped within fractions according to acrylamide and cross-linker concentration of the membrane gels *(8)*. Only proteins small enough to pass through the pore size of the membranes will move from the load fraction into the second fraction or further fractions. Using a low pH buffer and adding sample in fraction 2 will create proteins with an overall positive charge and migration to the cathode and fraction 1.

The rate of protein transfer decreases as the protein size approaches that of the separating membrane pores. Membrane pore sizes should be chosen either slightly smaller than the protein size to be retained or slightly larger than the protein to be transferred. Available membrane sizes are given in Table 2. The separation membranes are approximate in their separating sizes as retardation is relative to size as well as mobility.

1.1.1. Fractionation under Native Conditions

Native conditions preserve a proteins biological structure and conformation and are used for downstream enzymatic or ligand binding studies, conformation and association studies. Under native conditions, protein migration and mobility depends on both the size (of folded conformation) and net charge of the molecule, therefore, the pH of the buffer is important *(9)*. Higher mobility is achieved for more compact structures.

Native conditions are obtained by omitting reductants such as DTT, β-mercaptoethanol and denaturants such as urea from samples allowing inter and intra molecular bonds to remain intact.

Table 2
Pore sizes of separation membranes: The different charge densities and shapes of proteins affects their transfer through the membrane. The nominal molecular size of these membranes is an estimate only

Membrane nominal molecular weight cutoff (NMWCO)
1000
800
500
200
150
125
100
65
45
25
10
5

1.1.2. Fractionation Under Denaturing and/or Reducing Conditions

Proteins that have been dissociated into constituent polypeptides by urea, reducing agents, and detergents can be run under denaturing and reduced conditions. The addition of denaturants increases a proteins solubility especially membrane and associated proteins. Protein conformation in the denatured and reduced state should be linear with covalent thiol bonds reduced. Protein complexes are also broken down into their constituent protein partners. In this state, protein migration occurs in a similar manner to the classical 2-DE techniques.

1.1.3. Fractionation of Low Mass Proteins and Peptides

In the post-genomic era, the push for novel technology has also been driven by the search for biomarkers or targets for drug discovery for diagnostic purposes. The detection of early disease markers requires adequate sensitivity and specificity as biomarkers are often low mass proteins that are missed as a result of abundant proteins or inadequate resolution or capacity for low mass proteins and peptides *(10–12)*. The MF10 enables enrichment of proteins between 1 kDa and 10 kDa based on membrane pore sizes and these samples can be analysed directly via mass spectrometry.

1.2. Desalting Strategies

One of the major impediments to biomarker discovery is the high concentration of salts in readily accessible biological fluids such as urine, tears, and CSF. The transfer of proteins is inhibited by high concentrations of salts from biological sources or from upstream methods such as ion exchange chromatography. Electrodialysis using the MF10 allows salts to be removed. As salt levels decrease, the conductivity drops and the rate of protein transfer increases. The length of time taken for salt to be removed is dependent on the salt concentrations. Protein transfer is not affected by uncharged molecules such as 8 M Urea, 5% sucrose and non-ionic detergents *(13)*.

1.3. Depletion Strategies

Generally, enzymatically digested proteins resulting from samples dominated by high concentration of abundant proteins are detected in all salt fractions when using 2D-LC methods. This does not allow the distinction of lower abundance proteins because of the limit of range of detection of MS instrumentation. It is therefore beneficial to deplete or remove high abundance proteins from samples prior to further downstream analysis. The clustering of abundant proteins into fractions sequentially also provides enrichment for lower abundance proteins *(14)*. Depletion strategies may involve separations based on pH or size or a combination of both pH and size to effectively remove contaminant proteins from proteins of interest in complex biological samples *(1,15,16)*.

For electrophoretic protein separations the pH of the buffer must be kept constant to maintain the charge. For the rapid transfer of proteins, a buffer with a pH value well away from a contaminants pI is chosen so as to maximise the charge on the protein thereby increasing its rate of transfer (see Note 1). The pH of a buffer is a measurement of the H + ion concentration and given by the formula: pH= -log10 [H+]. The buffer streams of the MF10 are combined to increase the buffering capacity and there will be no net change in the pH. Buffering capacity, composition and ionic strength are dependent on the properties of the proteins. The pH of the buffer should be between the target protein pIs and the contaminant protein pI. For a 2-chambered pH separation there are four possible configurations seen in Table 3.

Table 3
Effect of buffering power and sample location on migration of proteins: Based on separating pore size of 1000 NMWCO and 2 chambered pH separation using Fraction 1 (F1) and Fraction 2 (F2)

Sample location	PH of buffer: contaminant pI	effect
F1 (near cathode)	pH<pI	Contaminating protein is positively charged. Migration of target proteins to F2. Contaminant remains in F1.
F1 (near cathode)	pH>pI	Contaminating protein is negatively charged. Migration of contaminating protein to F2. Target proteins remain in F1.
F2 (near anode)	pH<pI	Contaminating protein is positively charged. Migration of contaminating protein to F1. Target proteins remain in F2.
F2 (near anode)	pH>pI	Contaminating protein is negatively charged. Migration of target proteins to F1. Contaminant protein remains in F2.

For separations based on charge, large membrane pore sizes will allow the proteins to move freely and have minimal affect on protein transfer. Charge-based separations can be used in association with smaller pore size separating membranes. The smaller the pore size the lower the rate of transfer of lower mass proteins and increase transfer times. Therefore, relative mobilities can be increased by increasing the difference between the buffer pH and the protein pI (*see* Note 2).

2. Materials

2.1. Size Fractionations

2.1.1. Fractionation under Native Conditions at Low Protein Concentrations

1. Native buffer: 90 mM Tris (biochemical grade) and 10mM ε-amino-*n*-caproic acid (EACA) (Sigma) were dissolved in Ultrapure water. The pH should be 10.2. This buffer was used for both the circulating buffer and the sample buffer.
2. Protein 4 mix standards: 1 mg/ml total protein consisting of β-casein, carbonic anhydrase, bovine serum albumin and trypsin inhibitor was made up in Ultrapure

water. Five micrograms to 500 µg of the protein standard mix was diluted to a total volume of 140 µl using native buffer.

2.1.2. Fractionation under Denaturing and/or Reducing Conditions

1. Denaturing sample buffer: 4 M Urea (minimum assay 99.5%) (GE Healthcare) and 2% (3-[(3-cholomidopropyl)-dimethylammonio]-1-propane sulfonate (CHAPS)) (USB) was dissolved in native buffer and used as the sample buffer.
2. Protein 12 mix standards: Fifty micrograms of a 1 mg/ml total protein solution consisting of β-casein, carbonic anhydrase, bovine serum albumin, trypsin inhibitor, lysozyme C, transferrin, myoglobin, aldolase, β-lactoglobulin A/B, chymotrypsinogen A, ferritin and catalase in Ultrapure water was diluted to a total volume of 140µl using the denaturing sample buffer (see Note 3).
3. SDS-sample boiling buffer: 10% (w/v) sodium dodecyl sulphate, 87% (w/w) glycerol, 0.5 M Tris-HCl, pH 6.8, 0.1% (w/v) bromophenol blue and 2–5% (v/v) β-mercaptoethanol (98% min assay) was prepared in Ultrapure water.

2.1.3. Fractionation of Low Mass Proteins and Peptides

1. Millipore 1 kDa regenerated membranes (catalogue number: 13342) were used for low mass protein and peptide separations. These were prewashed in 3 changes of ultrapure water for 1 h.
2. TBP solution: Tributylphosphine stock was used at 200 mM (Bio-Rad).
3. IAA solution: α-iodoacetamide was prepared to 56 mg/ml.

2.2. Desalting Strategies

1. Desalting buffer: 1.8 M Tris/10 mM ε-amino-n-caproic acid at pH 10.2.
2. EDTA: Ethylenediaminetetraacetic acid was prepared to a final concentration of 1 mM.
3. TBE buffer: Tris-Borate-EDTA running buffer pH 8.0 for gel electrophoresis was diluted to 1 times concentrate.

2.3. Depletion Strategies

1. Charge fractionation buffer: 47 mM γ-aminobutyric acid (GABA) (minimum 99%) (Sigma) was titrated with 17.5 M acetic acid until pH 5.0. 4 M urea, 2% CHAPS.
2. Charge with Ampholine fractionation buffer: Charge fractionation buffer with 2% (v/v) Ampholine™ Preblended, pH 3.5–9.5 (GE Healthcare). This buffer was used as the sample buffer.
3. GABA buffer: 47 mM GABA/acetic acid, pH 5.0 was used as the circulating buffer.

4. Rehydration solution: 8 M urea, 2% (w/v) D-sorbitol, 0.3% (w/v) and 2.5% (v/v) ampholytes, pH 4–6.5.

3. Methods
3.1. Size Separations
3.1.1. Fractionation under Native Conditions at Low Protein Concentrations

With any electrophoretic prefractionation system, there is an optimum protein concentration working range in which efficient fractionations can occur. The described technique is for the fractionation of a mix of four standards under native conditions for as little as 5 µg of total protein. The resulting silver stained gel is shown in Fig. 3.

1. Native buffer was used as the circulating buffer and sample buffer.
2. Five micrograms of the protein 4 mix standards was made up in 140 µl of native buffer.
3. The MF10 two-chamber cartridge was assembled with polyacrylamide nominal molecular weight cut-off (NMWCO) membranes. The two end cartridges were fitted with 5 kD restriction membranes and the separation cartridge fitted with a 45 kD separation membrane. This will create 2 sets of 2 chambers parallel to each other.
4. 140 µl of the protein 4 mix standards was loaded using gel loading tips, into chamber 1 (closest to the cathode) of either set 1 or 2 and the remaining chambers filled with 140 µl native buffer.
5. 100 ml native buffer was circulated around the electrodes.
6. The instrument was run at 250V for 30 min to 2.5 h at 15 °C. The current was monitored during the separation.
7. Following the end of separation, the current was reversed for 2 min.
8. The fractionated sample was collected from the chambers using gel loading tips.
9. To collect any residual protein left in the instrument, 140 µl of PBS was loaded into the chambers and left for 5 min with no voltage applied. The PBS was collected and pooled with the corresponding fractions. This process was repeated for a total of 3 washes.
10. Final fractions were acetone precipitated by adding 4× sample volume of cold acetone and incubating the fractions at −20 °C for 1 h. The sample was centrifuged at 14,000g for 10 min and the acetone removed. The protein pellet was air-dried and resuspended in 20 µl Ultrapure water.
11. For SDS-PAGE, 2–5 µl of the fractions were mixed in a 1:1 volume ratio with 2× SDS sample boiling buffer and boiled for 5 min prior to loading on a 12.5% T SDS-PAGE gel.
12. Gels were silver stained.

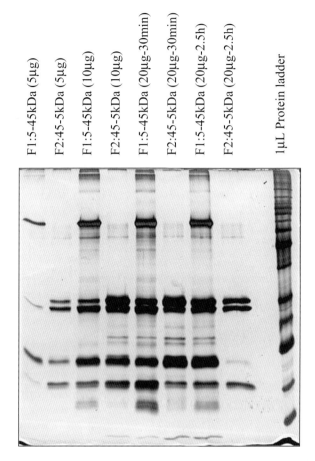

Fig. 3. Size fractionation of native protein standards consisting of BSA, β-casein, carbonic anhydrase and trypsin inhibitor. Different concentrations of total protein were tested on the MF10 instrument. The concentrations tested here were 5 μg, 10 μg and 20 μg in a 30-minute separation and 20 μg in a 2.5 hour separation. These samples were run on an SDS-PAGE gel. 5 μg fractions were loaded at 10 μL; 10 μg fractions were loaded at 10 μL; 20 μg fractions (30min and 2.5h) were loaded at 5 μL.

3.2. Fractionation Under Denaturing and/or Reducing Conditions

Separation under denaturing conditions is described using a mixture of 12 standards at a total amount of 50 μg (see Note 4).

1. Denaturing sample buffer was used as sample buffer and native buffer was used as the circulating buffer.
2. Fifty micrograms of the protein 12 mix standards was diluted to a total volume of 140 μl using the denaturing sample buffer (see Note 5).

3. The MF10 six-chamber cartridge was assembled with polyacrylamide NMWCO membranes. The two end cartridges were fitted with 5kD restriction membranes and the separation cartridges fitted with 1000, 500, 50, 45, and 25 kD separation membranes in decreasing pore size order from the cathode. This will create two sets of 6 chambers parallel to each other.
4. The protein 12 standard mix was loaded (140 μl) using gel loading tips, into chamber 1 (closest to the cathode) of either set 1 or 2 and the remaining chambers filled with 140 μl denaturing sample buffer.
5. 100 ml of native buffer was circulated around the electrodes.
6. The instrument was run at 250V for 2 h at 15 °C. The current was monitored during the separation.
7. Following the end of separation, the current was reversed for 2 min.
8. The fractionated sample was collected from the chambers using gel loading tips.
9. To collect any residual protein left in the instrument, 140 μl of PBS was loaded into the chambers and left for 5 minutes with no voltage applied. The PBS was collected and pooled with the corresponding fractions. This process was repeated for a total of 3 washes.
10. Final fractions were acetone precipitated by adding 4× sample volume of cold acetone and incubating the fractions at –20 °C for 1 hour. The sample was centrifuged at 14,000g for 10 minutes and the acetone removed. The protein pellet was air-dried and each fraction resuspended in 20 μl of denaturing storage solution.
11. For SDS-PAGE, 2–5 μl of the fractions were mixed in a 1:1 volume ratio with 2x SDS sample boiling buffer and boiled for 5 min prior to loading on a 12.5%T SDS-PAGE gel.
12. Gels were silver stained.

3.3. Fractionation of Low Mass Proteins and Peptides

Some separation technologies are limited to resolving proteins of molecular mass fractionation above 5kDa. Proteins or peptides in the mass range of 1-10 kDa can be enriched using the MF10 and analysed directly by mass spectrometry. Outlined here is a method to capture small proteins and peptides from complexes in biological samples. This is a method for enrichment of small plasma proteins.

1. Native buffer was prepared.
2. Plasma sample was diluted, reduced and alkylated. 200 μl plasma was diluted with 740 μl of native buffer and 60 μl of IAA solution and left at room temperature for 1 h. The reaction was quenched with 40 μl TBP solution at room temperature for 15 min.
3. The MF10 was fitted with Life Therapeutics polyacrylamide NMWCO membranes and the 1-kD Millipore cellulose membrane. The end closest to the cathode was bound by the 5-kD restriction membrane followed by the 10-kD

separating membrane, and then the 1-kD Millipore membrane at the anodic side of the unit (see Note 6).
4. 400 μl (4 mg) of reduced and alkylated plasma was loaded into chamber 1 and chamber 2 was loaded with 400 μl native buffer.
5. Native buffer was circulated around the electrodes and the instrument run at 250V for 35 min.
6. The low mass fraction 10-1 kDa was harvested from the anodic chamber and could be analysed using ESI or MALDI mass spectrometry.

3.4. Desalting strategies using Urine

Concentrating and desalting proteins for further downstream analysis can be achieved. The following method is for urine and results for 6 times concentrated urine are shown in Fig. 4.

1. Urine was collected and cellular debris pelleted at 2,500g for 15min.
2. The sample was buffered with 50 μl desalting buffer per 950 μl and EDTA was added to a final concentration of 1 mM to reduce protein aggregation.
3. The MF10 six separation chamber was bounded at either side by the 5 kD restriction membrane and interior chambers were fitted with the 1,000 kD separating membranes
4. All 6 chambers were loaded with 400 μl of the buffered urine samples.

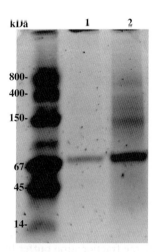

Fig. 4. Native profile of urine proteins six-fold concentrated in the MF10 as compared with unconcentrated urine. Centrifuged urine was concentrated for 35min in the MF10 using Tris/EACA/EDTA circulating buffer, 10 μl of unconcentrated (Lane 1) and 10 μl 6x concentrated (Lane 2) urine were separated by 4–20% native PAGE and stained with SYPRO Ruby.

5. Proteins were concentrated in the chamber closest to the anode, chamber 6, for 35min at 250V (see Note 7).
6. Concentrated proteins were harvested from Fraction 6, mixed with an equivalent volume of TBE loading buffer and analysed by native PAGE using 4–20% gradient gels and TBE buffer for 1 h 15min at 150V.
7. Gels were stained overnight with Sypro Ruby.

3.5. Depletion Strategies

Depletion strategies can be based on pH alone to retain a particular protein in one fraction while transferring remaining proteins to the second fraction and is based solely on the mobility of proteins and their pI in the buffered system. Either high pH or low pH of any value can be circulated to create a unique and specific migration process. Secondly, pH may not be sufficient to reduce sample complexity and thus a combination of pH and size can be used to create multiple fractions of contaminant depleted sample. The described depletion strategy uses a mix of 12 proteins in which Trypsin inhibitor with a pI of 4.5 (theoretical pI) has migrated from the remaining proteins into fraction 2 under denaturing conditions. Native conditions can also be employed. Results of silver stained gels are shown for charge separations in Figure 5a and charge and size fractionation in Figure 5b.

1. Charge with Ampholine fractionation buffer was used for charge-based fractionations alone and Charge fractionation buffer was used for size combined with charge-based separations under denaturing conditions as the sample buffer.
2. GABA circulating buffer was used as the circulating buffer for both modes of separation.
3. Fifty micrograms of the protein 12 standard mix was made up to 140 µl using the charge fractionation buffer (see Note 8).
4. The MicroFlow MF10 two-chamber cartridge was assembled with polyacrylamide nominal molecular weight cut-off (NMWCO) membranes. The 2 end cartridges were fitted with 5kD restriction membranes and the separation cartridge fitted with a 1,000 kD separation membrane for charge-based fractionations or a 45 kD separation membrane for a combination of charge and size-based fractionations. This configuration would yield two fractions.
5. The protein 12 standard mix was loaded using gel loading tips, into chamber 1 (closest to the cathode) of either set 1 or 2 and the remaining chambers filled with 140 µl of charge with Ampholine fractionation buffer for charge-based separations or charge fractionation buffer for a combination of charge and size-based separations.
6. 100 ml of GABA circulating buffer was circulated around the electrodes.
7. The instrument was run at 250V for 4 h at 15°C and the current was monitored during the separation.
8. Following the end of separation, the current was reversed for 2 min.

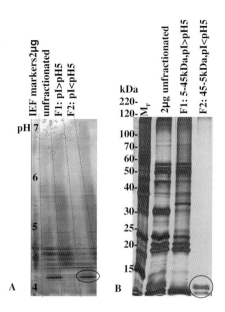

Fig. 5. A 45 kDa separation membrane was used to fractionate proteins according to size and a 47 mM GABA/acetic acid buffer, pH 5.0 was used for a pI-based separation. In charge and size-based fractionation there is migration of proteins with a pI < pH 5 to Fraction 2. Trypsin inhibitor protein fits this criterion with a pI of 4.5. This is observed in the silver stained IEF gel (A). This protein also has a mass less than 45 kDa. The SDS-PAGE gel (B) shows that size and charge fractionation restricted all proteins to Fraction 1 while Trypsin inhibitor protein migrated to Fraction 2.

9. The fractionated sample was collected from the chambers using gel loading tips.
10. To collect any residual protein left in the instrument, 140 µl of PBS was loaded into the chambers and left for 5 minutes with no voltage applied. The PBS was collected and pooled with the corresponding fractions. This process was repeated for a total of 3 washes.
11. Final fractions were acetone precipitated by adding 4× sample volume of cold acetone and incubating the fractions at −20 °C for 1 hour. The sample was centrifuged at 14,000g for 10 minutes and the acetone removed. The protein pellet was air-dried and resuspended in 20 µl of denaturing storage buffer.
12. For isoelectric focusing of the fractions, an (IEF) Immobiline(tm) DryPlate, pH 4-7 (GE) was rehydrated for 3 h in rehydration solution. 2 µl of each fraction was pipetted onto IEF sample application pieces and placed beside each other along the cathodic side of the IEF plate. The plate was focused using the Multiphor II electrophoresis unit (GE) at 15 °C for a total of 5.5 kVh.
13. The IEF plate was silver stained as per manufacturers recommendations.

4. Notes

1. Some caution should be taken when calculating pI from amino acid sequence alone as this will not take into account modifications.
2. High pH buffers are recommended as proteins are more stable in a high pH environment. If the target protein to be depleted is more abundant, move the other proteins.
3. The final concentration of the sample buffer must be the same as that of the circulating buffer. The sample buffer should be made at a higher concentration (2–10 times) than the circulating buffer so when it is added to the sample, it is diluted appropriately to match the concentrations of the circulating buffer. For example, a 10× more concentrated solution of 90 mM Tris/ 10 mM EACA buffer, pH 10.2 should be added at 14 µl to the sample and the rest made up with Ultrapure water for a total volume of 140 µl for separation. This will essentially dilute the 10× concentrated buffer to 1×, therefore matching the same concentration as the circulating buffer. Denaturing buffers added to the sample must be added at a higher concentration such that they become the required 4 M urea and 2% (w/v) CHAPS (and 2% (v/v) Ampholines ™ Preblended, pH 3.5–9.5 for charge-based only fractionations). A 2× buffer with added denaturants and detergents is sufficient
4. Reducing conditions can be applied by adding 10 mM DTT.
5. The final concentration of the sample buffer must be the same as that of the circulating buffer. The sample buffer should be made at a higher concentration (2 to 10 times) than the circulating buffer so when it is added to the sample, it is diluted appropriately to match the concentrations of the circulating buffer. For example, a 10× more concentrated solution of 90 mM Tris/ 10 mM EACA buffer, pH 10.2 should be added at 14 µl to the sample and the rest made up with Ultrapure water for a total volume of 140 µl for separation. This will essentially dilute the 10× concentrated buffer to 1×, therefore matching the same concentration as the circulating buffer. Denaturing buffers added to the sample must be added at a higher concentration such that they become the required 4 M urea and 2% (w/v) CHAPS (and 2% (v/v) Ampholines ™ Preblended, pH 3.5-9.5 for charge-based only fractionations). A 2× buffer with added denaturants and detergents is sufficient.
6. Shiny side of membrane should face into the chamber. There are only 3 chambers here allowing for an additional sealing gasket to be placed behind the Millipore membrane.
7. For 12 times concentration, the 6 chambers were reloaded with new membranes and the 400 µl of 6 times concentrated urine sample was loaded into chamber 6 and the remaining chambers loaded with unconcentrated urine. Sample was then further concentrated at 250V for another 35min.
8. The final concentration of the sample buffer must be the same as that of the circulating buffer.

References

1. Wasinger, V. C., Locke, V.L., Raftery, M.J., Larance, M., Rothemund, D., Liew, A., Bate, I., Guilhaus, M. 2005 Two-dimensional liquid chromatography/tandem mass spectrometry analysis of Gradiflow fractionated native human plasma. Proteomics 5, 3397–401.
2. Omenn GS, S. D., Adamski M, Blackwell TW, Menon R, Hermjakob H, Apweiler R, Haab BB, Simpson RJ, Eddes JS, Kapp EA, Moritz RL, Chan DW, Rai AJ, Admon A, Aebersold R, Eng J, Hancock WS, Hefta SA, Meyer H, Paik YK, Yoo JS, Ping P, Pounds J, Adkins J, Qian X, Wang R, Wasinger V, Wu CY, Zhao X, Zeng R, Archakov A, Tsugita A, Beer I, Pandey A, Pisano M, Andrews P, Tammen H, Speicher DW, Hanash SM. 2005 Overview of the HUPO Plasma Proteome Project: results from the pilot phase with 35 collaborating laboratories and multiple analytical groups, generating a core dataset of 3020 proteins and a publicly-available database. Proteomics 5, 3226–45.
3. Catzel D, L. H., Marquis CP, Gray PP, Van Dyk D, Mahler SM. 2003 Purification of recombinant human growth hormone from CHO cell culture supernatant by Gradiflow preparative electrophoresis technology. Protein Expr Purif. 32, 126–34.
4. Cheung, G. L., Thomas, T.M., Rylatt, D.B. 2003 Purification of antibody Fab and F(ab')2 fragments using Gradiflow technology. Protein Expr Purif. 32, 135–40.
5. Gee, S. C., Bate, I.M., Thomas, T.M., Rylatt, D.B. 2003 The purification of IgY from chicken egg yolk by preparative electrophoresis. Protein Expr Purif. 30, 151–55.
6. Righetti, P. G, Castagna. A., Antonioli, P., Boschetti, E. 2005 Prefractionation techniques in proteome analysis: the mining tools of the third millennium. Electrophoresis 26, 297–319.
7. Horvath Z. S, Corthals. G., Wrigley C.W., Margolis, J. 1994 Multifunctional apparatus for electrokinetic processing of proteins. Electrophoresis 15, 968-71.
8. Corthals, G. L., Margolis, J., Williams, K.L., Gooley, A.A. 1996 The role of pH and membrane porosity in preparative electrophoresis. Electrophoresis 17, 771–75.
9. Margolis, J., Corthals, G., Horvath, Z.S. 1995 Preparative reflux electrophoresis. Electrophoresis 16, 98–100.
10. Petricoin, E. E., Paweletz, C.P., Liotta, L.A. 2002 Clinical applications of proteomics: proteomic pattern diagnostics. J Mammary Gland Biol Neoplasia. 7, 443–40.
11. Nettikadan, S., Radke, K., Johnson, J., Xu, J., Lynch, M., Mosher, C., Henderson, E. 2006 Detection and quantification of protein biomarkers from fewer than ten cells. Mol Cell Proteomics 5, 895–901.
12. Tammen, H., Schorn, K., Selle, H., Hess, R., Neitz, S., Reiter, R., Schulz-Knappe, P. 2005 Identification of Peptide tumor markers in a tumor graft model in immunodeficient mice Comb Chem High Throughput Screen. 8, 783–88.
13. Rylatt, D. B., Napoli, M., Ogle, D., Gilbert, A., Lim, S., Nair, C.H. 1999 Electrophoretic transfer of proteins across polyacrylamide membranes. J chromatogr A 865, 145–53.

14. Li, R.X, Z. H., Li, S.J, Sheng, Q.H, Xia, Q.C, Zeng, R. 2005 Prefractionation of proteome by liquid isoelectric focusing prior to two-dimensional liquid chromatography mass spectrometric identification J Proteome Res. 4, 1256–64.
15. Catzel, D., Chin, D.Y., Stanton, P.G., Gray, P.P., Mahler, S.M. 2006 Fractionation of follicle stimulating hormone charge isoforms in their native form by preparative electrophoresis technology. J Biotechnol. 122, 73–85.
16. Rothemund, D. L., Locke, V.L., Liew, A., Thomas, T.T., Wasinger, V.C., Rylatt, D.B 2003 Depletion of the highly abundant protein albumin from human plasma using the Gradiflow Proteomics 3, 267–78.

22

Sample Prefractionation in Granulated Sephadex IEF Gels

Angelika Görg, Carsten Lück, and Walter Weiss

Summary

Prefractionation procedures not only aid in reducing sample complexity, but also permit loading of higher protein amounts within the separation range applied to two-dimensional electrophoresis (2-DE) gels and thus facilitate the detection of less abundant protein species. Hence we developed a simple, cheap, and fast prefractionation procedure based on flat-bed isoelectric focusing (IEF) in granulated Sephadex gels, containing chaotropes, zwitterionic detergents and carrier ampholytes. After IEF, up to ten Sephadex gel fractions alongside the pH gradient are obtained, and then applied directly onto the surface of the corresponding narrow pH range immobilized pH gradient (IPG) strips as first dimension of 2-DE. The major advantages of this technology are the highly efficient electrophoretic transfer of the prefractionated proteins from the Sephadex IEF fraction into the IPG strip without any sample dilution, and full compatibility with subsequent 2-DE, because the prefractionated samples have not to be eluted, concentrated or desalted, nor does the amount of the carrier ampholytes in the Sephadex fraction interfere with IEF in IPG strips. This sample prefractionation method has been successfully applied for the separation, detection and identification of low abundance proteins from pro- and eukaryotic samples.

Key Words: Isoelectric focusing; proteome; sample prefractionation; Sephadex; two-dimensional electrophoresis.

1. Introduction

Because of the limitations in sample loading capacity on the first dimension isoelectric focusing (IEF) gel, a whole cell lysate may not yield sufficient quantities of less abundant proteins to be displayed on a *single* two-dimensional electrophoresis (2-DE) gel. By replacing wide range pH gradients (e.g., IPG 3–11) by a number of overlapping (ultra)narrow (<1 pH unit) IPG strips, resolution can be considerably improved, and the number of different protein components that can be visualized on a 2-DE gel is substantially increased *(1,2)*. It has been demonstrated that direct application of total cell lysates to narrow pH range IPG strips works well for relatively low protein loads (<250 µg), permitting detection of the higher abundant "house-keeping" and structural proteins. However, when protein loads exceeding 500 µg (which are required for the detection of the less abundant proteins) are applied onto these gels, protein precipitation near the electrodes is observed. This phenomenon is the more pronounced the higher protein amounts of total cell extracts are loaded *(3,4)*. Obviously, proteins with pIs outside the pH interval precipitate at or near the electrodes, and many of the less abundant proteins, which would otherwise focus inside the pH interval used, are co-precipitated. This finding explains the paradoxical observation that the number of detectable protein spots on 2-D gels does not necessarily grow with higher protein loads, but may even decrease when protein loads exceeding 1.0 mg of unfractionated total protein extract are applied.

As a remedy, various sample prefractionation procedures have been proposed, because they permit increased loading of particular proteins while keeping the total protein load constant. These approaches are based on different principles, such as isolation of cell compartments or organelles, selective precipitation or sequential extraction of certain protein classes, or chromatographic and electrophoretic separation methods (reviewed by *(5,6)*). However, the major drawbacks of most of these prefractionation procedures include (*i*) the need for sophisticated instrumentation, (*ii*) dilution of the sample during or after the separation process, (*iii*) loss of proteins because of precipitation, and (*iv*) cross-contamination between different fractions.

One of the most sophisticated sample prefractionation devices is based on a multi-compartment electrolyzer with isoelectric membranes, initially developed by Righetti and co-workers *(7)* which has been down-scaled and simplified by Zuo and Speicher *(3)* to meet the requirements of proteome analysis. This procedure is particularly useful if the prefractionated proteins are subsequently applied onto corresponding narrow pH range IPG gels. It has been reported that this type of prefractionation allows higher protein loads (6–30-fold) on narrow IPG gels without protein precipitation and permits detection of less abundant proteins because major interfering proteins have been removed *(8)*.

As a simple, cheap, and fast alternative to the forementioned procedure, we have developed a prefractionation method for complex sample mixtures based on flat-bed isoelectric focusing (IEF) in granulated gels *(4)*, initially devised by Radola *(9,10)* for the separation of proteins under *native* conditions.

In contrast, IEF is now performed under *denaturing* conditions in flat-bed Sephadex gels containing chaotropes (urea or urea/thiourea mixtures), zwitterionic detergents, and carrier ampholytes *(4)*. After IEF, up to ten Sephadex gel fractions alongside the pH gradient are removed with a spatula and applied directly onto the surface of the corresponding narrow pH range immobilized pH gradient (IPG) strips as first dimension of 2-DE. The major advantages of this sample prefractionation technology include (*i*) a highly efficient electrophoretic transfer of the prefractionated proteins from the Sephadex IEF fraction into the IPG strip without any sample dilution, and (*ii*) full compatibility with subsequent IPG-IEF, because the prefactionated samples have not to be eluted, concentrated or desalted, nor does the amount of the carrier ampholytes in the Sephadex fraction interfere with IPG-IEF. This sample prefractionation method has been successfully applied for the separation, detection and identification of low abundance proteins from complex pro- and eukaryotic samples. Even when milligram-amounts of prefractionated sample were applied onto narrow pH range IPG strips, no protein precipitation at the electrodes occurred, and a considerably higher number of protein spots was detectable than with the unfractionated sample *(4,11)*.

2. Materials
2.1. Sample Preparation

1. Modified O'Farrell *(12)* lysis buffer for cell lysis and solubilization of hydrophophilic as well as moderately hydrophobic proteins: 9.5 M urea, 1.0% (w/v) dithiothreitol (DTT), 2.0% (w/v) 3-[(3-Cholamido propyl) dimethylammonio]-1-propanesulfonate, 2.0% (v/v) carrier ampholytes (pH range 3–10), 10 mM Pefabloc® proteinase inhibitor. To prepare 50 ml of urea lysis buffer, dissolve 30.0 g of urea (GE Healthcare/Amersham Biosciences, Freiburg, Germany) in deionized water (*see* Note 1) and adjust the volume to 50 ml. Add 0.5 g of Serdolit MB-1 mixed-bed ion exchanger resin (Serva, Heidelberg, Germany), stir for 10–15 min, and filter. Add 1.0 g CHAPS (Roche Diagnostics, Mannheim, Germany), 0.5 g DTT (Sigma-Aldrich, St. Louis, MO), and 1.0 ml of Pharmalyte pH range 3–10 (GE Healthcare/Amersham Biosciences) to 48 ml of the filtered urea solution. Store in aliquots at −70°C (*see* Notes 2 and 3). Immediately before use, add 1.0 mg Pefabloc$^{(r)}$ proteinase inhibitor (Merck, Darmstadt, Germany) per ml of lysis buffer.
2. Thiourea/urea lysis buffer *(13)* for cell lysis and solubilization of more hydrophobic proteins: 2 M thiourea, 7 M urea, 4% (w/v) CHAPS, 1% (w/v) DTT, 2% (v/v) carrier ampholytes (pH 3–10) and 10 mM Pefabloc$^{(r)}$ proteinase inhibitor. To prepare 50 ml of thiourea/urea lysis buffer, dissolve 22.0 g of urea (GE Healthcare) in deionized water, add 8.0 g of thiourea (Fluka/Sigma-Aldrich, Buchs, Switzerland) and adjust the volume to 50 ml with deionized water. Add 0.5 g of Serdolit MB-1 mixed bed ion exchange resin (Serva), stir for 10–15

min and filter. Add 2.0 g CHAPS (Roche), 1.0 ml of Pharmalyte pH 3–10 (GE Healthcare) and 0.5 g DTT (Sigma-Aldrich) to 48 ml of the filtrate and store in aliquots at $-70°C$ (*see* Notes 2 and 3). Immediately before use, add 1.0 mg Pefabloc® proteinase inhibitor (Merck) per milliliter of lysis buffer.

2.2. IEF in Granulated Sephadex Gels

1. Sephadex rehydration solution: 8 M urea, 1% (w/v) CHAPS, 25 mM dithiothreitol (DTT) or 200 mM hydroxyethyldisulfide (HED), and 4.0 % (v/v) carrier ampholytes. To prepare 50 ml of rehydration solution, dissolve 25.0 g of urea (GE Healthcare) in deionized water and adjust the volume to 50 ml. Add 0.5 g of Serdolit MB-1 mixed bed ion exchange resin (Serva), stir for 10 min and filter. Add 0.5 g CHAPS (Roche), 2.0 ml of carrier ampholytes (e.g., Pharmalyte pH 3–10; GE Healthcare) and 190 mg of DTT (Sigma-Aldrich) or, alternatively, 1.2 ml HED (2,2'-dithiodiethanol; Fluka or *DeStreak*™ *Reagent* ; GE Healthcare) to 48 ml of the filtrate. Store in aliquots at $-70°C$ (*see* Notes 2 and 3). For prefractionation of hydrophobic proteins, a combination of 6 M urea and 2 M thiourea instead of 8 M urea is favoured. Rehydration solution is also commercially available from GE Healthcare (*DeStreak*™ *Rehydration Solution*) (*see* Note 4).
2. Colored, low molecular mass isoelectric point (*pI*) markers on the basis of azo dyes *(14)*, aminomethylated sulfonephtaleins and nitrophenols with *pIs* 2.6, 3.3, 3.9, 4.7, 5.3, 5.7, 6.3, 7.5, 8.0, 8.6, and 9.0 were generously provided by Dr. Karel Slais (Institute of Analytical Chemistry, Brno, Czech Republic). A saturated solution of individual *pI* markers in deionized water was prepared and stored at 4 °C (for up to several months).
3. IPGstrip cover fluid (available from GE Healthcare)

2.3. IEF in Narrow pH Range IPG Strips

1. IPG Drystrip rehydration solution: 8 M urea, 1% (w/v) CHAPS, 25 mM dithiothreitol (DTT) or 200 mM hydroxyethyldisulfide (HED), and 0.5% (v/v) carrier ampholytes. To prepare 50 ml of rehydration solution, dissolve 25.0 g of urea (GE Healthcare) in deionized water and adjust the volume to 50 ml. Add 0.5 g of Serdolit MB-1 mixed bed ion exchange resin (Serva), stir for 10 min and filter. Add 0.5 g CHAPS (Roche), 0.25 ml of carrier ampholytes (e.g., Pharmalyte pH 3–10; GE Healthcare) and 190 mg of DTT (Sigma-Aldrich) or, alternatively, 1.2 ml HED (2,2'-dithiodiethanol; Fluka or *DeStreak*™ *Reagent* ; GE Healthcare) to 48 ml of the filtrate. Store in aliquots at $-70°C$ (*see* Notes 2 and 3). For prefractionation of hydrophobic proteins, we recommend a mixture of 6 M urea and 2 M thiourea instead of 8 M urea. Rehydration solution is also commercially available from GE Healthcare (*DeStreak*™ *Rehydration Solution*) (*see* Notes 4 and 5).
2. IPGstrip cover fluid (available from GE Healthcare)
3. Narrow pH range IPG Drystrips (typically one pH unit wide) (available e.g., from GE Healthcare or Bio-Rad, Hercules, CA)

3. Methods

3.1. Preparation of Samples

Briefly, cells or tissues are disrupted by techniques such as grinding in a liquid nitrogen-cooled mortar, shearing, homogenization etc. During -or immediately after- cell lysis, interfering substances such as proteolytic enzymes or high concentrations of salt ions must be inactivated or removed. Proteins are then solubilized, usually with sonication, in urea- or thiourea/urea-lysis buffer. As an example, solublization of mouse liver proteins is briefly described below (*see* Note 6).

1. Homogenize mouse liver (50 mg) in a morter with pestle under cooling with liquid nitrogen. Supend the resulting, deep-frozen powder in 1.0 ml urea or thiourea/urea lysis buffer. Sonicate briefly (3 × 20 sec) with cooling on ice. After 60 minutes, the protein extract is centrifuged (40,000g, 15 °C, 60 min) to remove any insoluble material (e.g., cell debris). Optimal protein concentration of the extract is between 5 and 10 mg/ml. The extract can be used immediately, or stored at −70 °C for several months.

3.2. IEF in Granulated Sephadex Gels

These instructions assume the use of a Multiphor II flatbed electrophoresis unit (GE Healthcare), but may be adaptable to other flat-bed IEF devices.

1. Prepare a Sephadex slurry by gently mixing (e.g., with a spatula) 210 mg of Sephadex G-100SF (GE Healthcare) and 3.0 ml of Sephadex rehydration solution (alternatively, DeStreak ™ Rehydration Solution may be used). Avoid trapping air bubbles, and let the Sephadex gel reswell at 20 °C for approximately 24–48 h (*see* Fig. 1A).
2. Insert a customized, tight-fitting template made from e.g., methacrylate or polycarbonate into the tray of the IPG DryStrip kit (GE Healthcare) and position on the cooling plate of the Multiphor II apparatus (*see* Fig. 2).
3. Add 1.5 ml of sample solution (protein concentration 5–10 mg/ml) and 5 µl of each colored pI marker solution to the Sephadex gel slurry (3.0 ml), and gently mix with a spatula (*see* Fig. 1A).
4. Pipet this mixture into the tray from step 2. Apply 4.5 ml of Sephadex-sample slurry into each individual lane of the template to form a 100 x 20 mm^2 wide and approximately 2.5 mm thick gel layer (*see* Fig. 1B). If the high viscosity of the Sephadex slurry prevents it from spreading evenly, gently tap the tray on the cooling plate of the Multiphor II device until the surface of the slurry is flat and devoid of air bubbles.
5. Soak two layers of 1-mm thick IEF electrode paper strips (e.g., GE Healthcare) in 10 mM sulfuric acid (anode) and 10 mM sodium hydroxide (cathode) and insert them in the notch at the anodic and cathodic ends of the tray, respectively. Cut off protruding parts so that the paper strips fit into the template, and apply platinum electrodes on the short side of the tray, so that the proteins are separated along the long axis.

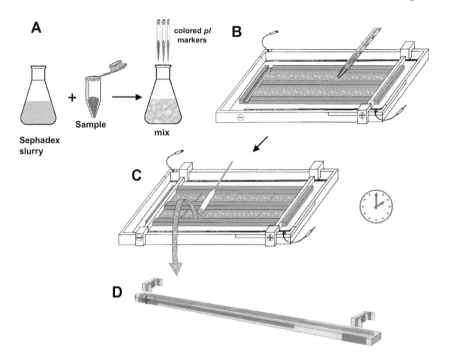

Fig. 1. Schematic presentation of the sample prefractionation procedure with Sephadex IEF before narrow pH range 2-DE gels. (**A**) Sephadex G-100SF is gently mixed with a solution containing urea or urea/thiourea, CHAPS, DTT (or HED), and carrier ampholytes and is then rehydrated for 24–48 h. Before IEF, the Sephadex slurry is mixed with the sample solution and colored *pI* markers. (**B**) The Sephadex slab gel is then poured by pipeting the Sephadex/sample mixture into the grooves of the template inserted into the IPG DryStrip kit. The surface of the Sephadex gel must be protected by a thin layer of Drystrip cover fluid to prevent desiccation of the gel. (**C**) After IEF, individual Sephadex fractions are removed with a spatula and applied onto rehydrated, narrow pH range IPG strips. (**D**) IPG-IEF is performed in the IPGphor instrument as described previously *(15,16)*. (Reproduced from *(4)* with permission from Wiley-VCH).

6. Cover the Sephadex gel with a thin layer of IPGstrip cover fluid (1.0 ml for 20 cm² of gel surface) to prevent water evaporation and desiccation of the gel during the IEF step.
7. Perform IEF at 20 °C. Typically, initial settings of 10 V/cm and 4 mA (per lane), and terminal settings of 100 V/cm are applied. For example, in case of a pH gradient 3–10 and a separation distance of 100 mm, IEF is run with an initial voltage of 150 V for 20 min, followed by 300 V for 30 min, 600 V for 60 min, and 1000 V for approximately 2 h until the proteins have focused. If the salt concentration of the sample exceeds 100 mM, the initial step (150V) should

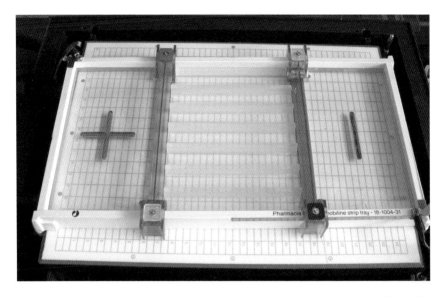

Fig. 2. Customized polycarbonate template inserted into the IPG DryStrip kit and placed on the cooling plate of the Multiphor II unit. Each groove of the template is 20 mm wide, 5 mm deep, and 100 mm long.

be extended for up to 3–4 h, and the IEF filter paper strips (where the salt has collected) should be replaced every 2 h.

8. After IEF, up to ten Sephadex fractions can be "harvested" by slicing the gel rectangular to the pH gradient with the help of a fractionation grid, or simply by removing the gel with a spatula (*See* Fig. 1C). Because of the problems of instability and irreproducibility of carrier ampholyte generated pH gradients, it is important to determine precisely the slope to the pH gradient and to define the exact positions where to "cut" and remove the individual Sephadex fractions to fit them to the corresponding narrow pH range IPG strips. The most convenient and reliable method is to incorporate synthetic, differentially colored low molecular mass *pI* markers *(14)* (which do not interfere with subsequent 2-DE) in the Sephadex-sample slurry (*see* Step 3).

9. The Sephadex fractions are either applied immediately onto rehydrated narrow pH range IPG strips (*see* Section **3.3**), or sealed in plastic sheets and stored at −70°C until further use.

3.3. IEF in Narrow pH Range IPG Strips

These instructions assume the use of the IPGphor electrofocusing unit with individual ("universal") or multiple *cup-loading* strip holders (Manifold™) available from GE Healthcare, but may be adaptable to similar devices from other manufacturers.

1. Rehydrate narrow pH range IPG strips with IPG Drystrip rehydration solution or DeStreak™ Rehydration Solution in an IPG DryStrip reswelling tray (e.g., GE Healthcare) according to the manufacturer's instructions. Then rinse the rehydrated IPG strips briefly with deionezed water, blot them between two sheets of moist filter paper and place them, gel-side up, and acidic ends facing towards the anode, in IPGphor strip holders. For better performance, insert moist filter paper pads (size: 4×4 mm^2) between the electrodes and the IPG strips during IEF.
2. Transfer the prefractionated proteins from Section 3.2 into the IPG gel by spreading the Sephadex fractions onto the surface (near the anode) of the corresponding narrow-range IPG gel strips (*see* Fig. 1D). Protect the surface of the IPG strips with 2 ml of IPG strip cover fluid. No protein elution from the Sephadex fraction before IPG-IEF was performed and, hence, no dilution of the fractionated sample happened.
3. Continue with IPG-IEF. Recommended settings and running conditions for IEF using the IPGphor are identical to those described in detail elsewhere *(15)*. Initial settings should be limited to 150 V, 300 V, and 600 V, respectively, for 60 min each to ensure efficient protein transfer from the Sephadex gel into the IPG matrix *(15,16)*. Then IEF should be continued with higher voltages 8,000 V to the steady state (approximately 150,000 Vh in case of 180 mm long and one pH unit wide IPG strips).
4. After termination of IPG-IEF, freeze the IPG strips at $-70°$C between two sheets of plastic film. Most of the Sephadex slurry sticks to the plastic film, whereas the other components are simply rinsed off during the IPG strip equilibration step

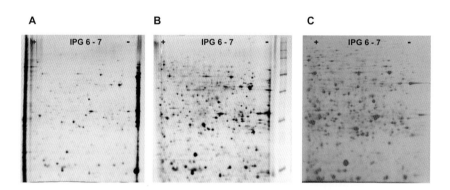

Fig. 3. Effect of sample prefractionation for high protein loads applied to 2-DE gels. First dimension: IEF in a narrow pH range IPG 6–7. Second dimension: SDS-PAGE (13%T, 2.6% C) Sample: mouse liver proteins. (**A**) unfractionated sample; protein load: 250 μg; silver stain; (**B**) sample prefractionated with Sephadex-IEF; protein load: 250 μg; silver stain; (**C**) sample prefractionated with Sephadex IEF; protein load: 1000 μg; Coomassie blue stain. (Reproduced from *(4)* with permission from Wiley-VCH).

before the second dimension of 2-DE. Then apply the equilibrated IPG strips onto SDS gels and continue with SDS-PAGE. An example of results (comparison of unfractionated and prefractionated samples) is presented in Fig. 3

4. Notes

1. All solutions should be prepared in water that has a resistivity of 18.2 MΩ/cm. This standard is referred to as "deionized water" in this text.
2. Urea and thiourea/urea lysis buffer should be freshly prepared. Alternatively, make 1 ml aliquots and store in Eppendorf vials at $-70°C$ for up to several months. Lysis buffer that has been thawed once should not be refrozen.
3. Never heat urea solutions above 37 °C. Otherwise, proteins in the sample solution may be carbamylated, resulting in charge artifacts ("spot trains").
4. The appearance of streaks that distort electrophoresis maps is a common problem, occurring particularly when running gels that contain regions *exceeding pH 7*. Increased sample load or using a narrower pH gradient worsen the problem. Hydroxyethyldisulfide (HED = 2,2'-dithiodiethanol = *DeStreak*™ reagent) *(17)* reduces nonspecific oxidation of proteins and streaking in the alkaline part of IEF gels. We recommend to incorporate HED in both Sephadex- and IPG Drystrip-rehydration solution. *DeStreak*™ *Rehydration Solution* that already contains urea, thiourea, CHAPS, and HED is available from GE Healthcare. The solution is ready for use after addition of the appropriate carrier ampholytes (e.g., Pharmalyte pH 3–10).
5. If proteins have been solubilized in thiourea/urea lysis buffer, thiourea should also be incorporated in the Sephadex- and IPG Drystrip-rehydration solutions.
6. This protocol can be easily adapted for other mammalian tissues. For more detailed information on sample preparation of mammalian cells or tissues, microorganisms and plant species s*ee* **other chapters in this book**.

Acknowledgements

The authors would like to thank Dr. Karel Slais (Institute of Analytical Chemistry, Brno, Czech Republic) for generously providing colored *pI* markers, and Mr. Andreas Klaus for excellent technical assistance.

References

1. Wildgruber, R., Harder, A., Obermaier, C., Boguth, G., Weiss, W., Fey, S.J., Larsen, P.M., and Görg, A. (2000) Towards higher resolution: Two-dimensional electrophoresis of *Saccharomyces cerevisiae* proteins using overlapping narrow immobilized pH gradients. *Electrophoresis* **21**, 2610–2616.
2. Westbrook, J.A., Yan, J.X., Wait, R., Welson, S.Y., and Dunn, M.J. (2001) Zooming-in on the proteome: Very narrow-range immobilized pH gradients reveal more protein species and isoforms. *Electrophoresis* **22**, 2865–2871.

3. Zuo, X., and Speicher, D.W. (2000) A method for global analysis of complex proteomes using sample prefractionation by solution isoelectrofocusing before two-dimensional electrophoresis. *Anal Biochem.* **10**, 266–278.
4. Görg, A., Boguth, G., Köpf, A., Reil, G., Parlar, H., and Weiss, W. (2002) Sample prefractionation with Sephadex isoelectric focusing prior to narrow pH range two-dimensional gels. *Proteomics* **2**, 1652–1657.
5. Righetti, P.G., Castagna, A., Antonioli, P., and Boschetti, E. (2005) Prefractionation techniques in proteome analysis: The mining tools of the third millennium. *Electrophoresis* **26**, 297–319.
6. Lescuyer, P., Hochstrasser, D.F., and Sanchez, J.-C. (2004) Comprehensive proteome analysis by chromatographic protein prefractionation. *Electrophoresis* **25**, 1125–1135.
7. Righetti, P.G., Wenisch, E., Jungbauer, A., Katinger, H., and Faupel, M. (1990) Preparative purification of human monoclonal antibody isoforms in a multicompartment electrolyser with immobiline membranes. *J. Chromatogr.* **500**, 681–696.
8. Zuo, X., and Speicher, D.W. (2002) Comprehensive analysis of complex proteomes using microscale solution isoelectrofocusing prior to narrow pH range two-dimensional electrophoresis. *Proteomics* **2**, 58–68.
9. Radola, B.J. (1973) Isoelectric focusing in layers of granulated gels. I. Thin-layer isoelectric focusing of proteins. *Biochim. Biophys. Acta* **295**, 412–428.
10. Radola, B.J. (1975) Isoelectric focusing in layers of granulated gels. II. Preparative isoelectric focusing. *Biochim. Biophys. Acta* **386**, 181–195.
11. Weiss, W., Lück, C., Reil, G., and Görg, A. (2005) Sample prefractionation with Sephadex IEF for enrichment of low abundance proteins and improvement of MALDI-MS spectra. *Mol. Cell. Proteomics* **4** (Suppl. 1), S346.
12. O'Farrell, P.H. (1975) High resolution two-dimensional electrophoresis of proteins. *J. Biol. Chem.* **250**, 4007–4021.
13. Rabilloud, T. (1998) Use of thiourea to increase the solubility of membrane proteins in two-dimensional electrophoresis. *Electrophoresis* **19**, 758–760.
14. Stastna, M., Travnicek, M., and Slais, K. (2005) New azo dyes as colored isoelectric point markers for isoelectric focusing in acidic pH region. *Electrophoresis* **26**, 53–59.
15. Görg, A., Obermaier, C., Boguth, G., Harder, A., Scheibe, B., Wildgruber, R., and Weiss, W. (2000) The current state of two-dimensional electrophoresis with immobilized pH gradients. *Electrophoresis* **21**, 1037–1053.
16. Görg, A, Weiss, W, and Dunn, M.J. (2004) Current two-dimensional electrophoresis technology for proteomics. *Proteomics* **4**, 3665–3685.
17. Olsson, I., Larsson, K., Palmgren, R., and Bjellqvist, B. (2002) Organic disulfides as a means to generate streak-free two-dimensional maps with narrow range basic immobilized pH gradient strips as first dimension. *Proteomics* **2**, 1630–1632.

23

Free-Flow Electrophoresis System for Plasma Proteomic Applications

Robert Wildgruber, Jizu Yi, Mikkel Nissum, Christoph Eckerskorn, and Gerhard Weber

Summary

This chapter describes the technology of free flow electrophoresis (FFE) and protocols to separate human plasma for proteome analysis. FFE is a highly versatile technology applied in the field of proteomics because of its continuous processing of sample and high resolution in separation of most kinds of charged or chargeable particles including ions, proteins peptides, organelles, and whole cells. FFE is carried out in an aqueous medium without inducing any solid matrix, such as acrylamide, so that it simplifies complex sample for the downstream analysis. Two FFE protocols are described to separate human plasma proteins under native and denaturing conditions. Plasma separated under native conditions was pooled into acidic-, alkaline-, and albumin- fractions that were furthered for gel-based analysis. Under denaturing condition plasma proteins were separated into 96 fractions. Each fraction can be supplied for in-solution digestion and further LC-MS/MS analysis. From a single FFE fraction 46 different proteins (protein family) have been identified, demonstrating FFE as a high efficient separation tool for human plasma proteome studies.

Key Words: Albumin depletion; free flow electrophoresis; isoelectric focusing; plasma protein separation; plasma proteomics.

1. Introduction

Free flow electrophoresis (FFE) is a well-established and highly versatile technology for the separation of a wide variety of charged or chargeable analytes including low-molecular weight organic compounds, peptides, proteins, protein complexes, membranes, organelles, and even whole cells. The separation is performed in aqueous media under either native or denaturing conditions.

From: *Methods in Molecular Biology, vol. 424: 2D PAGE: Sample Preparation and Fractionation, Volume 1*
Edited by: A. Posch © Humana Press, Totowa, NJ

The analyte is injected into a thin, laminar film of separation buffer which defines the electrophoretic mode like zone electrophoresis, isoelectric focusing or isotachophoresis, and is deflected by an electric field perpendicular to the flow direction. The absence of any types of solid matrix prevents unspecific adsorption of analytes to the separation matrix as well as precipitation of the analytes because of a concentration effect on a solid matrix. The FFE separation buffers continuously flow from the sample introduction through the complete separation process and are collected in single fractions after separation. These unique features of a continuous electrophoretic separation process are directly linked to high reproducible separations of plasma sample with duration of less than 20 min and with almost quantitative recovery in collected matrix-free fractionations.

The resolution of the separation is comparable to the high resolution of standard polyacrylamide electrophoresis (PAGE). The FFE technology ensures that separated samples are compatible with many kinds of downstream analyses such as concentration procedures (e.g., ultra filtration), separation techniques (1- and 2-dimensional electrophoresis), and analytical methods like matrix-assisted laser desorption ionization-time-of-flight mass spectrometry (MALDI-TOF-MS) or liquid chromatography (LC-MS). Because of the compatibility of FFE with almost any downstream technology, the combination of FFE with 1D- 2D-PAGE, MALDI-TOF-MS, and/or LC-MS/MS, ELISA as well as other activity tests, allows us to explore low abundant proteins as well as to proteins with an extreme physical-chemical properties, e.g., a very high or a very low molecular weight, high hydrophobicity, or an extreme acidic or alkaline isoelectric point (pI). With the advantage of the continuous process, the sample input can be adjusted individually from microgram to multi-gram levels, i.e., from analytical to highly preparative ranges, delivering recovery rates of more than 97%.

Isoelectric focusing in FFE allows for high-resolution fractionation of proteins with high recovery rates. The separation of sample can be continuously conducted in preparative quantities from less than 1.0 mg up to 50 mg protein/h. The pH gradient for the separation process is formed under constant voltage by the separation buffer with proprietary separation reagents, which consist of well-defined compounds with low molecular mass and specific pIs.

Any pH-gradient from 3 to 12, either in a wide or a narrow range, can be calculated and used for specific high resolution separations, depending on the sample and downstream application *(1)*.

The FFE system performs electrophoretic separations in patented and well-defined separation solutions without induction of a solid (e.g., polyacrylamide) matrix. It provides fluid-phase separation in three different operating modes: zone electrophoresis (ZE), which separates particles (cells, organelles and protein complexes) *(2)* by their net charge; isoelectric focusing (IEF), which separates proteins or peptides in a pH gradient by their isoelectric point

(3); and isotachophoresis, which separates analytes in a pH and conductivity step gradient by electrophoretic mobility. This enables the separation of cell organelles, proteins, and other charged or chargeable entities on a fast, preparative, and continuous basis (*see* **Fig. 1**).

The system allows researchers to reduce sample complexity using such subproteomics applications as cell or organelle purification in ZE mode, or enrich low-abundance proteins with high reproducibility in IEF.

The sample is applied using a peristaltic pump and is induced into a separation chamber consisting of two parallel plates. Under laminar flow, the sample is transported within a thin (0.4-mm) film of aqueous medium formed between the two plates. The plates are bordered by two electrodes that generate a high-voltage electric field perpendicular to the laminar flow. Charged particles (ions, peptides, proteins, organelles, membrane fragments, or whole cells) are deflected in this electric field, permitting subsequent separation and/or fractionation.

Three different operational modes can be used for FFE (*see* **Fig. 2**):

- IEF, which is used for the high-resolution fractionation of proteins; separation of quantities for analytical as well as preparative analysis
- ZE, which permits the rapid separation of organelles, membranes, and whole cells for the study of subproteomes
- Isotachophoresis, which offers an enhanced range of separation methods based on electrophoretic mobility.

Human plasma is a prolific source of potential disease and biological markers and is readily available in high amounts for clinical analysis *(4)*. The large dynamic range of proteins *(5)* present in plasma, as well as lipids, salts and the proteolytic activity limit their consistency, efficiency, sensitivity and resolution

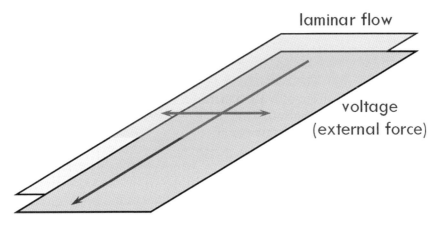

Fig. 1. Schematic overview of the FFE principle.

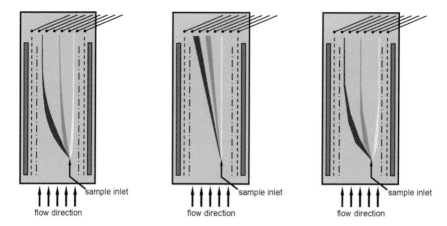

Fig. 2. FFE operating modes. (**a**) IEF mode for separation of proteins/peptides in a pH gradient, (**b**) zone electrophoresis for separation of particles like cells and organelles because of their net charge, and (**c**) isotachophoresis for separations of proteins and particles because of their electrophoretic mobility.

of many analytical techniques to identify potential biomarkers in plasma. As a consequence, the reduction of sample complexity is becoming widely accredited as a mandatory step in plasma proteome analysis. It is commonly accepted that proteins, e.g., albumin and IgGs should be depleted before in depth analysis of human plasma *(6)*, but protein losses caused by unspecific protein-protein and/or protein-matrix interactions limit the effectiveness and reproducibility of some depletion techniques.

As an alternative method to affinity depletion for HSA, we use free flow electrophoresis as a preceding separation step. The diluted plasma is directly injected under native conditions in a pH step gradient at pH 4.8 the pI of native albumin. Whereas the albumin is focused in one pool, all proteins below or above pH 4.8 are separated into a acidic or alkaline pool, respectively (see **Fig. 3**). This results in protein concentrations for the acidic pool of about 10%, for the albumin 60% and 30% alkaline pool. These pools can be used directly for top down proteomic approaches or ELISA testing, or for subsequent high resolution separations, giving access to low abundant proteins (*see* Fig. 4).

To access low abundant proteins it is either possible to separate plasma directly or use the pools of the native depletion FFE separation described before. Plasma proteins are separated into >70 fractions. Each fraction is supplied for in solution digestion and furthered for analysis by LC-MS/MS *(3)*. From a single fraction peptides are detected and fragmented for MS/MS analysis, followed by protein identification by search of MS/MS profiles against NCBInr. This study demonstrates that FFE can be used for large-scale sample preparation for

Free-Flow Electrophoresis System for Plasma Proteomic Applications 291

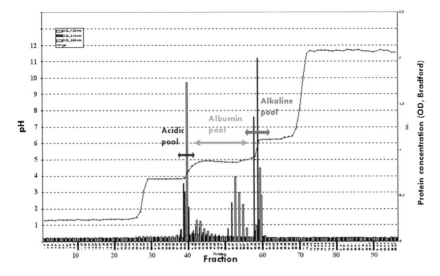

Fig. 3. pH gradient profile of native plasma depletion, showing acidic, albumin and alkaline pool of the FEE separation.

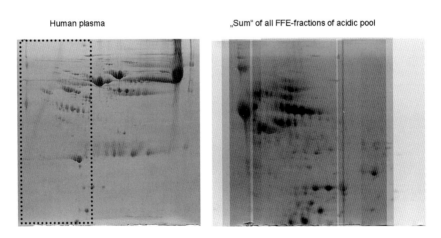

Fig. 4. Native depletion of HSA from human Plasma, Comparison of 2D electrophoretic protein patterns of unprocessed sample (left) and the summarized images as overlay of FFE processed samples (right). The marked area in the left area shows the corresponding region of the 2D gel which is corresponding to the pH range of the right image. A significant increase of protein spots of the FFE processed sample can be demonstrated.

high throughput proteomic analyses of plasma to raise new dimensions in the searching for biological markers.

2. Materials

2.1. Blood Collection and Plasma Preparation

1. BD™ P100 Blood Collection Tube (BD, Franklin Lakes, NJ) which includes EDTA and protease inhibitors.
2. A CellSep 6/720R centrifuge (SANYO, Japan) or a Sorvall RC3C Centrifuge.

2.2. Buffers for FFE-HSA depletion

1. Anodic stabilization medium 1 (I1): 10 % (w/w) glycerol, 100 mM H_2SO_4.
2. Anodic stabilization medium 2 (I2-I3): 10 % (w/w) glycerol, 200 mM Tris, 100 mM HAc.
3. Separation medium (I4): 10 % (w/w) glycerol, 10 mM Tris, 10 mM HAc.
4. Cathodic stabilization medium 2 (I5-I6): 10 % (w/w) glycerol, 100 mM Tris, 200 mM HAc.
5. Cathodic stabilization medium 1 (I7): 10 % (w/w) glycerol, 100 mM NaOH.
6. Counterflow medium: 10% (w/w) glycerol.
7. Anodic circuit electrolyte: 100 mM H_2SO_4.
8. Cathodic circuit electrolyte: 100 mM NaOH.

2.3. Buffers for denaturing FFE separation

1. Counterflow medium: weigh in 500 g H_2O dist. and add 315 g urea, 115 g thiourea and 34 g mannitol.
2. Separation buffer 1: 30.3 g of BD™ IEF buffer 1 (BD, Franklin Lakes, NJ), 69.7 g H_2O dist., 63.0 g urea, 23.0 g thiourea and 6.8 g mannitol.
3. Separation buffer 2: 100.0 g of BD™ FFE IEF buffer 2 (BD, Franklin Lakes, NJ), 100.0 g H_2Odist., 126.0 g urea, 46.0 g thiourea and 13.6 g mannitol.
4. Separation buffer 3: 37.0 g of BD™ FFE IEF buffer 3 (BD, Franklin Lakes, NJ), 63.0 g H_2Odist., 63.0 g urea, 23.0 g thiourea and 6.8 g mannitol.
5. Anodic stabilization medium: 3.0 g of H_2SO_4 (5M), 97.0 g H_2O dist., 63.0 g urea, 23.0 g thiourea and 6.8 g mannitol.
6. Cathodic stabilization medium: 30.0 g of NaOH (1M), 170.0 g H_2O dist, 26.0 g urea, 46.0 g thiourea and 13.6 g mannitol.
7. Anodic circuit electrolyte: 100 mM H_2SO_4.
8. Cathodic circuit electrolyte: 100 mM NaOH.

2.4. Post-FFE Sample Preparation for SDS-PAGE Analysis

1. XCell SureLock™ Mini-Cell (Invitrogen) and Power supply.
2. SDS buffer: add 2ml mercaptoethanol to 5 ml NuPAGE® LDS Sample Buffer (4×)

3. SDS Running buffer: Prepare 1X SDS Running Buffer by adding 50 ml 20× NuPAGE® MES buffer to 950 ml of deionized water.
4. SilverQuest® (Invitrogen) staining kit.

2.5. Post-FFE sample preparation for LC-MS/MS analysis

1. Membrane-based separator: Microcone YM-3 (Millipore, USA).
2. Washing solution 1: 30% acetonitrile, 0.1% TFA.
3. Washing solution 2: 100 mM Tris-HCl, pH 8.0.
4. Sample buffer: 100 mM Tris-HCl, pH 8.0, with 5 % acetonitrile, 100 mM NaCl.
5. Reducing agent: 200 mM DTT in 100 mM Tris-HCl, pH 8.0 (freshly made).
6. Alkylating agent: 200 mM iodoacetamide in 100 mM Tris-HCl, pH 8.0 (freshly made).
7. Trypsin solution: The protease for digestion is sequencing grade modified (methylated) trypsin of 20 ng/µl in a digestion buffer with 50 mM ammonium bicarbonate, pH 8.0, and 10% (v/v) acetonitrile (The solution was made freshly, and kept on ice until use).
8. C18 desalting spin column (Pierce Biotechnology, Inc, USA).

3. Methods

3.1. Preparation of plasma sample before FFE fractionation.

1. Proximately 8.5 ml of human blood is drawn directly into a BD™ P100 tube using a BD Vacutainer® Safety-Lok™ Blood Collection Set (BD, Franklin Lakes, NJ, USA). P100 tube with a lyophilized cocktail of protease inhibitors which inhibit proteolytic degradation of plasma proteins also (7).
2. Mix specimen with tube additives. For optimal mixing, slowly invert tube 8-10 times immediately after blood collection before centrifugation.
3. Blood tubes are spun immediately, latest ten min after collection in a CellSep centrifuge for 15 min at 2,500g and at room temperature. During the centrifuge, a cold temperature (4°C) should be avoided to minimize platelet activation.
4. Grasp the BD P100 Tube with one hand, placing the thumb under the Hemogard™ Closure. For added stability, place arm on solid surface. With the other hand, twist the Hemogard™ Closure while simultaneously pushing up with the thumb of the other hand only until the tube stopper is loosened.
5. Move thumb away before lifting closure. Do not use thumb to push closure off the tube. Caution: Any tube has the potential to crack. If the tube contains blood, an exposure hazard exists. To help prevent injury during removal, it is important that the thumb used to push upward on the closure be removed from contact with the tube as soon as the Hemogard™ Closure is loosened.
6. Lift closure off the tube. In the unlikely event of the plastic shield separating from the rubber stopper, do not reassemble the closure. Carefully remove stopper from the tube.

7. 0.1–1.0 ml plasma is removed from P100 tube into an Eppendorf tube for FFE analysis. Excess plasma is aliquot into storage tubes with 1.0 ml plasma per tube and stored in a -80°C freezer until further usage.

3.2. FFE Separation of Human Plasma

A native separation buffer (glycerol based) is used for HSA depletion while a denaturing separation medium (urea/thiourea based media) can be used for denaturing plasma protein separation.

A preparative IEF method is used for both HSA depletion and plasma separation. FFE is conducted at 10°C using the following conditions: the experiments run in a horizontal position of the separation chamber using a 0.4 mm spacer. A flow rate of ~60 g/h (inlet I1-7) is used in combination with a voltage of 1250 V which results in a current of ~20 mA. Samples are perfused into the separation chamber using the cathodic inlet at ~5 g/h.

The residence time in the separation chamber is approximately 20 min. Fractions are collected in polypropylene microtiter plates (MTP) numbered from 1 (anode) through 96 (cathode).

3.2.1. Native HSA Depletion

1. 100 µl of human plasma is diluted with 900 µl FFE separation buffer and applied to the sample holder on the FFE system (see Notes 1–4).
2. Clean interior surface of chamber: move separation chamber to an upright position and open the pairs of chamber clamps simultaneously use the following sequence for cleaning: water–isopropanol–petrolether–isopropanol–water (see Note 5).
3. Apply the spacer without covering the inlets; apply the electrode membrane soaked in glycerol/isopropanol, mix 1:1 and the filter paper for IEF with 0.6 mm thickness, soaked in dist. water onto the membrane.
4. Close separation chamber: lift the plexiglas side of the chamber to meet the mirrored side, first fasten central and adjacent clamps loosely (release chamber clamps by two turns each). Then fasten central and adjacent clamps tight, close top and bottom clamps, place fractionation plate on top of fractionation housing, finally make sure that all inlets of media and counterflow are free and are not covered by the spacer.
5. Fill separation chamber with degassed dist. water: open (lift) valves at the top of the chamber to degas the chamber, fasten wedge clamps (I1-I7) of the media pump and place tubes in reservoir filled with H_2O dist, switch on medium 1 pump on control board. When the chamber is totally flushed and free of air add counter flow by closing wedge clamps on pump.
6. Set up the electrode buffer, stabilization buffers and counter flow buffer and separation buffers using following configuration: the inlet 1 is connected to anodic stabilization medium 1, inlets 2 and 3 are connected with anodic stabilization medium 2, inlet 4 gets connected with separation medium, 5 and 6

are connected to cathodic stabilization medium 2, inlet 7 is connected with cathodic stabilization medium 1 and the counterflow inlets are connected with counterflow medium.
7. Connect anode (+) and cathode (–) electrolytes, connect plastic cover and start electrode pump.
8. Adjust the media flow rate to 60 ml/h and start the media flow, wait for 35 min.
9. Adjust the Voltage to 1250 V set the current to 50 mA and turn on high voltage, wait 15 min until the pH gradient builds up.
10. Set the sample flow rate to 500 µl/h and start the sample flow (*see* Note 6).
11. After 35 min the sample can be collected with a MTP or deep well plate, therefore place the MTP in the fraction collection device and remove the MTP from the device.

3.2.2. Denaturing Plasma Protein Separation

1. 100 µl of human plasma is diluted with 900µL FFE separation buffer and applied to the sample holder on the FFE system (*see* Notes 1–4).
2. Clean interior surface of chamber: move separation chamber to an upright position and open the pairs of chamber clamps simultaneously use the following sequence for cleaning: water–isopropanol–petrolether–isopropanol–water (*see* Note 5).
3. Apply the spacer (for IEF choose the spacer with a thickness of 0.4 mm) without covering the inlets, apply the electrode membrane soaked in glycerol/isopropanol, mix 1:1 and the filter paper for IEF with 0.6 mm thickness, soaked in dist. water onto the membrane.
4. Close separation chamber: lift the plexiglas side of the chamber to meet the mirrored side, first fasten central and adjacent clamps loosely (release chamber clamps by two turns each). Then fasten central and adjacent clamps tight, close top and bottom clamps, place fractionation plate on top of fractionation housing, finally make sure that all inlets of media and counterflow are free and are not covered by the spacer.
5. Fill separation chamber with degassed dist. water: open valves at the top of the chamber to degas the chamber, fasten wedge clamps (I1-I7) of the media pump and place tubes in reservoir filled with H_2O dist, switch on medium 1 pump on control board. When the chamber is totally flushed and free of air add counter flow by closing wedge clamps on pump.
6. Set up the electrode buffer, stabilization buffers and counter flow buffer and separation buffers using following configuration: connect inlet 1 with anodic stabilization medium, connect inlet 2 with separation medium 1, inlets 3, 4, and 5 are connected to separation medium 2, inlet 6 is connected to separation medium 3, inlet 7 is connected to cathodic stabilization medium and counterflow inlets are connected with counterflow medium.
7. Connect anode (+) and cathode (–) electrolytes, connect plastic cover and start electrode pump.
8. Adjust the media flow rate to 60 ml/h and start the media flow, wait for 35 min.

9. Adjust the voltage to 700V set the current to 50 mA and turn on high voltage, wait 15 min until the pH gradient builds up.
10. Set the sample flow rate to 500 μL/h and start the sample flow (*see* Note 6).
11. After 35 min the sample can be collected with a MTP or deep well plate, therefore place the MTP in the fraction collection device and remove the MTP from the device.

3.3. Post-FFE Gel-Based Analysis

The pH-values of the individual microtiter plate fractions are measured. Subsequently, the protein fractions get analysed by SDS-PAGE run on an XCell SureLock™ Mini-Cell in combination with precast NuPAGE® 4–12% Bis-Tris gels. Silver staining of the proteins is carried out using the SilverQuest kit (Invitrogen).

1. Take 10 μl of the separated FFE fraction and pipet into a 500 μl sample vial.
2. Add 5 μl SDS running buffer to the sample vial and mix sample and buffer.
3. Pipet 10 μl of the protein sample on the gel.
4. Repeat these steps with all fractions you want to analyze (usually every second or third fraction).
5. Close the separation cell and adjust the power supply to 200 V and 100 mA.
6. Start the electrophoresis and continue for 35 min.
7. After the run is finished take the gel out of the chamber, open the cassette and put the gel in a glass tray and silver stain the gel.

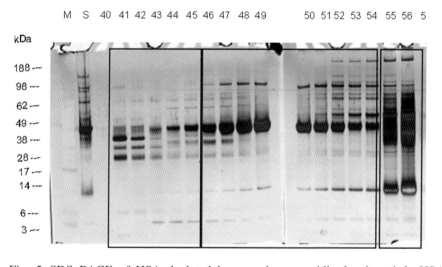

Fig. 5. SDS–PAGE of HSA depleted human plasma, acidic fractions left, HSA fractions middle and alkaline fractions right.

A typical example is shown in Fig. 5. After the free flow electrophoretic separation, the result is controlled by SDS-PAGE revealing the fractions used to be further analyzed (e.g., by 2-D electrophoresis, LC-MS/MS): the acidic, albumin or alkaline pools. Fig. 4 shows the comparison of 2D electropherograms of a crude plasma sample to the sum of the depleted and fractionated samples of an acidic pool which were subsequently analyzed. Laboratory results demonstrate an improvement in the resolution on 2D-PAGE by more than a factor of three, when compared to unfractionated samples.

3.4. Post-FFE LC-MS/MS analysis

After FFE separation of plasma under a denaturing condition with pH gradient from 3-10, samples are collected into 96 fractions. Each fraction is processed for washing and concentration, reduction, alkylation, in-solution trypsin digestion, and LC-MS/MS for protein identifications.

1. A collected fraction (300-500 µl) after FFE separation is loaded into a Microcone YM-3 (Millipore, USA), concentrated by centrifugation at 7,500g until it reached final volume of 30-50 µl.
2. The protein fraction is washed 3 times by adding 200 µl of washing solution 1 followed by spinning at 7,500g for 1 h or till 50-100 µl, and one time of 200 µl of washing solution 2 and spinning for 1 h. These washing steps can reduce the urea induced in FFE procedures and small molecules including peptides, which may interfere with the LC-MS/MS analysis.
3. The washed sample is collected by adding 100 µl of washing buffer 2, inverting Microcone to an Eppendorff tube, and spinning in a microcentrifuge for 10 sec.
4. The collected sample is reduced by adding 200 mM DTT until 10 mM of final concentration is reached, incubate for 1 h at room temperature (This step is skipped if the sample has been reduced before the FFE).
5. Add alkylating reagent (200 mM iodoacetamide) until 20 mM of final concentration, and allow the reaction at room temperate for 1 h. The alkylation reaction is stopped by addition of 20 mM DTT (final concentration). (This step is also skipped if the sample has been reduced and alkylated before the FFE).
6. Add trypsin solution till 1 ng/µl, and incubate for 16–18 h at 37°C.
7. The tryptic peptides are desalted by using C18 Spin Column, dried by using SpeedVac, and resuspended in 20 µl H_2O with 0.1% TFA, and 5% ACN.
8. The peptide samples are ready for LC-MS/MS analyses to identify proteins.

Usually, digested fractions analyzed by LC-MS/MS yielded hundreds of protein identifications per fraction. As one example, from fraction #57 (position at A8) after FFE separation under denaturing condition, total 455 peptides were detected in MS. These MS peaks combination with their MS/MS profiles were searched again NCBInr with Mascot (Matrix Science). Total 46 different proteins were identified (protein isoforms or precursors were

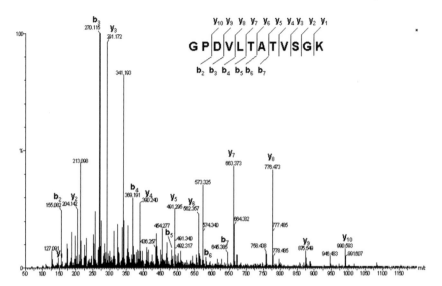

Fig. 6. The MS/MS profile of identified peptide with double charges (M/z: 572.806) in the digested fraction A8. The peptide peaks were mapped to sequence of GPDVL-TATVSGK with MOWSE score of 53. The matched b and y ions are labeled above their peaks.

accounted as a single protein), with at lease one peptide having a Mowse score higher than 50. **Fig. 6** shows, as an example, the MS/MS profile of a peptide. This peptide was mapped a sequence of G P D V L T A T V S G K, which was found in 6 proteins: (1) GP120 peptide, gi|1041907, an inter-alpha-trypsin inhibitor heavy chain-related protein found in plasma (8), (2) PK-120 precursor, gi|1402590, (3) Inter-alpha-trypsin inhibitor family heavy chain-trelated protein, gi|1483187, or gi|4096840, or inter-alpha-trypsin inhibitor heavy chain H4 precursor, gi|13432192 (9), (4) PRO1851, gi|7770149, (5) hypothetical protein, gi|51476525 (6) Inter-alpha (globulin) inhibitor H4 (plasma kallikrein-sensitive glycoprotein) variant, gi|62088356. These six proteins are related in their sequences. Our upcoming bioinformatics analyses might reveal how they are related in their functions.

4. Notes

1. Turbidity of protein samples is an indication of insufficient solubility and/or protein precipitation. Protein samples may not be turbid; otherwise the resolution of the electrophoretic separation will be poor, turbid protein samples have to be cleared by centrifugation before the FFE separation. Whereas organelle samples are turbid by nature and must not be cleared before the FFE separation!

2. The salt concentration of samples may not exceed 25 mM. If the conductivity of the sample because of the salt concentration is too high, it has to be desalted or diluted to reach <25mM of total salt concentration.
3. Generally, the sample should be as similar to the separation media as possible in behalf of the chemical and physical properties like density, conductivity and viscosity. Therefore we dilute or dissolve the sample at least 1+3 with the separation medium.
4. Additionally, tolerable additives for sample preparations can be: Urea up to 8 M; 0.1–1% of detergents like CHAPS, CHAPSO, Digitonin, Dodecyl-β-D-maltoside, Octyl-β-D-glucoside, Triton-X-114 (IEF); up to 50 mM DTT.
5. The system has to be cleaned thoroughly prior use, usually the separation chamber is wiped with water, isopropanol and petrolether and again with isopropanol and water to make sure any organic and anorganic contamination is cleaned.
5. Protein sample is usually applied on the cathodic inlet, only few samples work better when applied at the middle or anodic inlet, this can be checked for each sample. Plasma on the depletion gradient for example is applied in the middle. Typical sample application is between 200 µl and 750 µl/h.

References

1. Weber G. and Bocek P. (1998) Recent developments in preparative free flow isoelectric focusing. *Electrophoresis*. 19, 1649–53.
2. Zischka H., Weber G., Weber P.J., Posch A., Braun R.J., Buhringer D., Schneider U., Nissum M., Meitinger T., Ueffing M. and Eckerskorn C. (2003) Improved proteome analysis of *Saccharomyces cerevisiae* mitochondria by free-flow electrophoresis. *Proteomics*. 3(6), 906–16.
3. Hoffmann P., Ji H., Moritz R.L., Connolly L.M., Frecklington D.F., Layton M.J., Eddes J.S., Simpson R.J. (2001) Continuous free-flow electrophoresis separation of cytosolic proteins from the human colon carcinoma cell line LIM 1215: a non two-dimensional gel electrophoresis-based proteome analysis strategy. *Proteomics*. 1, 807–18.
4. Omenn, G.S., (2004) The Human Proteome Organization Plasma Proteome Project pilot phase: reference specimens, technology platform comparisons, and standardized data submissions and analyses. *Proteomics*. 4, 1235–1240.
5. Anderson, N. L. and Anderson, N.G. (2002) The human plasma proteome: history, character, and diagnostic prospects. *Mol. Cell Proteomics*. 1, 845–866.
6. Omen, G.S., States, D. J., Adamski, M., Blackwell, T.W., Hannash, S. M., et al., (2005) Overview of the HUPO Plasma Proteome Project: results from the pilot phase with 35 collaborating laboratories and multiple analytical groups, generating a core dataset of 3020 proteins and a publicly-available database. *Proteomics*. 5, 3226–3245.
7. Rai, A. J., Gelfand C.A., Haywood, B. C., Warunek, D., Yi, J., Chan, D. W., et al. (2005) HUPO Plasma Proteome Project specimen collection and handling: towards

the standardization of parameters for plasma proteome samples. *Proteomics.* 5. 3262–3277.
8. Choi-Muira, et al., (1995) Purification and characterization of a novel glycoprotein which has significant homology to heavy chains of inter-alpha-trypsin inhibitor family from human plasma. *J. Biochem.* 117 (2), 400–407.
9. Saguchi, M., Tobe, T., Hashimoto, K., Sano, Y., and Tomita, M. (1995) Isolation and characterization of the human inter-alpha-trypsin inhibitor family heavy chain-related protein (IHRP) gene (ITIHL1). *J. Biochem.* 117 (1), 14–18.

24

Protein Fractionation by Preparative Electrophoresis

Michael Fountoulakis and Ploumisti Dimitraki

Summary

Preparative electrophoresis is a protein fractionation approach useful for the enrichment of low-abundance gene products. Preparative electrophoresis is usually performed in the PrepCell apparatus. Proteins are separated according to their size in a cylindrical gel in the presence of an ionic detergent. The method is particularly efficient for the enrichment of low-molecular-mass gene products. Preparative electrophoresis can be followed by proteomic analysis, and the proteins eluted from the preparative gel can be separated by two-dimensional gel electrophoresis and identified by mass spectrometry.

Key Words: Low-abundance proteins; low-molecular-mass proteins; preparative electrophoresis; protein enrichment; proteomics.

1. Introduction

Low-abundance gene products are thought to be involved in important biological processes. Therefore, the identification of low-abundance proteins is essential for a thorough proteomic analysis and represents a scientific challenge. Proteomics is a powerful technology which studies the identity and expression levels of proteins. Changes of protein expression levels as well as of their modifications that occur in various disorders may be informative of the pathogenesis of these disorders and could result in the identification of potential drug targets and disease markers. Detection of low-abundance proteins is achieved by enrichment from crude extracts by biochemical protein-enriching approaches like selective fractionation, chromatography, or electrophoretic procedures *(1–3)*. The original protein mixture is separated into less complex fractions each containing a lower number of total proteins in comparison with the starting material. Thus, fractionation increases the likelihood of detecting low-abundance proteins.

Electrophoretic techniques represent an important approach for protein enrichment (4,5). There are two major types of preparative electrophoresis: (i) those based on the isoelectrofocusing (IEF) principle where the proteins are separated according to their charge and (ii) those based on the gel-electrophoresis principle where separation depends on the protein size. The IEF techniques include gel-free, classical methodologies, like Rotofor (6), free flow electrophoresis (7) and more sophisticated approaches, like multicompartment electrolyzers with isoelectric membranes (8) and isoelectric beads (9,10).

In the gel-based, preparative electrophoresis approach, the proteins are usually separated in cylindrical gels in the presence of ionic detergents. Preparative electrophoresis is useful for the detection of low-abundance proteins. We have previously applied this approach for the fractionation of rat and mouse brain proteins to develop the methodology for the detection of low-abundance gene products possibly involved in neurological diseases (3,11) and of liver proteins which are potential targets of toxic agents (12). Fig. 1 shows the pathway followed for the proteomic analysis of mouse brain proteins fractionated by preparative electrophoresis. An example of enrichment of plasma proteins is shown in Fig. 2. In addition, preparative electrophoresis has the advantage that it allows the enrichment of low-molecular-mass proteins which can not be enriched by other approaches. Thus, almost half of the proteins identified in the fractions collected from the preparative electrophoresis analysis of mouse brain have masses lower than 30 kDa. The proteins with

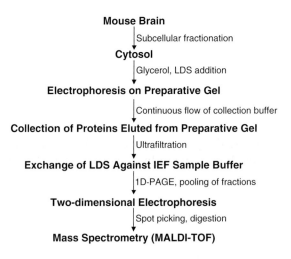

Fig. 1. Pathway of proteomic analysis of mouse brain proteins fractionated by preparative electrophoresis.

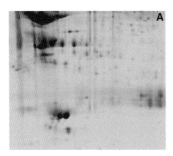 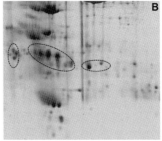

Fig. 2. Two-dimensional gel analysis of plasma proteins, partially depleted of albumin, without fractionation (**A**) and of fractions eluted from the preparative electrophoresis gel (**B**). Corresponding regions of the 2-D gels are shown. The presence of additional spots after preparative electrophoresis representing enriched proteins can be seen (circled in **B**). Certain spots in (**B**) appear to be weaker in comparison with the corresponding spots in the starting material (**A**). This happens because the largest amount of these proteins has been eluted in other fractions eluted from the preparative gel which are not shown here. The gels were stained with Coomassie blue.

molecular masses lower than 20 kDa represented 15% of total proteins, whereas in brain proteomes not fractionated by preparative electrophoresis such gene products represent about 8% of total proteins *(11)*. Preparative electrophoresis is often the method of choice to separate protein isoforms to homogeneity for crystallization purposes *(13)*.

We perform preparative gel electrophoresis with the PrepCell apparatus (Bio-Rad), following the instructions of the supplier and as described *(3,11,13)*. The principle of the method is that proteins are fractionated by continuous elution electrophoresis in the presence of an ionic detergent in a cylindrical gel. During electrophoresis, the proteins are separated into ring-shaped bands along the gel, the individual bands (rings) migrate off the bottom of the gel in a collection chamber and a dialysis membrane underneath the elution frit traps proteins within the chamber. The eluted proteins are then drawn with the elution buffer and are collected in a fraction collector. To assure that molecules migrate in compact parallel bands, the gel temperature is kept constant by cooling and circulating the running buffer and/or by performing the electrophoresis at 4°C.

2. Materials
2.1. Protein Sample Buffer
1. The protein sample to be fractionated (about 10–50 mg) should be prepared in a low-salt buffer, for example 50 mM Tris-HCl, pH 6.8. Glycerol and lithium

dodecyl sulfate (LDS) are added before electrophoresis to a final concentration of 25% and 1%, respectively (*see* Note 1).

2.2. Preparative Electrophoresis Apparatus

1. Preparative electrophoresis is performed in the PrepCell apparatus, Model 491 (Bio-Rad).
2. The PrepCell apparatus needs to be equipped with two peristaltic pumps with a liquid supply capacity of 100 ml/min and 100 ml/h, respectively, and a fraction collector (*see* Note 2).

2.3. Preparative Polyacrylamide Gel

2.3.1. Separation Gel

1. The volume of the acrylamide solution necessary for the preparation of the separation gel depends on the gel dimensions. For a 6-cm high gel, 100 ml of acrylamide solution should be prepared. The concentrations of the ingredients are shown in Table 1.
2. To prepare 100 ml of 11% acrylamide solution, mix 36.7 ml of stock acrylamide/PDAsolution (acrylamide/piperazine-di-acrylamide (PDA) solution 37.5:1, w/v) and 25 ml of 1.5 M Tris-HCl, pH 8.8, and bring to 100 ml with Milli-Q water. Degas by filtering through a 0.22 µm filtration device, add 250 µl of 10% APS and 25 µl of TEMED, mix and quickly proceed with gel pouring.

2.3.2. Stacking Gel:

1. Usually a 2.5 cm high stacking gel (4%) is prepared for which 50 ml of acrylamide solution are required. The final concentrations of the ingredients are shown in Table 2.
2. To prepare 50 ml of 4% acrylamide solution for the stacking gel, mix 6.65 ml of stock acrylamide/PDA solution (37.5:1, w/v) and 12.5 ml of 0.5 M Tris-HCl, pH 6.8, and bring to 50 ml with Milli-Q water. Degas by filtering through a 0.22 µm filtration device, add 250 µl of 10% APS (w/v) and 50 µl of TEMED, mix and quickly proceed with gel pouring.

Table 1
Reagents for preparation of the separation gel

Acrylamide/PDA solution (37.5:1, w/v)	depends on the gel strength
1.5 M Tris-HCl, pH 8.8	375 mM
TEMED (N,N,N',N'-tetramethylethylenediamine)	0.65 ml/L
10% Ammoniumpersulfate (APS, w/v)	3.5 ml/L
Water (Milli-Q)	to fill-up to the required volume

Table 2
Reagents for preparation of the stacking gel

Acrylamide/PDA solution (37.5:1, w/v)	4%
0.5 M Tris-HCl, pH 6.8	125 mM
TEMED (N,N,N',N'-tetramethylethylenediamine)	1.0 ml/L
10% Ammoniumpersulfate (APS, w/v)	5.0 ml/L
Water (Milli-Q)	to fill-up to the required volume

2.4. Running and Elution Buffer

1. LDS-gel running buffer A: 25 mM Tris-base, 198 mM glycine, 0.1% LDS (w/v). To prepare 3 L of running buffer, dissolve 9.09 g of Tris-base, 44.59 g of glycine and 3 g of lithium dodecyl sulfate (LDS) in 2 L of water and bring the final volume to 3 L with Milli-Q water.
2. Elution buffer B: 25 mM Tris, 198 mM glycine, 0.1% CHAPS (w/v). To prepare 600 ml of elution buffer, dissolve 1.82 g of Tris-base, 8.92 g of glycine and 600 mg of CHAPS in 500 ml Milli-Q of water and bring to 600 ml with Milli-Q water.

2.5. Water-Saturated Isobutanol

Prepare 10 ml of water-saturated isobutanol by mixing 5 ml of isobutanol with 5 ml of Milli-Q water and shake well for 5 s. Leave solution over night and use the upper phase for gel casting.

2.6. Analysis of the Fractions from the Preparative Gel by SDS-PAGE

1. Standard SDS-electrophoresis equipment is required.
2. SDS-gels of large format or minigels can be purchased or prepared following the instructions for the preparation of the preparative gels (*see* Sections 2.3.1 and 2.3.2). For the preparation of the analytical separation and stacking gels use the reagents listed in Tables 1 and 2, respectively.
3. SDS-PAGE running buffer: 25 mM Tris-base, 198 mM glycine, 0.1% SDS. To prepare 3 L of running buffer, dissolve 9.09 g of Tris-base, 44.59 g of glycine and 3 g of SDS in 2 L of Milli-Q water and bring the final volume to 3 L with Milli-Q water.

3. Methods

3.1. Preparation of Protein Sample

The protein sample to be fractionated by preparative electrophoresis is prepared according to the protocols, experience and the interests of the individual laboratory. Sample preparation eventually requires subcellular fractionation (i.e. cytosol, mitochondria, membrane preparation) and protein concentration by ultrafiltration or trichloroacetic acid (TCA) precipitation before the preparative electrophoresis *(14)*. The protein amount (usually 10–50 mg) to be applied on the preparative gel should be in a volume of about 4–5 ml in 50 mM Tris-HCl, pH 6.8, 25% glycerol, 1% LDS (*see* Notes 3–5).

3.2. Fractionation by Preparative Electrophoresis

We perform preparative electrophoresis using the PrepCell Model 491 (Bio-Rad) (*see* Note 6). The proteins are electrophoresed in a cylindrical gel which consists of a lower separation gel and an upper stacking gel. The separation gel is about 6 cm long and has an acrylamide concentration of 11%. The stacking gel is about 2.5 cm long and has an acrylamide concentration of 4% (*see* Note 7). Here we describe a general method for performing preparative electrophoresis under denaturing conditions (*see* Notes 8 and 9).

3.2.1. Preparation of the Preparative Gel

1. Use the device parts to form the gel tube assembly which is required to pour the preparative gel following the instructions of the apparatus supplier. The system consists of an outer cylinder, the cooling core (which functions as the inner cylinder) and a gel casting stand. Place carefully the gel tube assembly on the casting stand and check for liquid leakage. The gel is poured in the space between the two cylinders.
2. Connect the tubing from one outlet of the cooling coil to the peristaltic pump supplying running buffer and start with circulating water or buffer of a temperature of about 18°C through the cooling core (*see* Note 10).
3. Prepare 100 ml of acrylamide solution for the separation gel (*see* Section 2.3.1 and Table 1). Transfer the solution to the space between the two cylinders with a pipet and pour carefully to avoid formation of bubbles.
4. Pour enough acrylamide solution to form a gel of height about 6.5 cm (*see* Note 11).
5. Overlay the gel with 3 ml of water-saturated isobutanol. After 2 h, carefully remove the isobutanol, wash the gel surface twice with 5 ml of Milli-Q water and overlay the gel with 5 ml of Milli-Q water.
6. Transfer the system to the cold room (4°C) and allow the gel to polymerize overnight.
7. Next morning, transfer the gel to room temperature and remove the water from the gel surface.
8. Prepare 50 ml of acrylamide solution for stacking gel (*see* Section 2.3.2 and Table 2).
9. Pour the acrylamide solution on top of the separation gel and overlay with 3 ml of water- saturated isobutanol.
10. Allow polymerizing for 2 h at room temperature, remove the isobutanol and wash twice with 5 ml of Milli-Q water.

3.2.2. Protein Fractionation

1. Prepare the protein solution, about 4 ml, in 50 mM of Tris-HCl, pH 6.8, 25% glycerol, 1% LDS, traces of bromophenol blue. Add 200 µl of β-mercaptoethanol and allow staying at room temperature for 10 min (*see* Note 3).
2. Assemble the apparatus following the instructions of the supplier. Disconnect the gel tube assembly with the polymerized gel from the casting stand, assemble the elution chamber and place the gel tube assembly on the elution chamber base (*see* Note 12).
3. Fill the lower tank of the apparatus with 2.5 L of running buffer A and place the gel tube assembly with connected elution chamber into the lower tank. The level of the running buffer should be high enough to cover the gel. Avoid the formation of air bubbles under the base of the gel.
4. Pour running buffer A on top of the gel and fill the cylinder.
5. Place the upper tank of the apparatus and fill the inner part of it with running buffer A (500 ml).
6. Fill the outer space of the upper tank with 600 ml of elution buffer B and connect the elution buffer tank with the base of the gel (elution chamber) using the tubing provided with the apparatus.
7. Flush air bubbles from the elution buffer supply system (*see* Note 13).
8. Transfer the apparatus to the cold room (4°C) (*see* Note 14).
9. Connect the tanks of the running buffer A to each other and to the circulating, peristaltic pump.
10. Apply the protein sample with a syringe on top of the stacking gel.
11. Place the lid of the apparatus and start electrophoresis at 250 V.
12. Turn on the running buffer A pump and adjust the flow rate to about 100 ml/min (*see* Note 2).
13. When the bromophenol blue front reaches the bottom of the (separation) gel (after about 4–5 h, depending on the length of the gel), connect the tubing from the cooling core (center of the gel tube assembly) to the pump supplying the elution buffer B and to a fraction collector and adjust the flow rate to 30 ml/h (*see* Notes 15 and 16).
14. Collect 5-ml fractions and continue with the collection of about 80 fractions (*see* Note 17).

3.3. Analysis of the Fractions from the Preparative Gel by SDS-PAGE

The fractions collected from the preparative electrophoresis gel are analyzed by one-dimensional SDS-PAGE. In the consecutive fractions, proteins with gradually increasing molecular masses are contained. If the goal of the experiment is the detection of low-molecular-mass proteins present in the starting protein sample, Tricine gels instead of Laemli gels *(15)* are used for SDS-PAGE analysis (*see* Note 18).

1. Analyze each second fraction from the preparative gel by 12% SDS-PAGE using 25 μl of protein solution. Stain the gels with Coomassie blue.
2. On the basis of the SDS-PAGE, select the fractions to be used for further experiments or further analysis.
3. Determine the protein concentration in the pools using the Coomassie blue method *(16)* (*see* Note 19). See also other chapters in this book.

3.4. Proteomic Analysis

The fractions collected from the preparative gel can be further analyzed by two-dimensional gels *(17)* and the proteins can be identified by mass spectrometry, for example by the peptide mass fingerprint approach using matrix-assisted laser desorption ionization time-of-flight mass spectrometry (MALDI-TOF-MS) *(18)*. The proteomic analysis (protein separation by 2-D gels and protein identification by MALDI-TOF-MS *(19,20)*) is optional and it is not described here. Here we describe the preparation of the samples which are recovered from the preparative electrophoresis for the 2-D gel electrophoresis analysis.

Because the samples are diluted with elution buffer and the proteins are in an ionic detergent (LDS), which is incompatible with isoelectric focusing (the first step of the 2-D electrophoresis), exchange of the buffer with simultaneous protein concentration are necessary before the proteomic analysis. Furthermore, for the proteomic analysis, the fractions from the preparative electrophoresis should be combined to form a lower number of pools for the convenient analysis by 2-D gels. The fractions are pooled on the basis of the SDS-PAGE profile to usually form ten to twelve pools. The pools are further analyzed by 2-D gel electrophoresis on broad range 3-10 immobilized pH gradient (IPG) strips. In the 2-D gels, carrying consecutive pools, the spots are forming zones with gradually increasing molecular masses (Fig. 3).

1. Prepare 10-12 pools with the fractions collected from the preparative electrophoresis following the SDS-PAGE profile (*see* Note 20).
2. Concentrate the content of each of the pools to about 200 μl by ultrafiltration at 2000*g*.
3. Dilute the concentrate(s), which contain ionic detergent, 10-fold with isoelectric focusing sample buffer (20 mM Tris-HCl, pH 7.5, 8 M urea, 4% CHAPS) and concentrate to about 200 μl by ultrafiltration at 2000*g*.
4. Repeat the dilution-concentration process for a second time. The concentrate can be used for 2-D gel electrophoresis (add IPG buffer, which is not included in the sample buffer, to 1% final concentration). Apply the protein sample at the acidic and basic ends of rehydrated IPG strips and proceed with the further proteomic analysis *(20)*.

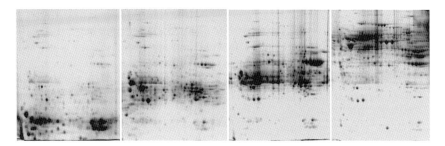

Fig. 3. Two-dimensional gel analysis of consecutive pools of rat brain cytosolic proteins eluted from the PrepCell. The gels were stained with colloidal Coomassie blue. The increasing average molecular mass of the proteins in the consecutive fractions can be seen.

4. Notes

1. Reagents for the preparation of polyacrylamide gels should be kept at 4°C. Acrylamide/PDA solutions can be purchased ready-made. Acrylamide is a potent neurotoxin and purchase of the reagent in solution should be preferred to avoid exposure to dust. Wear gloves and protective clothes whenever working with acrylamide and also during the whole procedure of the preparative electrophoresis.
2. One pump is used for the circulation of the running buffer at about 100 ml/min and the other for the constant supply of the elution buffer preferentially at 30 ml/h.
3. The protein sample applied on the preparative gel should have a protein concentration of about 10 mg/ml so that the final volume not to be larger than 5 ml if a total protein amount of 50 mg will be applied. If a larger total protein amount will be applied, the protein concentration can be brought up to about 20 mg/ml. Allow sufficient time for the sample preparation, in particular if prefractionation is required. The protein sample should be ready when polymerization of the stacking gel has finished. Start with preparative electrophoresis early enough to have sufficient time to control the elution of the bromophenol blue front and to connect the elution buffer tubing to the fraction collector (approximately 5 h from the start of electrophoresis).
4. Protein determination is usually performed by the Coomassie blue method *(16)*.
5. The protein sample contains 25% glycerol for a quick and efficient settlement on top of the gel. The sample also contains 1% LDS for the efficient disruption of protein complexes. LDS is used instead of SDS because the former can be easier removed from the proteins in comparison with SDS and it does not interfere with 2-D gel electrophoresis, whereas SDS can not be completely removed and often produces horizontal streaking in the gels *(21)*.
6. The PrepCell apparatus is assembled following the instructions of the supplier. Read carefully the compatibility of the instrument and the safety guidelines in

the manual. The technically most critical step of the whole procedure is the assembly of the device parts to pour the gel and the assembly of the elution chamber. It is recommended to assemble the apparatus, prepare a gel and run an experiment with a less precious protein sample for the first time.

7. The dimensions of the preparative gel (6-cm-long separation and 2.5-cm-long stacking gel) have been selected so that an efficient separation of proteins with average molecular mass values can be achieved and simultaneously the gel volume is not very large. On a gel of such dimensions, about 50 mg of total protein can be applied. The same gel dimensions can be used for a protein amount up to about 100 mg. Larger gels may be result in a more efficient separation of the low-molecular mass proteins but the protein losses will be larger. Carrying out experiments to find out the optimal acrylamide concentration for the preparative gel is laborious and time consuming. We recommend to start with the conditions described which can be applied for the separation of the majority of proteins and only if the goal of the particular experiment is different then to try to optimize the conditions.

8. Two gel tube assemblies are provided with the PrepCell instrument. We usually use the lager tube of 37 mm inner diameter (ID). The use of the smaller tube of 28 mm ID is similar to the described here and it is recommended for small protein amounts (about 10 mg or less).

9. In the analytical 2-D electrophoresis process the proteins are separated in the second dimension according to their masses; however, the number of detected proteins is relatively low because of the small volume (usually about 200 μl, containing up to 1 mg of protein) which is applied in an analytical process. On the other hand, in the preparative electrophoresis, one can start with large volumes and a large protein amount (about 50–100 mg or more). The proteins of the fractions collected from the preparative electrophoresis gel, which contain a large quantity of proteins with similar masses, can be subsequently separated by 2-D electrophoresis and the separation will result in the detection of a larger number of gene products which could not have been detected before.

10. The gel tube assembly should have been cooled to 4°C before pouring the gel. To avoid overheating and to obtain a homogeneous gel, during polymerization, water or buffer of a temperature of 18°C should be circulated through the cooling core.

11. Pour more acrylamide solution in the cylinder than necessary to form a 6-cm gel (the gel should originally have a height of about 6.5 cm). The gel shrinks during polymerization.

12. When not in use, the ultrafiltration membrane necessary for the formation of the elution chamber should be kept at 4°C between the two sponges (which are used to assemble the elution chamber) in elution buffer containing 0.02% sodium azide.

13. The elution chamber and the tubing must be free of air bubbles before start the elution. Connect a 50 ml syringe to the elution buffer outlet tubing at the top of the cooling core. Gently suck elution buffer until no more air bubbles are coming into the syringe. Then, fill the syringe with elution buffer, connect to

the full with elution buffer outlet on top of the cooling core and gently push buffer into the elution chamber. Air trapped in the system is released into the elution buffer tank. Degas elution buffer before use.

14. The electrophoresis is preferentially performed at 4°C. The gel should be additionally cooled by circulating running buffer in the cooling core. The running buffer can be cooled by passing it through a metallic coil immersed in an ice-water bath.

15. During electrophoresis, the bromophenol blue front forms a very thin, horizontal disk which migrates slowly to the lower end of the gel. Shapes different from the thin, horizontal disk are signs that the gel has not been prepared properly. Collect the bromophenol blue front as well because it can contain small size proteins.

16. The proteins as they are eluted from the preparative gel are drawn with the elution buffer which contains 0.1% CHAPS to reduce precipitation. Elution buffers of different composition can be used as well. The particular buffer described here should be used if a proteomic analysis will follow the preparative electrophoresis.

17. The volume collected in each fraction depends on the goal of the experiment. Collection of 5 ml fractions is convenient for SDS-PAGE analysis and protein content determination and the information is usually not affected. Protein elution can be followed with the use of an UV recorder. However, the elution buffer absorbs strongly and the recording of the protein absorbance is not satisfactory. Alternatively, protein elution can be followed by determination of the protein content in each second fraction using the Coomassie blue method *(16)*. The protein concentration can be plotted versus the fraction number. Fig. 4 shows the elution profile of mouse brain proteins from the preparative gel.

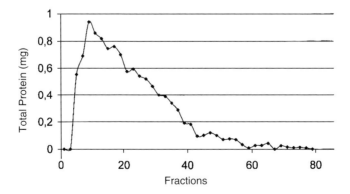

Fig. 4. Elution profile of mouse brain proteins from the preparative gel. The protein concentration was determined with the Coomassie blue method *(16)* and the values were plotted against the fraction numbers.

18. Preparative electrophoresis can be accordingly modified so that a selective enrichment of small size proteins (less than 15 kDa) is achieved (see also Note 7).
19. Approximately 13 mg of total protein are recovered in all fractions from the 50 mg applied on the preparative gel.
20. Fractions can be pooled on the basis of the protein distribution revealed by SDS-PAGE and the protein content of the fractions. The volume of the protein solution in the various pools can be different. For the pool formation, important are the protein distribution profile in the fractions and the protein amount. Fractions containing proteins with similar masses are usually pooled together, so that consecutive pools include proteins with higher average masses (see rat brain protein distribution in Fig. 3).

References

1. Fountoulakis, M. and Takács, B. (2002) Enrichment and proteomic analysis of low-abundance bacterial proteins. *Methods Enzymol.* **358**, 288–306.
2. Lescuyer, P., Hochstrasser, D. F., and Sanchez, J.-C. (2004) Comprehensive proteome analysis by chromatographic protein prefractionation. *Electrophoresis* **25**, 1125–1135.
3. Fountoulakis, M. and Juranville, J.-F. (2003) Enrichment of low-abundance brain proteins by preparative electrophoresis. *Anal. Biochem.* **313**, 267–282.
4. Righetti, P. G., Castagna, A., Antonioli, P., and Boschetti, E. (2005) Prefractionation techniques in proteome analysis: The mining tools of the third millennium. *Electrophoresis* 2005, **26**, 297–319.
5. Garbis, S., Lubec, G., and Fountoulakis M. (2005) Limitations of current proteomics technologies. *J. Chromatogr. A* **1077**, 1–18.
6. Mesa, C., Dembic, Z. Garotta, G., and Fountoulakis, M. (1995). Interferon γ receptor extracellular domain-IgG fusion protein was produced in Chinese hamster ovary cells as a mixture of glycoforms. *J. Interferon Cytokine Res.* **15**, 309–315.
7. Zischka, H., Weber, G., Weber, P. J., Posch, A., Braun, R. J., Buhringer, D., Schneider, U., Nissum, M., Meitinger, T., Ueffing, M., and Eckerskorn, C. (2003) Improved proteome analysis of Saccharomyces cerevisiae mitochondria by free-flow electrophoresis. *Proteomics*, **3**, 906–916.
8. Pedersen, S. K., Harry, J. L., Sebastian, L., Baker, J., Traini, M. D., McCarthy, J. T., Manoharan, A., Wilkins, M. R., Gooley, A. A., Righetti, P. G., Packer, N. H., Williams, K. L., and Herbert, B. R., Unseen proteome: mining below the tip of the iceberg to find low abundance and membrane proteins. (2003) *J. Proteome Res.*, **2**, 303–311.
9. Fortis, F., Girot, P., Brieau, O., Boschetti, E., Castagna, A., and Righetti, P. G. (2005) Amphoteric, buffering chromatographic beads for proteome prefractionation. I: theoretical model. *Proteomics* **5**, 620–628.
10. Fortis, F., Girot, P., Brieau, O., Castagna, A., Righetti, P. G., and Boschetti, E. (2005) Isoelectric beads for proteome pre-fractionation. II: experimental evaluation in a multicompartment electrolyzer. *Proteomics* **5**, 629–638.

11. Xixi, E., Dimitraki, P., Vougas, K., Kossida, S., Lubec, G., and Fountoulakis, M. (2006) Proteomic analysis of the mouse brain following protein enrichment by preparative electrophoresis. *Electrophoresis* **27**, 1424–1431.
12. Fountoulakis, M., Juranville, J.-F., Tsangaris, G., and Suter, L. (2004) Fractionation of liver proteins by preparative electrophoresis. *Amino Acids* **26**, 27–36.
13. Fountoulakis, M., Takács-di Lorenzo, E., Juranville, J,-F., and Manneberg, M. (1993). Purification of interferon γ-interferon γ receptor complexes by preparative electrophoresis on native gels. *Anal. Biochem.* **208**, 270–276.
14. Jiang, L., He, L., and Fountoulakis, M. (2004) Comparison of protein precipitation methods for sample preparation prior to proteomic analysis. *J. Chromatogr. A* **1023**, 317–320.
15. Schaegger, H. and von Jagow, G. (1987) Tricine-sodium dodecyl sulfate-polyacrylamide gel electrophoresis for the separation of proteins in the range from 1 to 100 kDa. *Anal. Biochem.* **166**, 368–379.
16. Bradford, M. (1976) A rapid and sensitive method for the quantitation of microgram quantities of protein utilizing the principle of protein-dye binding. *Anal. Biochem.* **72**, 248–254.
17. Langen, H., Roeder, D., Juranville, J.-F., and Fountoulakis, M. (1997) Effect of the protein application mode and the acrylamide concentration on the resolution of protein spots separated by two-dimensional gel electrophoresis. *Electrophoresis* **18**, 2085–2090.
18. Lahm, H. W. and Langen, H. (2000) Mass spectrometry: a tool for the identification of proteins separated by gels. *Electrophoresis* **21**, 2105–2114.
19. Fountoulakis M. (2005) Analysis of membrane proteins by two-dimensional gels. In The Proteomics Protocols Handbook (Walker, J. M., Ed.), Humana Press, Totowa, New Jersey, p. 133–144.
20. Fountoulakis M. (2004) Application of proteomics technologies in the investigation of the brain. *Mass Spectrom. Rev.* **23**, 231–258.
21. Fountoulakis, M. and Takács, B. (2001) Effect of strong detergents and chaotropes on the protein detection in two-dimensional gels. *Electrophoresis* **22**, 1593–1602.

V

Enrichment Strategies for Organelles, Multiprotein Complexes, and Specific Protein Classes

25

Isolation of Endocitic Organelles by Density Gradient Centrifugation

Mariana Eça Guimarães de Araújo, Lukas Alfons Huber, and Taras Stasyk

Summary

Advanced prefractionation strategies, in combination with highly sensitive and accurate mass spectrometers provide powerful means to detect and analyze low abundant proteins on the subcellular and organelle-specific level. Among enrichment techniques, subcellular fractionation has become the most commonly used. Its application gives access to less complex subproteomes and organelle constituents, facilitating downstream analysis. Furthermore, subcellular fractionation allows the identification of proteins that shuttle between different subcellular compartments in a stimulus dependent manner. As a paradigm of subcellular organelle isolation, we describe here endosomal purification protocols, based on differential centrifugation in continuous and discontinuous sucrose gradients. Described methods can be easily modified to isolate other organelles and are compatible with subsequent organelle- and functional organelle proteome analyses by, e.g., two-dimensional gel electrophoresis.

Key Words: Endosomes; gradient centrifugation; homogenization; organelles; proteomics; subcellular fractionation.

1. Introduction

Recent progress in proteomics technology has enabled the comprehensive profiling of enriched organelle fractions, resulting in the identification of hundreds of proteins *(1,2)*. Subcellular fractionation, the first step among enrichment techniques for proteomic studies, is a flexible approach that greatly reduces sample complexity. Such complexity reduction increases the number of less abundant proteins that can be subsequently analyzed *(3,4)*. Furthermore, subcellular fractionation can be efficiently combined with two-dimensional electrophoresis (2-DE) followed by mass spectrometry as well as with gel-independent techniques.

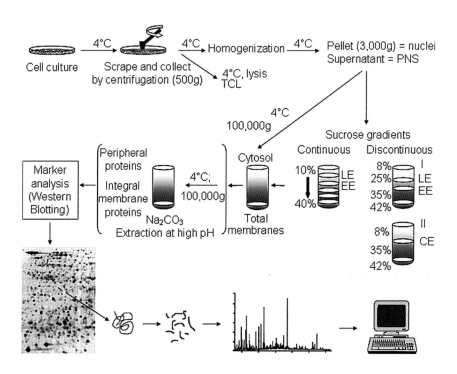

Fig. 1. Schematic outline of subcellular fractionation by density gradient centrifugation and subsequent organelle proteome analysis. Cultured cells are scraped and sedimented by centrifugation. This pellet, when lysed with appropriate buffer at 4°C, gives rise to a fraction designated as total cell lysate (TCL). Alternatively, pelleted cells can be homogenized and centrifuged to separate nuclei from post nuclear supernatant (PNS). The PNS contains the cell organelles in suspension. These, can be purified by continuous or discontinuous gradient centrifugation. Individual organelles (e.g., late [LE], early [EE] or crude [CE] endosomes) are then retrieved from these gradients at different positions/interfaces. Cytosol is readily obtained from the PNS by ultracentrifugation. The total membranes sedimented during cytosol preparation, or in fact any of the organelle fractions obtained by gradient centrifugation, can be subjected to further downstream treatments like sodium carbonate extraction. In this specific case, peripheral membrane proteins will be separated from integral/transmembrane proteins. Upon assessing the quality of the fractionation, samples of interest can be analyzed by 2-DE and mass spectrometry methods (see also Fig. 3).

Subcellular fractionation is of special importance for the analysis of intracellular organelles and multi-protein complexes. The method consists of two major steps *(5)*: (i) homogenization and (ii) fractionation of the homogenate to separate the different organelles (*see* Fig. 1). Cells are collected by centrifugation and homogenized. After homogenization, an additional centrifugation step separates the post-nuclear supernatant (PNS) from unbroken cells, cellular debris and nuclei. The latter can be purified from the pellet for further studies by adapting the protocol described in *(6)*. The PNS contains the cell organelles in suspension, which can be subsequently separated by gradient ultracentrifugation. The position of membrane particles in density gradients is determined mainly by the ratio of their lipid to protein content, e.g., mitochondrial and endoplasmic reticulum membranes have a high density because of high protein content, whereas endosomes have lipid-rich membranes and therefore low density.

Among different techniques for organelle isolation, centrifugation is the most effective one. Discontinuous gradients as well as step gradients can be applied for the separation of organelles and membranes (*see* Fig. 1). For better resolution, equilibrium separation in a continuous gradient is the method of choice. After centrifugation to equilibrium, membranes distribute throughout the entire gradient according to their specific densities. A drawback of continuous gradients is the low enrichment of organelles, resulting in rather diluted fractions.

In this chapter we describe state of the art endocitic organelle isolation protocols by gradient centrifugation. The methods described can be integrated in more refined prefractionation approaches (e.g., used in combination with chromatography separation), and can be easily modified to isolate other subcellular organelles. As an example, if purifying mitochondria, an initial adjustment of the sucrose concentrations on a continuous gradient to 25–40% would provide a better separation between mitochondria and other heavy membrane fractions like the endoplasmic reticulum. In brief, the chapter has been written in such a way which provides the reader with robust basic building blocks from which to start and design their personal strategy.

2. Materials
2.1. Standard Homogenization Protocol

1. Homogenization buffer (HB): 250 mM sucrose, 3 mM imidazole, pH 7.4, ddH$_2$O. Add a cocktail of protease inhibitors (10 μg/ml aprotinin, 1 μg/ml pepstatin, 10 μg/ml leupeptin, and 1 mM Pefabloc) and phosphatase inhibitors (1 mM Na$_3$OV$_4$, 5 mM Na$_4$P$_2$O$_7$, 50 mM NaF, 10 mM β-glycerophosphate) immediately before starting the experiment.

2. Homogenization buffer plus (HB$^+$): HB containing 1 mM EDTA, 0.03 mM cycloheximide.
3. Phosphate buffered saline (PBS): 137 mM NaCl, 2.7 mM KCl, 1.5 mM KH$_2$PO$_4$, 6.5 mM Na$_2$HPO$_4$, 1 mM CaCl$_2$, 1 mM MgCl$_2$.
4. PBS$^+$: PBS containing 0.5× protease inhibitors (5 µg/ml aprotinin, 0.5 µg/ml pepstatin, 5 µg/ml leupeptin, and 0.5 mM Pefabloc).

2.2. Homogenization Protocol for Cells Requiring Hypotonic Shock

1. Homogenization buffer A (HBA): 3 mM imidazole pH 7.4, 1 mM EDTA, ddH$_2$O, protease and phosphatase inhibitors.
2. Homogenization buffer B (HBB): 500 mM sucrose, 3 mM imidazole, pH 7.4, 1 mM EDTA, ddH$_2$O, 0.06 mM cycloheximide, protease, and phosphatase inhibitors.
3. Phosphate buffered saline (PBS): 137 mM NaCl, 2.7 mM KCl, 1.5 mM KH$_2$PO$_4$, 6.5 mM Na$_2$HPO$_4$, 1 mM CaCl$_2$, 1 mM MgCl$_2$.
4 PBS$^+$: PBS containing 0.5× protease inhibitors (5 µg/ml aprotinin, 0.5 µg/ml pepstatin, 5 µg/ml leupeptin, and 0.5 mM Pefabloc).

2.3. Latency Measurement

2.3.1. Fluid Phase Internalization of Horseradish Peroxidase (HRP) on Dishes

1. High activity Horseradish peroxidase (HRP) 1050 U/mg, lyophilized powder from Serva.
2. Phosphate buffered saline (PBS): 137 mM NaCl, 2.7 mM KCl, 1.5 mM KH$_2$PO$_4$, 6.5 mM Na$_2$HPO$_4$, 1 mM CaCl$_2$, 1 mM MgCl$_2$.
3. PBS/BSA: PBS containing 5 mg/ml BSA.
4. Internalization Medium: Modified Eagles Medium (MEM), 10 mM Hepes, pH 7.4, 5 mM D-glucose.

2.3.2. Determination of Horseradish Peroxidase (HRP) Specific Activity Present in Each Subcellular Fraction.

1. Homogenization Buffer plus (HB$^+$): 250 mM sucrose, 3 mM imidazole, pH 7.4, 1 mM EDTA, 0.03 mM cycloheximide, protease inhibitors.
2. O-dianisidine. This is a very toxic and light sensitive reagent. Keep the reagent light protected and perform the reactions in the dark.
3. HRP Reagent Solution: 0.003% H$_2$O$_2$, 50 mM phosphate buffer, pH 5.0, 0.1% TritonX-100.
4. 1 mM KCN.
5. Spectrophotometer.

2.4. Subcellular Fractionation using a Step Gradient

High percentage sucrose solutions take some time to dissolve and should, therefore, be prepared the day before the experiment. The concentration of all sucrose solutions must be controlled using a refractometer.

1. Homogenization buffer (HB): 250 mM sucrose, 3 mM imidazole, pH 7.4, 1 mM EDTA, ddH$_2$O, protease and phosphatase inhibitors.
2. 25% (w/w) sucrose solution in ddH$_2$O: 0.806 M sucrose, 3 mM imidazole, pH 7.4, 1 mM EDTA, ddH$_2$O. The refractive index of this solution should be 1.3723 at 20 °C.
3. 35% (w/w) sucrose solution in ddH$_2$O: 1.177 M sucrose, 3mM imidazole, pH 7.4, 1 mM EDTA, ddH$_2$O. The refractive index of this solution should be 1.3904 at 20 °C.
4. 62% (w/w) sucrose solution in ddH$_2$O: 2.351 M sucrose, 3 mM imidazole, pH 7.4, 1 mM EDTA, ddH$_2$O. The refractive index of this solution should be 1.4463 at 20 °C.

2.5. Subcellular Fractionation using a Continuous Gradient

Please take into consideration the information provided in Section 2.4 of this chapter concerning the preparation of sucrose solutions.

1. Homogenization buffer (HB): 250 mM sucrose, 3 mM imidazole, pH 7.4, 1 mM EDTA, ddH$_2$O, protease and phosphatase inhibitors.
2. 10% (w/w) sucrose solution in ddH$_2$O: 0.3 M sucrose, 3 mM imidazole, pH 7.4, 1 mM EDTA, ddH$_2$O. The refractive index of this solution should be 1.3479 at 20 °C.
3. 40% (w/w) sucrose solution in ddH$_2$O: 1.375 M sucrose, 3 mM imidazole, pH 7.4, 1 mM EDTA, ddH$_2$O. The refractive index of this solution should be 1.3997 at 20 °C.
4. When possible, access to both a gradient forming device (e.g., gradient master, BioComp) and a fraction collector (e.g., Auto density flow, Labconco) is of advantage.

2.6. Organelle Sample Preparation for 2-DE Analysis.

1. 3 mM imidazole pH 7.4, 1 mM EDTA.
2. Micro BCA Protein Assay Reagent Kit containing: Micro Reagent A (MA), Micro Reagent B (MB), Micro Reagent C (MC).
3. Working Reagent (mix 4.8 ml MB with 0.2 ml MC (green precipitate is formed). Add 5 ml MA (solution should get clear again). This solution should be prepared freshly and is stable up to 1 day at RT.
4. BSA standard solution: Dilute the 2 mg/ml BSA stock solution provided with the Micro BCA Protein Assay Reagent Kit to 0.1 µg/µl in Homogenization Buffer (*see* Section 2.4 of this chapter). Store the solution in the freezer.

5. Methanol and chloroform.
6. 2DE Sample buffer: 7M urea, 2M thiourea, 4% CHAPS, 30 mM Tris, 65 mM DTT, 2% carrier ampholytes, buffered to pH 8.5 with HCl. Aliquots can be stored at $-20\,°C$.

3. Methods

3.1. Standard Homogenization Protocol

For mechanically breaking the plasma membrane we describe here a needle/syringe based protocol, which does not require specialized equipment, is easy to handle and gives reproducible results. Passage cells in such a way that they are approximately 80–90% confluent at the start of the experiment (*see* Note 1). The subsequent harvesting, homogenization and fractionation steps should be performed at $4\,°C$. Keep all solutions on ice.

1. Place the dish on the cold plate (metal plate on ice). Wash cells three times with ice cold PBS.
2. Add PBS^+ (3–5 ml per 15-cm dish) and scrape the cells in one continuous sheet using a rubber policeman and turning the plate 360° in one continuous movement while keeping the scraper with slight pressure always in the fluid (tilt a bit). Remove the sheet of cells from the scraper by fast vertical shaking within the fluid without touching the dish. Scrape the centre of the filter with a single movement as well. Transfer the cells to 15-ml tubes using a transfer pipet. Centrifuge at $200g$ for 5 min at $4\,°C$.
3. Aspirate PBS^+ and add three times the pellet volume ($3V$) of homogenization buffer (HB). Slightly resuspend pellet using a cut pipet tip. The aim is to resuspend the cell pellet and it is not necessary to remove all existing cell clumps.
4. Centrifuge at $1,300g$ for 10 min at $4\,°C$, aspirate the buffer.
5. Resuspend the cells gently with HB^+ buffer (3–5 times the pellet volume) using a cut blue 1,000-μl tip. Avoid formation of air bubbles while pipeting (*see* Note 2). Verify that no big cell aggregates are visible, it is important to disrupt them efficiently before homogenization. Split the suspension into several 15-ml tubes (1 ml/tube).
6. Attach a 22-G needle to a 1 ml syringe. The syringe should be prewashed with ice cold HB^+ buffer (*see* Note 3) and make sure that all air bubbles are removed from the needle.
7. Pass the suspension through the syringe 3–10 times with some force but still gently, with the tip of the needle being in contact with the tube. Monitor the homogenization by phase contrast microscopy. Please note that the number of strokes is highly dependent on the manual force of the operator and should, therefore, be carefully controlled. Homogenization is best achieved at a ratio of 70–80% clear nuclei to intact cells (*see* Note 4).
8. Add 700 μl HB buffer to each 1 ml of homogenized cells. Centrifuge at $2,000g$ for 10 min at $4\,°C$.

9. Carefully retrieve the post nuclear supernatant (PNS) and the nuclear pellet. Leave an aliquot of both for later analysis. The PNS is now ready for fractionation. For additional analysis, nuclei can be purified from the pellet fraction by adapting the protocol described in (6).

3.2. Homogenization Protocol for Cells Requiring Hypotonic Shock

Certain cells are harder to homogenize either because they tend to round up when scraped and/or because they do not form strong intercellular contacts. These cells need to be swollen before homogenization. The hypotonic shock renders the cells more fragile and, therefore, extra care should be taken when using this protocol. The presence of sucrose in the homogenization buffer B is necessary to assure membrane integrity.

1. Perform steps 1 and 2 of the standard homogenization protocol.
2. Rebuffer the cells using three times the pellet volume of homogenization buffer A (HBA) in such a way that the entire pellet floats up. Centrifuge 10 min at $200g$, 4 °C and aspirate the buffer.
3. Resuspend the cells gently in HBA (0.5–1 × the pellet volume) using a cut blue 1,000 µl tip. Check for the absence of cellular clumps. Let the cells swell on ice for additional 10–20 min.
4. Add an equal volume of homogenization buffer B (HBB) as of HBA used. Carefully and slowly agitate the tube until no visible smears remain. Split the suspension into several 15-ml tubes (1 ml/tube maximum).
5. Perform steps 6–9 of the standard homogenization protocol (Section 3.1).

3.3. Latency Measurement

The latency measurement allows the quantification of disrupted endosomal organelles during homogenization using internalized phase markers such as horseradish peroxidase (HRP) (5). In brief, the protocol consists of an internalization step followed by thorough wash steps to remove any nonspecifically adsorbed marker. Upon homogenization, HRP amounts in the different fractions are quantified in a spectrophotometer at 455 nm (a brown product is formed upon the reaction of the peroxidase with o-dianisidine and H_2O_2).

3.3.1. Fluid Phase Internalization of Horseradish Peroxidase (HRP) on Dishes

1. Use a metal plate to cool dishes to ice temperature using a box filled with crushed ice. Assure that the dishes are set completely flat.
2. Wash dishes three times with PBS for 2–3 min on a shaking platform.
3. Place the dishes for 2–3 sec on a 37 °C water bath and replace the PBS with warmed up internalization medium containing 1.8 mg/ml HRP (*see* Note 5).

4. Incubate dishes on a rocking platform at 37 °C for the desired time (*see* Note 6). A water bath can also be used.
5. Cool the dishes again and wash them three times with PBS/BSA (5 min per wash).
6. Wash twice with PBS. The cells are ready for homogenization.
7. Homogenize cells according to the standard homogenization protocol or the protocol for cells requiring hypotonic shock. Collect aliquots of the nuclear pellet and PNS.
8. The remaining PNS can then be used to purify the fraction of interest (e.g., perform a step gradient and collect interphases to analyze the integrity of late and early endosomes).

3.3.2. Determination of Horseradish Peroxidase (HRP) Specific Activity Present in Each Subcellular Fraction.

1. Dilute the PNS in HB$^+$ and centrifuge 30 minutes at 260,000g, 4 °C.
2. Collect the supernatant. Resuspend the pellet in HB$^+$.
3. Using clean glassware prepare the HRP reagent solution by mixing 12 ml of 0.5 M phosphate buffer pH 5.0 and 6 ml of 2% TritonX-100 with 100.8 ml of ddH$_2$O. Gently dissolve 13 mg of *o*-dianisidine in this buffer and finally add 1.2 ml of 0.3% H$_2$O$_2$). Do not stir the solution. Be careful when weighing o-dianisidine, it is a carcinogenic and a highly toxic substance.
4. Prepare blanks and standards diluting 1–10 ng of HRP in HB$^+$.
5. Dilute samples, blanks and standards 1:10 into the solution previously prepared as explained in Section 3.2.3. Quickly mix and place the plate in the dark at room temperature. Record the time.
6. When the mixture turns brownish, stop the reaction by adding 0.1 times the reaction volume (0.1V) of 1 mM KCN. Read the absorbance at 455nm. Results are expressed as OD (optical density) units/min. Make a standard curve using the concentrations and the absorbance values obtained for the standards, and calculate the amount of HRP present in the PNS and in each sample. The reaction is in the linear range up to an OD of about 1.5.
7. Calculate the peroxidase specific activity present in each subcellular fraction. Values should be expressed as ng of HRP per µg of total protein. Calculate the fraction's enrichment as its HRP activity per HRP activity of the PNS. Determine the yield of each fraction as percentage of its specific activity per total activity present in the PNS.

3.4. Subcellular Fractionation using a Step Gradient

During centrifugation, organelles sediment (or float) until they reach their isopycnic position within the density gradient. Depending on their content, protein/lipid ratio, shape and size, different organelles will have different densities. The degree of separation during gradient centrifugation will also depend on the nature of the medium. Although sucrose is the most common,

Isolation of Endocitic Organelles by Density Gradient Centrifugation 325

there are many other alternatives, e.g., Ficoll, Percoll, Nycodenz or Metrizamide *(5)*. Based on this knowledge the step gradient protocol was designed so that different organelles are enriched at different interphases between cushions of sucrose solutions. For instance, late endosomes and lysosomes will be found floating on the interface between 25% sucrose and HB (appr. 8.6% sucrose), early endosomes and carrier vesicles will be enriched at the 35%/25% interphase and finally, heavy membranes like Golgi and ER membranes will be found floating at the interphase between 40.6% and 35% *(3,4)*.

Before starting, add protease inhibitors and cycloheximide to the sucrose solutions. Cycloheximide inhibits the release of ribosomes from the ER; the high protein content makes this organelle heavier and, therefore, easier so separate from other lighter organelles *(4)*.

1. Adjust the sucrose concentration of the PNS to 40.6% by adding 62% sucrose (usually 1:1.2 V/V), (*see* Note 7). Control the sucrose concentration of the solution using a refractometer. Add HB^+ or 62% sucrose if further adjustment is required.
2. Load the diluted PNS on the bottom of an ultracentrifuge tube. Which tube to use will depend on the volume of the PNS and the ultracentrifuge rotor of choice.
3. Overlay sequentially with 1.5 V of 35% sucrose, 1 V of 25% sucrose and finally, fill up the rest of the tube with HB^+. Mark the interphases with a waterproofed pen.
4. Mount the tubes on the appropriate rotor and centrifuge at $210,000g$ at $4\,°C$ for 1.5 h. It is very important to balance the tubes with an identical gradient.
5. After centrifugation, one should be able to detect a milky band of membrane particles at each interphase. Collect the different interphases. It is easier to collect an interface with the top cushion still remaining.
6. Fractions should be snap-frozen in liquid nitrogen and stored at $-80\,°C$. Upon storage, thaw up the samples as rapidly as possible before use or analysis.
7. Using any of the standard methods to measure protein concentration, calculate yield and enrichment of each fraction relative to the initial PNS.
8. Prepare a gradient profile. In brief, load equal volumes of each of the fractions collected on an SDS-PAGE. Use antibodies against well established organelle markers to determine the degree of separation achieved *(4)*, *see* also Fig. 2.

3.5. Subcellular Fractionation using a Continuous Gradient

Continuous gradients provide better resolution than step gradients at the expense of yield. After centrifugation to equilibrium the organelles are distributed throughout the gradient according to their densities. The presence of subdomains within an organelle can be responsible for small differences in density, resulting in comigration of organelles in several adjacent fractions. This effect also increases the final volume required to collect a specific organelle.

For this protocol it is of advantage to have access to both a gradient forming device (e.g., gradient master, BioComp) and a fraction collector (e.g., Auto

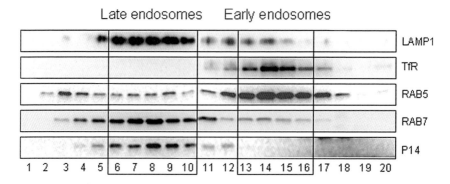

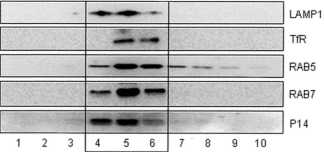

Fig. 2. Western blotting of gradient fractions with markers specific to late endosomes (LAMP1, Rab7, p14) and early endosomes (TfR, Rab5). Late and early endosomes from murine EpH4 cells are well separated in continuous sucrose gradient (10% to 40%, corresponding to fractions from 1 [top] to 20 [bottom]). Crude endosomal fraction (containing early and late endosomes) can be efficiently obtained by differential centrifugation in discontinuous sucrose gradient (consisting of three steps of 8%, 35%, 42% sucrose).

density flow, Labconco). Such instruments can largely improve handling and reproducibility. As mentioned for the step gradient protocol, add protease inhibitors and cycloheximide to the sucrose solutions before starting. Work strictly on ice or in the cold room and remember to keep a small aliquot of the PNS as a reference sample. Warm up the 10% and 40% sucrose solutions to room temperature before starting.

1. Draw a mark for 50% of the volume of the tubes using the small metal block provided with the gradient master.
2. Fill half the tube with 10% solution. Using a metal syringe (also from the gradient master equipment), load the 40% sucrose underneath the 10% solution until the interphase between both is found exactly at the 50% volume mark.

3. Carefully close the tubes with the long black lids (*see* Note 8). Level the gradient master and perform the adequate run (*see* instruction manual of the equipment).
4. Remove the caps with care and place the gradients on ice. Once cooled, the gradient can be overloaded with the prepared PNS. Avoid disturbing the existing gradient when performing this step. Spin gradients overnight at 210,000g, 4°C.
5. Using the centering jaws, fix the centrifuged tube to the fraction collector. Turn the instrument on until the probe tip reaches the surface of the fluid.
6. Collect the fractions. Calculate the number of drops to place in each 1.5 ml precooled tube considering the total volume of the gradient prepared. The lightest fraction is labeled as fraction 1.
7. Fractions should be frozen in liquid nitrogen and stored at −80 °C. Upon storage, thaw the samples as rapidly as possible before use.
8. Analyze the fractions as explained for the step gradient.

3.6. Organelle Sample Preparation for 2-DE Analysis.

1. For 2-DE analysis organelles/membranes fractions from several gradients are pulled together and diluted 1:1 in 3 mM imidazole, pH 7.4, 1 mM EDTA so as to reduce the sucrose concentration *(4)*.
2. Obtained dilutions are centrifuged for 1 hour at 100,000g and stored on ice. Supernatant is removed, and the membrane pellet resuspended in the same buffer (*see* Note 9).
3. To determine protein concentrations in much diluted organelle fractions (as well as in a limited amount of organelle protein pellet) we recommend the use of Micro BCA Protein Assay reagent Kit, available from Pierce (Rockford, IL) (*see* Note 10). Since the assay is very sensitive always wear gloves to avoid contamination with proteins such as skin keratins.
4. Prepare standards in a 96-well plate in a final volume of 150 µl bidistilled water: A1=blank; B1=0.1µg; C1=0.3µg; D1=0.6µg; E1=1µg; F1=3µg ; G1=6µg; H1=10µg.
5. Dilute appropriate volume (usually 1–2 µl PNS, 10 µl heavy membranes, 50 µl early endosomes, 100 µl late endosomes) of your samples in a final volume of 150 µl as well, filling the plate in vertical direction.
6. Add 150 µl of working reagent to each well, using a multichannel pipet, and mix thoroughly.
7. Cover the plate with the lid and incubate for 2 h to over night at 37 °C in an incubator (longer incubation increases the sensitivity of the assay).
8. Cool down to room temperature and read the absorbance at 562 nm with the use of the spectrophotometer (do not start more than 20 samples at the same time. The reaction is slowly ongoing!).
9. Export the data to a Microsoft Excel worksheet and calculate the protein concentration of the samples.
10. Proteins are then precipitated by chloroform-methanol (Wessel/Fluegge method) *(7)*. Add 4V methanol followed by 1V of chloroform to 20 µg of protein sample

to be analyzed by 2-D gel electrophoresis. Vortex and check that there is only one phase.
11. Add 3V of distilled water and vortex thoroughly.
12. Centrifuge 1 minute at 16,000 g, room temperature. Remove upper organic phase with drawn out Pasteur pipet without disturbing the interphase.
13. Add at least 3V of methanol, vortex thoroughly and centrifuge at 16,000g for 2 minutes.
14. Carefully remove supernatant with drawn out Pasteur pipet without disturbing the pellet.
15. Air dry the pellet with care (overdrying makes the pellet very difficult to solubilize).
16. The pellet is then directly solubilized in sample buffer and used for 2-DE

Using the presented strategy, it is possible to avoid the negative effect of high sucrose concentrations in the efficiency of protein precipitation. Furthermore, lipids are readily removed from the sample, thereby making it fully compatible with subsequent 2-D gel electrophoresis *(8)*. Protein precipitation before two-dimensional gel electrophoresis is of particular importance when the gel is to be analyzed for phosphoprotein content. Lipid depletion minimizes background staining because of phospholipids and other cell constituents *(9)*.

Please note that membrane proteins require special handling because of their hydrophobicity. A combination of 4% CHAPS, 2 M thiourea, 7 M urea, 2% carrier ampholytes is used in the sample buffer to increase the solubility of integral and peripheral membrane proteins *(10)*.

3.7. Concluding Remarks

Subcellular fractionation is an essential step among prefractionation techniques in proteomics research. The approach described above simultaneously allows the reduction of the complexity of the protein sample and the enrichment of the organelles. 2D-maps of several subcellular fractions purified by density gradient centrifugation are presented in Fig. 3, as an example of successful application of the method described here. Several hundreds of protein species could be specifically enriched in certain organelle/subcellular fraction *(1)*, thereby making possible detection of proteins of lower abundance and organelle-specific ones and subsequent identification of proteins by mass spectrometry. Analysis of sub-proteomes of purified organelles and different subcellular fractions allows tracking of proteins that shuttle between different compartments. Though the high purity of isolated organelles is essential for comprehensive analysis of total organelle proteomes, it is important to point out that complete purification is hardly possible. Similar physical properties shared

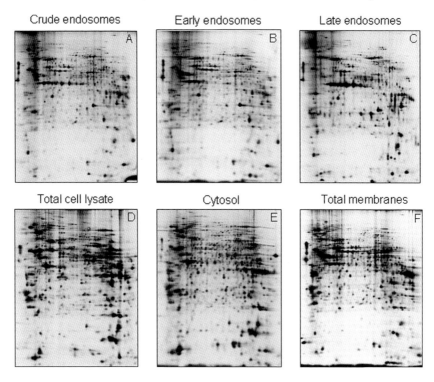

Fig. 3. Two-dimensional differential gel electrophoresis (DIGE) of subcellular fractions purified from murine EpH4 cells by sucrose gradient centrifugation. Crude endosomal fraction (**A**) was collected from the interface between 8% and 35% sucrose cushions in discontinuous gradient, consisting of three steps of 8%, 35%, 42% sucrose. Late (**B**) and early endosomes (**C**) were efficiently separated in similar discontinuous gradient with an additional step of 25% sucrose. Total cell lysate (**D**) was obtained by direct lysis of cell pellet; cytosol (**E**) and total membranes (**F**) were obtained from PNS by ultracentrifugation as described in Fig. 1. Organelle and other subcellular proteomes are enriched with specific proteins and are of reduced protein sample complexity.

by different subcellular compartments determine their comigration in conventional gradients and, therefore, considerable cross-contamination of isolated subcellular fractions. As the size and density of the organelles differ, the method needs to be optimized for every type of cells/tissue. Despite that, the enrichment of specific organelles or certain subcellular fractions is beneficial for the detection of low abundant proteins and tracking of their changes after cell stimulation.

4. Notes

1. In certain cell types the passage number influences the homogenization result. Use always cells with a known passage number. Confluence of cells on the plate is another critical parameter (cell/cell interactions must be broken during homogenization). Check that they have reached at least 80% confluence before starting. Finally, care should be taken that plates are set flat on the shelf of the incubator, so that a proper and even monolayer can be formed.
2. Homogenization is more efficient at high density of cells 20–30% v/v, therefore keep volume of HB buffer low.
3. Fill the syringe with buffer and blow out all air-bubbles trapped in the needle before starting. The presence of air bubbles in the needle generates unpredictable forces leading to bad homogenization and or extensive nuclear breakage.
4. Cytoplasmic aggregates are often found during homogenization because of the presence of cytoskeletal elements that entrap organelles *(11)*. These aggregates should be resuspended as their presence might result in loss of yield due to sedimentation. In over-homogenized cells on the other hand, organelles will break releasing their content. As a consequence, the cytosolic fraction will be contaminated with typical organelle/membrane bound material. Also, nuclear breakage releases DNA and acts as a potential source for aggregates. Therefore, it is crucial to access the quality of the homogenization (phase contrast microscopy or latency measurement).
5. HRP is a relatively stable enzyme and the assay is very sensitive. Avoid cross-contaminating samples or standards.
6. Different endocytic compartments require specific internalization times e.g., 5 min for early endosomes, 10 min followed by 30 min chase for late endosomes. If chase is required, place the dishes on ice and wash them 2–3 times with precooled PBS/BSA immediately after internalization. Change the PBS/BSA solution to internalization medium containing 2 mg/ml BSA and incubate the dishes at 37°C for the necessary time. Proceed to step 5 of the fluid phase internalization protocol (Section 25.3.3.1 of this chapter).
7. Please note that endosomes are very fragile under these conditions. Ensure the solution is mixed both thoroughly and gently. Confirm the homogeneity of the solution under a lamp (no smears should be visible).
8. Be sure to remove all existing air bubbles as they will interfere with gradient formation. When closing the tubes with the back lids, it is common that a bit of 10% solution superfluous through the existing lid hole. Remove this excess of solution before starting the gradient master run.
9. Sedimentation of organelles by ultracentrifugation efficiently removes traces of cytosolic proteins from the organelle fractions of interest. This is of particular importance when analyzing light fractions obtained from continuous gradients (late endosomes) in which the PNS was loaded on the top of gradient.
10. The color reaction is based on the Biuret reaction (production of Cu^{1+} ions because of the reaction of proteins with Cu^{2+} in alkaline medium). BCA (Bicinchoninic acid), in its water-soluble Na-salt form, is a sensitive, highly specific

reagent for Cu^{1+}. The purple reaction product of two molecules BCA with one Cu^{1+} ion is water-soluble as well and exhibits a strong absorbance at 562 nm. When performing this method on a plate reader, absolvance can be read at 540–590 nm.

Acknowledgments

Work in the Huber laboratory is supported by the Austrian Proteomics Platform (APP) within the Austrian Genome Program (GEN-AU), Vienna, Austria and by the Special research Program "Cell Proliferation and Cell Death in Tumors" (SFB021, Austrian Science Fund).

References

1. Stasyk T, Huber LA. (2004) Zooming in: fractionation strategies in proteomics. *Proteomics* **4**, 3704–16.
2. Huber LA. (2003) Is proteomics heading in the wrong direction? *Nat Rev Mol Cell Biol.* **4**, 74–80.
3. Pasquali C, Fialka I, Huber LA. (1999) Subcellular fractionation, electromigration analysis and mapping of organelles. *J Chromatogr B Biomed Sci Appl.* **722**, 89–102.
4. Fialka I., Pasquali C., Lottspeich F., Ahorn H., Huber L.A. (1997) Subcellular fractionation of polarized epithelial cells and identification of organelle-specific proteins by two-dimensional gel electrophoresis. *Electrophoresis* **18**, 2582–2590.
5. Huber LA, Pfaller K, Vietor I. (2003) Organelle proteomics: implications for subcellular fractionation in proteomics. *Circ Res.* **92**, 962–8.
6. Dignam J.D., Lebovitz,R.M. and Roeder,R.G. (1983) Accurate transcription initiation by RNA polymerase II in a soluble extract from isolated mammalian nuclei. *Nucleic Acids Res.*, **11**,1475–1489.
7. Wessel D., Flügge U.I. (1984) A method for the quantitative recovery of protein in dilute solution in the presence of detergents and lipids. *Anal. Biochem.* **138**, 141–143.
8. Stasyk T., Hellman U., Souchelnytskyi S. (2001) Optimizing sample preparation for 2-D electrophoresis. *Life Science News* **9**, 8–11.
9. Stasyk T, Morandell S, Bakry R., Feuerstein I., Huck C.W, Stecher G, Bonn GK, Huber LA. (2005) Quantitative detection of phosphoproteins by combination of two-dimensional difference gel electrophoresis and phosphospecific fluorescent staining. *Electrophoresis* **26**, 2850–4.
10. Pasquali C., Fialka I., Huber L.A. (1997) Preparative two-dimensional gel electrophoresis of membrane proteins. *Electrophoresis* **18**, 2573–81.
11. Gruenberg, J. and Howell K.E. (1988) Fusion in the endocytic pathway reconstituted in a cell-free system using immuno-isolated fractions. *Prog. Clin. Biol Res.* **270**, 317–331.

26

Isolation of Highly Pure Rat Liver Mitochondria with the Aid of Zone-Electrophoresis in a Free Flow Device (ZE-FFE)

Hans Zischka, Josef Lichtmannegger, Nora Jaegemann, Luise Jennen, Daniela Hamöller, Evamaria Huber, Axel Walch, Karl H. Summer, and Martin Göttlicher

Summary

This protocol describes the purification of mitochondria from rat liver with the aid of zone electrophoresis in a free flow device (ZE-FFE). Starting from liver homogenate, cell debris and nuclei are removed by low speed centrifugation. A crude mitochondrial fraction is obtained by medium speed centrifugation and is further purified by washing followed by a Nycodenz® gradient centrifugation. Lysosomes and microsomes are located at the upper parts of the gradient, whereas mitochondria are found in the medium part of the gradient. A subsequent purification step with ZE-FFE efficiently removes remaining lysosomes and microsomes and, importantly, damaged mitochondrial structures. The resulting purified mitochondria can be concentrated by centrifugation and used for further experiments. Finally, possible modifications of this protocol with respect to the isolation of pure lysosomes are discussed.

Key Words: rat liver mitochondria; free flow electrophoresis; lysosomes; zone electrophoresis.

1. Introduction

This protocol describes the isolation and purification of mitochondria from rat liver by a combination of centrifugation steps and zone electrophoresis in a free flow device (ZE-FFE). The latter is based on the pioneering work of Kurt Hannig, Hans-G. Heidrich and Roland Stahn, who set the standards for free flow electrophoresis (FFE) of subcellular particles almost forty years ago *(1–4)*. Originally Stahn et al. described the electrophoretic purification of rat liver lysosomes from mitochondria, showing that these two organelles markedly co-sediment in the preceding differential centrifugation steps *(3)*. This cross contamination, also observed by others *(5)*, may lead to substantial structural impairment of the mitochondria, e.g., by proteases originating from damaged lysosomes *(3)*. As Hannig and Heidrich described, a further difficulty is the persistent contamination of rat liver mitochondria with microsomes (and

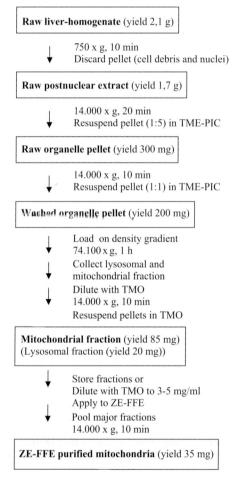

Fig. 1. Flow chart of the purification of rat liver mitochondria. Typical yields of the individual samples are given as determined by their protein amount and normalized to 10 g wet weight of liver tissue.

peroxisomes), which are difficult to separate electrophoretically because of their similar surface charge *(1,2)*.

Based on these findings we modified Stahn's protocol with the aim to obtain mitochondria with a high degree of purity and intactness (Fig. 1). Thereto a density gradient centrifugation step was implemented *(6,7)* to remove a large part of the contaminating organelles from the mitochondria (Fig. 2) preceding the subsequent ZE-FFE purification (Fig. 3). In fact, the isolated lysosomal (-microsomal) fraction from the density gradient typically deflects two fractions further towards the anode compared to the mitochondrial fraction (Fig. 3B). To monitor the effectiveness of the isolation and purification procedure immuno-analysis against proteins with a specific subcellular localization was employed (Fig. 4). Rat liver mitochondria purified by density gradient centrifugation are deemed to be "pure" and are considered to be the "gold standard" in mitochondrial research *(8)*. They are, however, although to a minor extent, still contaminated with lysosomes and microsomes (Fig. 4, lane 5). An even bigger drawback is the occasional presence of structurally damaged mitochondria in such preparations, presumably arising from the mechanical stress of the tissue homogenization, which inevitably challenges mitochondrial stability. It is known, for example, that mitoplasts (i.e., mitochondria lacking outer membranes) can be generated from mitochondria by extensive homogenization *(2)*. In fact,

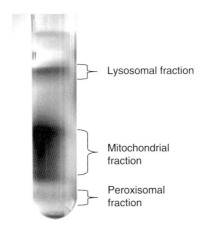

Fig. 2. Nycodenz® density step gradient separation of the washed organelle pellet (Fig. 1) with the following composition from top to bottom: 2 ml 24%, 2 ml 27%, 3 ml 28%, 1 ml 33%, and 1 ml 40% w/v Nycodenz® in 10 mM Tris/HCl, pH 7.4. Lysosomes (and microsomes) can be collected from the upper parts of this gradient, mitochondria are found in the medium part and peroxisomal activities in the lower part of the gradient.

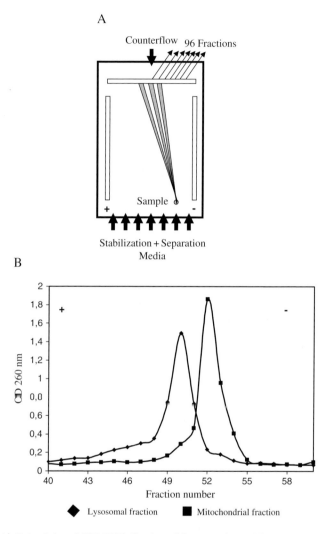

Fig. 3. (**A**) Principle of ZE-FFE displayed in top view. The separation chamber is constantly flushed with separation (TMO) and stabilization (RSO) media. At the end side of the chamber these media are blocked by the counterflow (CF) and deviated via the fraction collector into a 96-well plate. The electrical field is oriented perpendicular to this buffer stream. The prefractionated mitochondria are injected via the sample inlet into the electrophoresis chamber, transported with the laminar buffer flow and deflect towards the anode because they are negatively charged at neutral pH *(14)*. Optical density measurements of the 96 collected fractions allow monitoring of the separation. The major peak in these separation profiles indicates the mitochondria-containing fraction from which ZE-FFE purified mitochondria can be collected by centrifugation. (**B**) Typical ZE-FFE separation profiles of a mitochondrial and a lysosomal fraction. A difference in deflection between the organelle fractions can be observed with mitochondria being collected more at the cathodal side (anode +; cathode −).

Isolation of Highly Pure Rat Liver Mitochondria

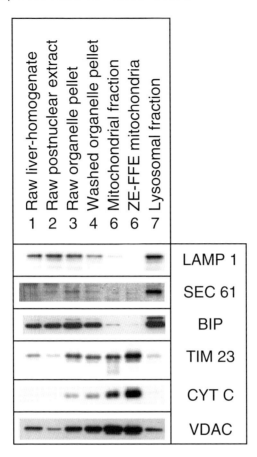

Fig. 4. Immunoblotting analysis of the obtained samples of the isolation and purification procedure (outlined in Fig. 1). LAMP 1 was used as validation for lysosomes, SEC 61 and BIP as validation for microsomes and TIM 23, CYT C and VDAC as validation for mitochondrial compartments (i.e., inner membrane, inter membrane space and outer membrane). Protein load in each lane was 25 µg.

such structures were still observed (with varying extent) on electron micrographs although the analyzed mitochondria were purified by density gradient centrifugation (Fig. 5A). A further challenge of mitochondrial integrity might arise from the co-sedimentation of lysosomes with mitochondria in differential centrifugation steps preceding the density gradient. Clearly, damaged lysosomes may release a whole "cocktail" of degradation enzymes, which may only be partially inactivated (e.g., by applied inhibitors or altered pH). Consequently mitochondrial preparations containing severely damaged mitochondria

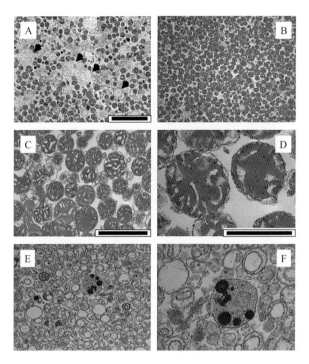

Fig. 5. Electron microscopy analysis: (**A**) Mitochondrial fraction from the density gradient containing mitochondria and some mitoplasts (arrows, bar equals 5 μm). (**B–D**) ZE-FFE purified mitochondria (bar equals 2 μm in **C** and 1 μm in **D**, respectively). (**E–F**) lysosomal fraction.

are obtained (Fig. 6), demonstrating that they are not efficiently removed by the density gradient. Thereto the subsequent ZE-FFE purification step is implemented, which integrates a further separation and purification parameter, i.e., the electrophoretic mobility *(1,9)*. Given a sufficient difference in this parameter, contaminants will deflect differently from intact mitochondria in the electric field and, in consequence, highly "pure" mitochondria are obtained (Fig. 5B-D).

This final ZE-FFE purification step specifically aids in: (i) the removal of residual Nycodenz® in density gradient purified mitochondria, which might disturb subsequent analyses; (ii) a further removal of contaminating lysosomes and microsomes (Fig. 4, sample 6); (iii) the removal of mitoplasts (cp. Fig. 5A and 5B); (iv) the removal of severely damaged mitochondria which have been proteolytically attacked; (Fig. 6).

Isolation of Highly Pure Rat Liver Mitochondria

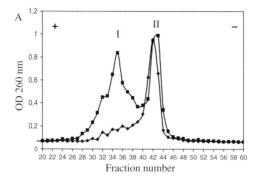

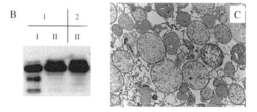

Fig. 6. (**A**) ZE-FFE separation profiles of two mitochondrial preparations, whereby preparation 1 was obtained by more vigorous resuspension of the organelle pellets compared to preparation 2 and omission of protease inhibitor in these resuspension steps. Preparation 1 displayed two major peaks (I and II) in contrast to preparation 2 where only one major peak (II) was evident (anode +; cathode −). (**B**) TIM 23 immuno-blotting analysis of the separated ZE-FFE peaks I and II of both preparations reveals marked proteolytic degradation signals of this inner membrane protein in peak I. (**C**) Electron micrograph of ZE-FFE purified mitochondria from peak I of preparation 1. Significant structural alterations and damage of the mitochondria are visible (magnification ×20,000).

2. Materials

2.1. Isolation of Rat Liver Mitochondria

1. 10 mM Tris-HCl, pH 7.4: To prepare 1 L, 1.21 g Tris is dissolved in 850 ml deionized H_2O, adjusted to pH 7.4 with HCl and adjusted to a total volume of 1 L.
2. 0.9% (w/v) NaCl solution. Stored at 4 °C.
3. 5× electrode solution (5× ELO): 500 mM acetic acid (HAc), 500 mM triethanolamine (TEA) pH 7.4. To prepare 1 L, 28.6 ml HAc and 66.65 ml TEA are added to 800 ml ultra pure H_2O, adjusted to pH 7.4 with 10 M KOH and adjusted to a total volume of 1 L and stored at 4 °C.
4. 0.5 M EDTA (ethylenediaminetetraacetic acid tetrasodium salt): 38.02 g are dissolved in 200 ml deionized H_2O.

5. Protease inhibitor solution (PIC): e.g., prepared with protease inhibitor Complete™ tablets from Roche Diagnostics GmbH, Mannheim, Germany. To prepare 10× concentrated aliquots one tablet is dissolved in 5 ml deionized H_2O and aliquots are stored at −20 °C.
6. Isolation buffer (TME): 10 mM HAc, 10 mM TEA, 1 mM EDTA, 0.28 M sucrose, pH 7.4. To prepare 1 L, 95.8 g sucrose are dissolved in 700 ml deionized H_2O, 20 ml 5× ELO and 2 ml EDTA solution are added, adjusted to a total volume of 1 L and stored at 4 °C.
7. Isolation buffer with protease inhibitor (TME-PIC): Two tablets protease inhibitor Complete™ (Roche Diagnostics GmbH, Mannheim, Germany) are dissolved in 100 ml TME and the buffer is stored at 4 °C.
8. Nycodenz® solutions for the density step gradient. To prepare the 24%, 27%, 28%, 33%, and 40% w/v Nycodenz® solutions 12 g, 13.5 g, 14 g, 16.5 g, or 20 g Nycodenz® (Axis-Shield PoC, Oslo, Norway) are dissolved in 50 ml 10 mM Tris-HCl, pH 7.4. Aliquots of these solutions can be stored at 4 °C for several weeks.

The Nycodenz® density step gradient is prepared essentially as described by Klein et al. *(6,7)*. To avoid strong perturbations the gradient is formed from the bottom using a peristaltic pump (*see* Note 1).

1. Nycodenz® solutions with increasing w/v concentration are poured from the bottom to give the following composition: 2 ml 24%, 2 ml 27%, 3 ml 28%, 1 ml 33%, and 1 ml 40% w/v Nycodenz® in 10 mM Tris-HCl, pH 7.4. A larger volume of the 28% Nycodenz® fraction is used, because the major part of the washed organelle pellet (Sample 4 in Fig. 1) are mitochondria, which gather in this part of the gradient (Fig. 2).
2. Typically six tubes are prepared for the purification of mitochondria from one rat liver. The gradients could be prepared during Steps 3.1.6. and 3.1.7. and are stored on ice until use.

2.2. Purification of Rat Liver Mitochondria by ZE-FFE and FFE Setup (see Notes 2–4)

1. ZE-FFE electrode solution (ELO): 100 mM HAc, 100 mM TEA, pH 7.4. To prepare 1 L, 200 ml 5× ELO are added to 800 ml ultra pure H_2O and stored at 4 °C.
2. ZE-FFE stabilization medium (RSO): 100 mM HAc, 100 mM TEA, 0.28 M sucrose, pH 7.4. To prepare 1 L, 95.8 g sucrose are dissolved in 600 ml ultra pure H_2O, 200 ml 5× ELO are added, adjusted to a total volume of 1 L and stored at 4 °C.
3. ZE-FFE separation medium (TMO): 10 mM HAc, 10 mM TEA, 0.28 M sucrose, pH 7.4. To prepare 1 L, 95.8 g sucrose are dissolved in 700 ml ultra pure H_2O, 100 ml ELO solution are added, adjusted to a total volume of 1 L and stored at 4 °C.
4. ZE-FFE counterflow medium (CF): 0.28 M sucrose. 95.8 g sucrose are dissolved in ultra pure H_2O, adjusted to a total volume of 1 L and stored at 4 °C.

The elaboration of the present protocol was done with the "OCTOPUS" FFE apparatus in its latest version (Dr. Weber GmbH, Kirchheim, Germany), but

Isolation of Highly Pure Rat Liver Mitochondria 341

should be applicable to more recently released instruments, because of a similar work mode schematically displayed in Fig. 3A. The electrophoresis chamber measures 500 mm in length, 100 mm in width and 0.5 mm in thickness. The chamber can be tilted and is run in horizontal mode. Separated fractions are collected via a pulse free splitting device using the counterflow technique *(10)*. Further, stabilization medium (RSO) is used to avoid the penetration of electrode products across the electrophoresis membranes *(10)*.

Setup of the FFE is done according to the manufacturer's instructions on a daily basis which allows a preseparation check up of the system parts and avoids abrasion of the silicone seals by prolonged pressure. The back plate is routinely inspected for breakage of glass or ruptures in the backing foil and the media pump are regularly checked and calibrated. Calcinations of the cooling system are carefully avoided to ensure appropriate cooling. In addition new tubes are used on a regular basis (i.e., after 6 months or approximately 200 work-hours). This avoids abrasion of the tubes because of peristaltic pump action and avoids precipitation and bacteria growth in the tubes because of residual buffer. The setup is done during the density gradient centrifugation (i.e., Step 3.1.9 in the protocol) and is completed if all 96 fractionation tubes are enabled and no entrapped air bubbles are visible in the separation chamber and the sample inlet tubing. The cooling unit of the apparatus is turned on and set to 5 °C to start precooling the system.

2.3. Monitoring of the Isolation and Purification of Rat Liver Mitochondria

Standard equipment is used for SDS-PAGE and subsequent immunoblotting analysis. Ready made gels may be used for SDS-PAGE or gels may be casted according to instructions provided by the instrument manufacturers. For the immunoblotting analysis PVDF membranes are used.

A list of the in this study selected validation markers for the immunoblotting analysis (Fig. 4) of the obtained samples of the isolation and purification procedure (as outlined in Fig. 1) is given in Table 1. Appropriate other markers may be alternatively used.

3. Methods
3.1. Isolation of Rat Liver Mitochondria

The flow chart in Fig. 1 summarizes the major steps of the isolation procedure.

1. Mitochondria are isolated from one rat liver. The animal is killed by exsanguination from the lower abdominal vein under light ether anesthesia and the liver is flushed via the portal vein with ice cold 0.9% NaCl to wash out residual blood.

Table 1
List of selected validation markers for the immunoblotting analysis (Fig. 4) of the obtained samples of the isolation and purification procedure (as outlined in Fig. 1).

Immunoreactivity of the primary antibody	Abbreviation	Subcellular Localization		Applied Dilution	Company	Host
Lysosomal associated membrane protein 1	LAMP 1	Lysosomes	Mono-clonal	(1:1000)	BD Biosciences	Mouse
Protein transport protein SEC61 alpha	SEC 61	Microsomes	Poly-clonal	(1:1000)	Acris	Rabbit
Binding protein / glucose regulated protein 78	BIP/GRP 78	Microsomes	Mono-clonal	(1:1000)	BD Biosciences	Mouse
Translocase of the Inner Membrane	TIM 23	Mitochondria (inner membrane)	Mono-clonal	(1:1000)	BD Biosciences	Mouse
Cytochrome C	CYT C	Mitochondria (inter membrane)	Poly-clonal	(1:1000)	Cell Signaling	Rabbit
Voltage dependent anion channel	VDAC	Mitochondria (outer membrane)	Poly-clonal	(1:1000)	Acris	Rabbit

Isolation of Highly Pure Rat Liver Mitochondria

2. The liver is removed from the animal, immersed in ice cold TME, the wet weight is determined and the liver is cut up into small pieces with scissors. The minced tissue is washed twice with TME. Subsequently given quantities and buffer volumes refer to 10 g of liver wet weight and may be up- or downscaled accordingly.
3. The minced liver tissue is resuspended 1:3 (w/v) in TME (i.e., 30 ml) and homogenized with seven to ten strokes at 500 rpm in a Teflon-glass homogenizer (*see* Note 5).
4. The homogenate is diluted 1:10 (w/v) with TME (i.e., add 100 ml) and filtered through nylon gauze (200 μm).
5. To remove cell debris and nuclei the filtered raw liver homogenate (Fig. 1) is spun at 750g for 10 min at 4 °C.
6. From the supernatant (the raw postnuclear extract, Fig. 1) a crude organelle fraction containing mitochondria, lysosomes, peroxisomes and also some microsomes is pelleted by centrifugation at 14,000g for 20 min at 4 °C.
7. The raw organelle pellet is carefully resuspended 1:5 (w/v) in TME-PIC (i.e., 50 ml) (*see* Note 6) and repelleted by centrifugation at 14,000g for 10 min at 4 °C (*see* Note 7).
8. The washed organelle pellet is carefully resuspended 1:1 (w/v) in TME-PIC (i.e., 10–12 ml).
9. 2 ml of the resuspended sample is loaded on each Nycodenz® gradient and centrifuged at 74,100g_{max} for 1 h at 4 °C in a swing out rotor.
10. The lysosomal and mitochondrial fractions (*see* Fig. 2) are collected with a Pasteur glass pipet. The lysosomal fraction is diluted at least 1:2 and the mitochondrial fraction at least 1:4 with TME.
11. The density gradient purified lysosomes and mitochondria are pelleted at 14,000g for 10 min at 4 °C.
12. If the isolation procedure is completed at this level, organelles may be resuspended in 2–3 ml TME-PIC, snap-frozen with liquid nitrogen and stored frozen for further analyses. If mitochondria shall be further purified by ZE-FFE, they should be resuspended in TMO at a concentration of 3–5 mg/ml (*see* Note 8).

3.2. Purification of Rat Liver Mitochondria by ZE-FFE.

1. The correctly setup and precooled FFE is connected to the diverse media: ELO to the electrode chambers, RSO to the media tubes 1 and 7, TMO to the media tubes 2–6 and CF to media tube 8. Care should be taken to avoid the introduction of any air bubbles in the separation chamber. The chamber is flushed with media for at least 10 min at a delivery rate of 300–400 ml/h.
2. The voltage is set to 750 V and switched on (*see* Note 9). After approximately 10 min the current should reach a stable value (typically between 80–120 mA).
3. The mitochondrial preparations isolated above are used for the ZE-FFE purification. Mitochondrial samples in TMO at a concentration of 3–5 mg/ml are injected in the ZE-FFE apparatus via the sample inlet pump. Lower concentrations will decrease the yield and may hamper a visual inspection during the

separation; higher concentrations may negatively influence the separation because of agglomeration of the sample which should be carefully avoided.

4. The sample is injected in the separation chamber at a rate of 1–2 ml/h. The solution should enter the laminar media stream without clotting or agglomeration and a major line should be visible that deflects towards the anode (*see* Note 10).
5. The separated fractions are collected in 96-well plates (deep-well plates) as soon as the sample reaches the end of the separation chamber. A stable sample deflection pattern should be observed in the separation chamber ensuring that constant separation conditions exist during the run. Separation profiles (Fig. 3B) can be measured by means of a 96-well plate reader (*see* Note 11).
6. The remaining sample is processed (*see* Note 12) and according to the associated separation profiles corresponding main peaks from the deep well plates are pooled. The ZE-FFE purified mitochondria are collected by centrifugation (14,000g, 10 min, 4 °C). Organelles may be resuspended in small volumes of TME-PIC, snap-frozen with liquid nitrogen and stored frozen for further analyses.

3.3. Monitoring the Mitochondrial Isolation and Purification Procedure

3.3.1. SDS-PAGE and Immunoblotting

The effectiveness of the mitochondrial isolation and purification is monitored by SDS-PAGE followed by immunoblotting analysis against proteins with specific subcellular localization (Fig. 4). A detailed description of these techniques is beyond the scope of this chapter and only a brief summary of the individual steps is given here. The reader is kindly referred to the instructions provided by the manufacturers of the gel electrophoresis and immunoblotting equipment, as well as to the articles of Laemmli *(11)* for SDS-PAGE and Towbin et al. *(12)* for immunoblotting analysis, respectively.

1. SDS-PAGE gels are casted and polymerized according to standard recipes. 12% T gels are routinely used. For the detection of proteins with a molecular weight below 20 kDa 15% T gels are used.
2. Equal amounts of the protein extracts (typically 25 μg per lane) of the samples taken from the individual purification steps (Fig. 1) are separated by SDS-PAGE.
3. After the SDS-PAGE proteins are electrophoretically transferred to PVDF membranes using standard blotting equipment. To assess for correct transfer to the PVDF membranes SDS-PAGE separated proteins are subsequently stained with Ponceau S.
4. PVDF membranes are destained, unspecific binding sites are blocked, incubated e.g. with the primary antibody solutions given in Table 1 and subsequently developed using the appropriate secondary antibodies and for example ECL staining.

As result, with the progress of the isolation and purification steps, a significant increase in the amount of mitochondrial proteins (TIM 23, CYT C, VDAC) is paralleled by a decrease in the amount of lysosomal (LAMP 1) or microsomal (SEC 61, BIP) proteins (Fig. 4).

3.3.2. Electron Microscopy

The homogeneity and overall structural integrity of the ZE-FFE purified mitochondria is assessed by electron microscopy analysis of ultra thin slices of pelleted mitochondrial preparations (Fig. 5). A detailed description of this analysis is beyond the scope of this chapter and the reader is kindly referred to the instructions provided by the original literature, e.g., the article of Lewis and Shute *(13)*.

4. Notes

1. Ultra-Clear™ tubes (16×76 mm, Beckman Instruments, Inc., CA, USA) may be used as this facilitates the visual inspection and fractionation after the density gradient centrifugation.
2. Buffers and solutions used for ZE-FFE should be prepared with ultra pure H_2O. In addition it is advisable to precool the FFE buffers and to keep them on ice during the separation to avoid strong temperature gradients in the separation chamber.
3. New electrophoresis membranes must be reswollen by overnight incubation in a 1:1 mixture of glycerol/isopropanol. Membranes should always be wet and handled with care. The wet membranes should be assembled in correct orientation. The smooth side (i.e., the membrane) should face the electrode chamber seal; the rough side (i.e., the paper backing) should face the filter paper strip. In ZE-FFE 0.5 mm spacers should be used and 0.8-mm filter paper strips.
4. ZE-FFE is a support and carrier free electrophoresis. This implies that any disturbance of the laminar buffer stream through the separation device has to be carefully avoided. It is therefore strongly recommended to control the flow profile after the setup with the "stripes test" as described by the instrument's manufacturer on a daily basis. If this test fails, a consequent unsatisfactory separation is very likely and remedies should be taken before starting with the separation.
5. Forced homogenization should be avoided, because mitoplasts (i.e., mitochondria without outer membranes) are generated by such treatments *(2)*.
6. Resuspension of the crude mitochondrial pellets obtained by differential centrifugation should be done carefully to avoid lysosomal and consequent mitochondrial damage (*see* also Fig. 6). As further precautionary measure it is advisable to resuspend the pellet in TME-PIC to inhibit released proteolytic activities. It may also be possible to use BSA (e.g., 0.1–1% w/v) as protective measure. One should keep in mind however, that the presence of high amounts of added proteins may subsequently alter the electrophoretic deflection of the organelles because of surface binding *(3)*.

7. If the aim is to purify lysosomes from rat liver (*see* also below), the washing step 3.1.7 should be repeated two or three more times to remove microsomes as completely as possible *(2)*.
8. It is advisable to estimate the duration of a subsequent ZE-FFE purification of the isolated mitochondria and to limit the amount of resuspended organelles to a quantity which can be processed in a reasonable time. An upper limit may be 10 mg of mitochondrial sample applied to ZE-FFE per hour.
9. Important safety note: the electrical current occurring during separation is sufficient enough to cause severe bodily harm. It is strongly advised that, during an electrophoresis run, only visual inspection should take place. In case of problems, power should be switched off first!
10. If the introduced sample contains an unusually high amount of broken and severely damaged organelles, large scale clotting might occur and the whole procedure will fail. It is therefore important to handle the mitochondria with care and samples should be kept on ice during the manipulations. Especially resuspension of pelleted samples should be done carefully and strong shear forces should be avoided. If clotting is visible in the starting sample for ZE-FFE a short spin (e.g., 500*g* for 3 min) can remove larger aggregates. The introduction of air bubbles, e.g., when media are exchanged, should be avoided during the run.
11. The progress of the separation should be monitored during the run. A fast and direct way is to repeatedly transfer equal volumes (e.g., 200 µl) of the collected fractions to a 96-well plate with a multichannel pipet and measure the optical density with an appropriate reader. Routine measurements at 260 nm can be done in appropriate UV-transparent 96-well plates (e.g., Microtest™ 96-well Clear Plates, BD Falcon). Other parameters might be determined as well, e.g., the protein content or enzymatic activities like succinate dehydrogenase for mitochondria or acid phosphatase for lysosomes *(7)*.
12. To ensure a stable separation pattern over the whole sample application time, the sample volume as well as the media should be regularly checked during the separation run. The fractionation plate should be inspected for blocked tubes during the run and the introduction of air bubbles should be strictly avoided.
13. As can be seen in Fig. 5B–D mitochondria with a very high degree of purity can be isolated by this protocol. Fig. 5E–F show that lysosomes may be isolated parallel to the mitochondrial fraction by the current protocol. However, as can be seen in Fig. 4 (lane 7), such preparations do contain microsomes and, to a minor extent, mitochondria or submitochondrial structures like outer mitochondrial membranes. The herein described workflow, however, is a flexible approach which can be modified and adapted towards the isolation and purification of lysosomes. Thereto the washing step 3.1.7 in the protocol should be repeated twice before the density gradient centrifugation to remove microsomes more efficiently and to proceed with the density gradient centrifugation as described *(6,7)*. As alternative Stahn et al. have omitted the density gradient (Section 3.1.9) and directly applied the thoroughly washed (at least 3–4 times) crude lysosomal

pellet to ZE-FFE *(3)*. Their approach takes advantage of the more negative surface charge of lysosomes compared to mitochondria, which consequently migrate more towards the anode in ZE-FFE (*see* also Section 3.2).

Acknowledgements

The authors would like to thank Drs. E. E. Rojo and G. Greif for critical reading of the manuscript. HZ would especially like to thank Dr. M. Ueffing for his support.

References

1. Hannig, K. and Heidrich, H. G. (1974) The use of continuous preparative free-flow electrophoresis for dissociating cell fractions and isolation of membranous components. Meth. Enzymol. 31, 746–61.
2. Heidrich, H. G., Stahn, R., and Hannig, K. (1970) The surface charge of rat liver mitochondria and their membranes. Clarification of some controversies concerning mitochondrial structure. J. Cell Biol. 46(1), 137–50.
3. Stahn, R., Maier, K. P., and Hannig, K. (1970) A new method for the preparation of rat liver lysosomes. Separation of cell organelles of rat liver by carrier-free continuous electrophoresis. J. Cell Biol. 46(3), 576–91.
4. Hannig, K. and Wrba, H. (1964) Isolation of Vital Tumor Cells by Carrier-Free Electrophoresis. Z. Naturforsch. B. 19, 860.
5. Murayama K., Fujimura T., Morita, M., and Shindo, N. (2001) One-step subcellular fractionation of rat liver tissue using a Nycodenz density gradient prepared by freezing-thawing and two-dimensional sodium dodecyl sulfate electrophoresis profiles of the main fraction of organelles. Electrophoresis 22(14), 2872–80.
6. Klein, D., Lichtmannegger, J., Heinzmann, U., and Summer, K. H. (1998) Association of copper to metallothionein in hepatic lysosomes of Long-Evans cinnamon (LEC) rats during the development of hepatitis. Eur. J. Clin. Invest. 28(4), 302–10.
7. Klein, D., Lichtmannegger, J., Heinzmann, U., Muller-Hocker, J., Michaelsen, S., and Summer, K. H. (2000) Dissolution of copper-rich granules in hepatic lysosomes by D-penicillamine prevents the development of fulminant hepatitis in Long-Evans cinnamon rats. J. Hepatol. 32(2), 193–201.
8. Fuller, K.M. and Arriaga, E.A. (2004) Capillary electrophoresis monitors changes in the electrophoretic behavior of mitochondrial preparations. J. Chromatogr. B 806(2), 151–9.
9. Zischka, H., Weber, G., Weber, P. J., Posch, A., Braun, R. J., Buhringer, D., Schneider, U., Nissum, M., Meitinger, T., Ueffing, M., and Eckerskorn, C. (2003) Improved proteome analysis of *Saccharomyces cerevisiae* mitochondria by free-flow electrophoresis. Proteomics 3(6), 906–16.
10. Weber, G. and Bocek, P. (1996) Optimized continuous flow electrophoresis. Electrophoresis 17(12), 1906–10.

11. Laemmli, U. K. (1970) Cleavage of structural proteins during the assembly of the head of bacteriophage T4. Nature 227(5259), 680–5.
12. Towbin, H., Staehelin, T., and Gordon, J. (1979) Electrophoretic transfer of proteins from polyacrylamide gels to nitrocellulose sheets: procedure and some applications. Proc. Natl. Acad. Sci. U S A 76(9), 4350–4.
13. Lewis, P. R. and Shute, C.C. (1969) An electron-microscopic study of cholinesterase distribution in the rat adrenal medulla. J. Microsc. 89(2), 181–93.
14. Ericson, I. (1974) Determination of the isoelectric point of rat liver mitochondria by cross-partition. Biochim. Biophys. Acta 356(1), 100–7.

27

Isolation of Proteins and Protein Complexes by Immunoprecipitation

Barbara Kaboord and Maria Perr

Summary

Immunoprecipitation (IP) uses the specificity of antibodies to isolate target proteins (antigens) out of complex sample mixtures. Three different approaches for performing IP will be discussed; traditional (classical) method, oriented affinity method and direct affinity method. The traditional method of incubating the IP antibody with the sample and sequentially binding to Protein A or G agarose beads (resin) facilitates the most efficient target antigen recovery. However, this approach results in the target protein becoming contaminated with the IP antibody that can interfere with downstream analyses. The orientated affinity method uses Protein A or G beads to serve as an anchor to which the IP antibody is crosslinked thereby preventing the antibody from co-eluting with the target protein. Similarly, the direct affinity method also immobilizes the IP antibody except in this case it is directly attached to a chemically activated support. Both methods prevent co-elution of the IP antibody enablings reuse of the immunomatrix. All three approaches have unique advantages and can also be used for co-immunoprecipitation to study protein:protein interactions and investigate the functional proteome.

Key Words: Aldehyde activated support; antibody contamination; co-immunoprecipitation; immune complex; immunoaffinity support; immunoprecipitation; Protein A; Protein G; protein-protein interactions; protein purification.

1. Introduction

Many scientists rely on immunoprecipitation as a method for isolating small amounts of antigen or target protein from complex samples such as cell lysates (*1,2*), serum (*3–5*) and tissue homogenates (*6,7*). IP can be used to evaluate the differential expression of a protein and to characterize the protein's molecular

From: *Methods in Molecular Biology, vol. 424: 2D PAGE: Sample Preparation and Fractionation, Volume 1*
Edited by: A. Posch © Humana Press, Totowa, NJ

weight, posttranslational modifications, and interacting ligands. The antibodies used may be polyclonal or monoclonal and may recognize the protein of interest, a particular posttranslational modification *(8–10)*, or an epitope tag *(11,12)* if the protein is overexpressed.

The traditional (or classical) method for performing an IP requires incubating the antibody first with the sample containing the target protein (antigen). After the antigen-antibody complex is formed, it is bound to Protein A or Protein G beads (typically cross-linked agarose) via the Fc region of the IP antibody. The beads and the sample are centrifuged to pellet the captured immune complex, the supernatant discarded and the beads washed and centrifuged again to remove any unbound proteins *(13)*. For high-throughput applications, immobilized Protein A or G magnetic beads can be used to facilitate separation of the isolated immune complex from the remainder of the sample. The resin pellet is then exposed to denaturing conditions or low-pH conditions to dissociate the complex. The released proteins are typically analyzed via one-dimensional (1D) *(14,15)* or two-dimensional gel (2D) *(16)* electrophoresis followed by mass spectrometry *(15,16)* or immunoblotting *(17,18)*.

Both the antibody and target protein are released in the final step of the traditional IP technique. This can present a problem for 1D or 2D analysis of the target protein when the molecular weight is similar to the heavy chain or light chain of the IP antibody *(19)*. For this reason alternative methods have been developed to prevent the antibody from contaminating the target protein including crosslinking the IP antibody to Protein A or Protein G beads *(20,21)* or directly coupling the antibody to an activated support to create an immunoaffinity resin *(22–24)*.

Each of these IP strategies has also become an effective tool in studying protein:protein interactions *(19,25–28)*. Co-immunoprecipitation (Co-IP) uses the antigen–antibody complex to isolate unknown proteins bound to the antigen ("bait-prey" complex). Co-IP is a valuable *in vitro* tool to verify receptor-ligand or enzyme-substrate interactions, to identify multiprotein complex formation and to confirm yeast two-hybrid results *(11,12)*. As in IP, identifying and characterizing these unknown interacting proteins can be hampered by the presence of contaminating antibody heavy and light chains. Therefore, the advantages of the antibody immobilization methods outlined here also readily apply to Co-IP experiments. All three strategies for performing IP and Co-IP studies discussed in this review can be found in a variety of commercially available kits.

2. Materials (see Note 1)

1. Protein G agarose (50% slurry) from Thermo Fisher Scientific (Rockford, IL). Resin capacity is typically 15 mg antibody bound per milliliter of resin bed.
2. Dulbecco's phosphate-buffered saline (PBS); 8 mM NaHPO, mMKHPO, 140 mM NaCl, 10 mM KCl, pH 7.4.

3. Tris-buffered Saline (TBS): 25 mM Tris-HCl, pH 7.2, 150 mM NaCl.
 4. Elution Buffer: 0.1 M glycine, pH 2.8, (pH adjusted with HCl).
 5. Neutralization Buffer: 1.0 M Tris-HCl, pH 9.5.
 6. Disuccinimidyl suberate (DSS) from Thermo Fisher Scientific (Rockford, IL): Dissolve 2 mg DSS in 150 μl DMSO or DMF immediately before use. Once reconstituted, the DSS must be used immediately. DSS is moisture sensitive and must be stored desiccated.
 7. AminoLink® Plus Coupling Gel (50% slurry) from Thermo Fisher Scientific (Rockford, IL). Store at 4 °C.
 8. Sodium cyanoborohydride: 5.0 M. Sodium cyanoborohydride is highly toxic and should be prepared and used in a fume hood.
 9. Spin Columns (such as the Handee™ Spin Columns) from Thermo Fisher Scientific (Rockford, IL). Volume capacity per column is 900 μl.
 10. Microcentrifuge tubes.
 11. 5X SDS-PAGE Sample Buffer: 0.2 M Tris-HCl, pH 6.8, 10% (w/v) SDS, 10 mM DTT, 20% glycerol, 0.05% (w/v) bromophenol blue.

3. Methods

The basic procedure for any immunoprecipitation (IP) or coimmunoprecipitation (CoIP) involves extracting antigen from cells or tissue samples in an appropriate lysis buffer, incubating the resulting lysate with antibody to allow formation of an immune complex, and precipitating or capturing those complexes on a solid phase support, typically immobilized Protein A or Protein G crosslinked agarose beads. The decision to use Protein A or Protein G depends upon the source and class of the antibody used to IP and the overall cost. Immobilized Protein A is generally less expensive than Protein G however Protein G binds a wider variety of antibody classes (Table 1).

Three immunoprecipitation procedures will be described, varying by the type of affinity matrix used to capture the target antigen(s). Traditional or classical IP involves the use of Protein A or G beads to capture the antibody:antigen complex. Protein A or G binds to the F region of most antibody classes and most species. This enables efficient, unhindered binding of the antigen to the variable sites of the antibody molecule (Fig. 1A). Problems with the traditional method include contamination of the target protein with the IP antibody, the destruction of costly IP antibody and sample loss incurred when pipeting/removing supernatants from small resin pellets. Contamination is most problematic if the antigen or its interacting partner has similar molecular weights to the antibody heavy and light chains because these bands will co-migrate during SDS-PAGE run under reducing conditions.

Alternatively, the IP antibody can be covalently crosslinked to the immobilized Protein A or G resin (oriented affinity method) before addition of the

Table 1
Comparison of antibody-binding characteristics for Protein A and Protein G

Species	Antibody class	Protein A	Protein G
Human	Total IgG	S	S
	IgG_1	S	S
	IgG_2	S	S
	IgG_3	W	S
	IgG_4	S	S
	IgM	W	NB
	IgD	NB	NB
	IgA	W	NB
	Fab	W	W
	ScFv	W	NB
Mouse	Total IgG	S	S
	IgM	NB	NB
	IgG_1	W	M
	IgG_{2a}	S	S
	IgG_{2b}	S	S
	IgG_3	S	S
Rat	Total IgG	W	M
	IgG_1	W	M
	IgG_{2a}	NB	S
	IgG_{2b}	NB	W
	IgG_{2c}	S	S
Cow	Total IgG	W	S
	IgG1	W	S
	IgG2	S	S
Goat	Total IgG	W	S
	IgG_1	W	S
	IgG_2	S	S
Sheep	Total IgG	W	S
	IgG_1	W	S
	IgG_2	S	S
Horse	Total IgG	W	S
	IgG(ab)	W	NB
	IgG(c)	W	NB
	IgG(T)	NB	S
Rabbit	Total IgG	S	S
Guinea Pig	Total IgG	S	W

Table 2
(Continued)

Species	Antibody class	Protein A	Protein G
Pig	Total IgG	S	W
Dog	Total IgG	S	W
Cat	Total IgG	S	W
Chicken	Total IgY	NB	NB

Legend: W = weak binding NB = no binding M = medium binding S = strong binding

antigen-containing sample (lysate, serum, or homogenate). Like the traditional IP method, this strategy has the advantage of orienting the capture antibody via the Fc region for sterically unhindered antigen binding (Fig. 1B). Before sample addition, the antibody is crosslinked to Protein A or G using a common bifunctional crosslinker such as DSS resulting in covalent attachment of the antibody to Protein A or G. Because the native antibody is retained on the resin, there is an added benefit of being able to reuse the antibody-Protein A or G resin, conserving valuable antibody (Fig. 2). The one caveat of this oriented affinity method is that DSS will react with any exposed amine, including any lysines

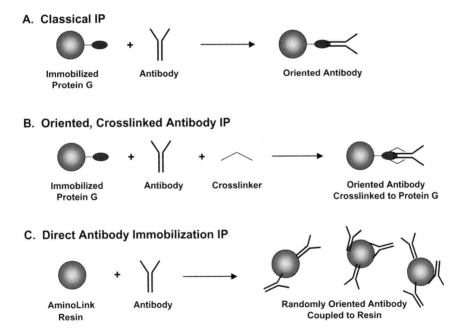

Fig. 1. Strategies for preparing immunoprecipitation matrixes.

that may be in the antigen-binding site of the antibody. These side reactions can be minimized with careful titration of the DSS crosslinker, however, they cannot be completely eliminated. Therefore, a decrease in overall antigen-binding efficiency may be observed under certain conditions (*see* Note 2).

Another way to prevent antibody heavy and light chains from contaminating the eluted fraction is to directly immobilize the antibody to an activated beaded support, eliminating the need for Protein A or G altogether (Fig. 1C). This method is especially beneficial when using antibodies that do not bind strongly to Protein A or G such as chicken antibodies and mouse IgG_1 (Fig. 3). This direct affinity strategy, however, results in a random orientation of the antibody on the beads potentially causing a reduction in antibody binding efficiency under certain conditions (*see* Note 2). However, like the oriented affinity method, the eluted fraction is not contaminated with antibody heavy and light chains, and the resin may be reused several times.

Choosing an IP strategy will depend upon a variety of factors including IP antibody class, the host species in which it was raised and the purity of

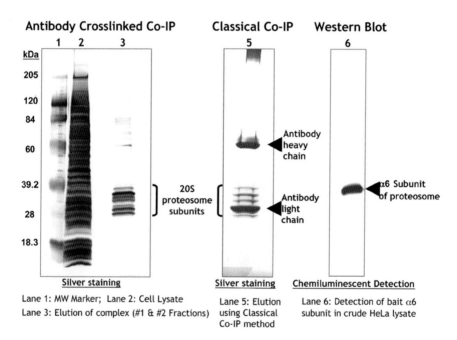

Fig. 2. Comparison of Co-IP results obtained with the Classical and the Crosslinked Oriented Antibody approaches. Mouse monoclonal preproteosome subunit α6 antibody (100 μl) and 100 μl Protein G resin were used for both Co-IPs. HeLa ($\sim 5 \times 10^7$ cells) lysate was precleared with Protein G agarose and the Co-IP was performed at 4 °C overnight. (Adapted from *(19)*.)

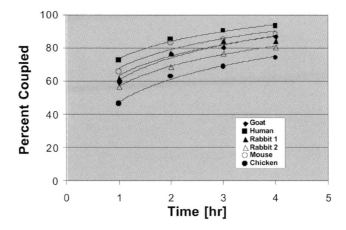

Fig. 3. Antibody coupling efficiency using the direct antibody immobilization method. Purified antibody (200 μg) from various species was coupled to 200 μl of AminoLink® Coupling Gel (settled gel) at room temperature for 4 h. For the chicken antibody, 500 μg was used. (Adapted from *(19)*.)

the antibody chosen along with its availability and cost. These factors will be discussed in further detail at the beginning of each subsection.

Regardless of the IP method chosen, each method can be made more efficient through the use of spin columns to separate the bound and free fractions *(29)*. Typically, IPs and Co-IPs are performed using a "batch" process with the sample and resin in microcentrifuge tubes. Careful pipeting is required to remove the supernatant, wash volumes and eluted fraction from the resin pellet to prevent aspiration of the small resin bed volume. Also, if some of the supernatant is left behind, the unbound protein can be spread over more fractions. Using spin columns helps eliminate these pipeting errors. Commercially available spin columns (0.5–1.0 ml size) have frits that retain the resin beads and any bound proteins while the flow through, wash, and elution fractions are collected into standard microcentrifuge tubes. Although each of these strategies can be adapted to the standard technique of pelleting resin directly in microcentrifuge tubes, the procedures described herein will use the spin column format with cell lysate as the sample.

3.1. Immobilized Protein A or G Equilibration (Classical or Traditional IP)

Traditionally, the immune complex is formed by adding antibody directly to the lysate (sample) followed by capture of the complexes onto immobilized Protein A or G (*see* Note 3). Alternatively, the antibody can be bound first to the Protein A or G resin and then exposed to the source of antigen. The latter sequence is especially useful when crude serum, ascites or hybridoma culture supernatant is the source of the primary antibody because this purifies the IgG fraction and removes contaminating proteins before adding the antigen-containing lysate. Elution of the bound antigen will result in co-elution of the IP antibody, preventing reuse of the antibody. However, the Protein A or G resin can be reused if desired.

1. Gently swirl the Protein G resin to obtain an even suspension. Dispense 20–50 μl of the immobilized Protein G (50% slurry) into a spin column. Centrifuge at 3,000–5,000g for 1 min to remove the storage solution (*see* Note 4).
2. Wash the resin by adding 0.4 ml PBS, gently invert several times to mix. Centrifuge and discard wash.
3. Add 5–20 μg purified antibody to a protein sample or lysate. Incubate the lysate (sample) with the antibody for 2 h to overnight at 4 °C with mixing (*see* Note 5).
4. Proceed to Section 2.4.

3.2. Crosslinking Antibody to Protein A or G (Oriented Affinity Method)

Crosslinking the antibody to Protein A or G beads results in a permanent affinity support with the antibody properly oriented to bind the target antigen(s). A common homobifunctional crosslinker such as disuccinimidyl suberate (DSS) reacts with amine groups present on both the antibody and the Protein A or G molecules to covalently link the molecules, preventing loss of the antibody during antigen elution. This method is a good choice when the amount of antibody is in short supply or expensive because the antibody:Protein A or G resin can be reused several times. This approach is preferred more than the direct antibody immobilization strategy when the antibody contains gelatin or BSA as a carrier protein because it still uses the affinity purification characteristics of Protein A or G.

1. Gently swirl the immobilized Protein G resin to evenly suspend the slurry. Dispense 0.4 ml of immobilized Protein G (50% slurry) into a spin column (*see* Note 6).
2. Centrifuge the resin at 3,000–5,000g for 1 min (*see* Note 4). Discard the flow through.
3. Wash the resin 2–3 times with 0.4 ml Dulbecco's PBS, being careful to fully resuspend the resin during each wash before centrifugation (*see* Note 7).
4. Incubate 100–200 μg of monoclonal or polyclonal antibody (*see* Note 5) in a volume of 0.3–0.4 ml with the 0.2 ml bed volume of washed protein G agarose at room temperature for 1 h.

5. Wash the resin three times with 0.5 ml PBS to remove unbound antibody. For each wash invert the tube 5–10 times before centrifugation at 3,000g for 1 min. Discard the washes (*see* Note 8).
6. Resuspend the resin in 0.4 ml PBS. Add 0.1 ml of 13 mg/ml DSS crosslinker freshly prepared in DMSO or DMF (*see* Note 9).
7. Incubate the crosslinker with antibody-Protein G resin for 1 h at room temperature.
8. Remove uncrosslinked antibody by washing 4 times with 0.5 ml 0.1 M glycine, pH 2.8. Glycine will also serve to inactivate any unreacted DSS crosslinker (*see* Note 10).
9. Immediately re-equilibrate resin by washing 3 times with 0.5 ml TBS.
10. Store resin as a 50% slurry in TBS containing 0.02% sodium azide at 4 °C.
11. Determine the antibody immobilization efficiency by reading the A_{280} of the flow through and first wash. Because Protein G is in molar excess over antibody, there should be very little A_{280} signal in the flow through or wash.
12. Use 20–100 µl of the DSS-crosslinked antibody resin slurry for each immunoprecipitation reaction (10–50 µl of settled gel). Centrifuge resin and wash with PBS to remove the storage buffer.
13. Proceed to Section 2.4.

3.3. Direct Antibody Immobilization to Resin (Direct Affinity Method)

In this method, the antibody is coupled directly to the agarose beads, eliminating the need for a crosslinker or Protein A or G. The direct strategy is more universal because it is not limited to species that interact strongly with Protein A or G. Because any amine-containing molecule is immobilized, this method works best with antibodies that are affinity-purified and free of carrier proteins. Proteins in crude preparations (serum, ascites, or hybridoma supernatant) will also be coupled unless the IgG fraction is isolated first (*see* Note 11). Because all surface-accessible amines are reactive, the antibody is captured in a variety of orientations. However, despite the random antibody orientation, the binding efficiency of this method is usually higher than the DSS crosslinking method, because there is no diffusible crosslinker used that could inactivate essential lysines in the antigen recognition sites. Like the DSS method, the immobilized antibody resin may be reused several times, conserving scarce or expensive antibody.

1. Gently swirl the AminoLink® Plus Coupling Gel to evenly suspend the slurry. Dispense 0.4 ml of AminoLink® Plus (50% slurry) into a spin column.
2. Centrifuge the resin at 3,000–5,000g for 1 min (*see* Note 4). Discard the flow through.
3. Wash the resin twice with 0.4 ml PBS, being careful to fully resuspend the resin during each wash before centrifugation.

4. Dilute 100–400 μg of the affinity-purified monoclonal or polyclonal antibody (*see* Note 5) in 400 μl PBS (*see* Notes 12 and 13).
5. Add the diluted antibody to the tube or spin column containing the washed AminoLink® Plus coupling gel.
6. In a fume hood, add 1 μl 5 M sodium cyanoborohydride for every 100 μl of diluted affinity-purified antibody being coupled. Close the cap of the tube and invert five times.
7. Incubate at room temperature for 4 h to overnight with mixing/inversion.
8. Remove any uncoupled antibody by washing the resin with 0.4 ml PBS. Invert several times to mix and centrifuge at 3,000–5,000g for 1 min. Remove the flow through and save.
9. Quench any unreacted sites on the AminoLink® Plus resin by adding 0.5 ml 1M Tris-HCl, pH 7.4. Invert the tube 10 times to mix. Centrifuge at 3,000–5,000g for 1 min. Discard the buffer supernatant or flow through.
10. Add 0.4 ml 1M Tris-HCl, pH 7.4, to the resin. In a fume hood add 4 μl 5 M sodium cyanoborohydride. Close the cap and invert five times to mix. Incubate for 30 min with gentle end-over-end mixing. Centrifuge and discard the supernatant or flow through.
11. Wash the resin 6 times with 0.4 ml 1M NaCl.
12. Wash the resin 3 times with 0.4 ml PBS. Store the antibody-coupled resin in 0.2 ml PBS containing 0.02% sodium azide at 4 °C (50% slurry).
13. Determine the antibody coupling efficiency by reading the A_{280} of the flow through and first wash. There should be very little A_{280} signal in the flow through or wash.
14. Use 20–100 μl of antibody-coupled resin slurry for each immunoprecipitation reaction.

3.4. Capture and Elution of the Immune Complex

1. For classical immunoprecipitation, add the immune complex prepared in Section 2.1.3 to the spin column containing the equilibrated Protein G and cap the tube. To the crosslinked or coupled antibody resins (Section 2.2.10 or 2.3.12), add 0.5–1.0 mg protein lysate to 10–50 μl of the prepared settled resin for each IP reaction (*see* Note 14). Incubate for 2 h to overnight at 4 °C. Centrifuge the tube and discard the flow-through or save it to analyze capture efficiency.
2. Add 0.5 ml of PBS into the spin column. Cap the tube and invert it 5–10 times. Centrifuge the tube and discard flow-through or save it in a separate tube for analysis. Repeat this step a least two additional times (*see* Note 15).
3. Place 10 μl 1M Tris-HCl, pH 9.5, into the microfuge tubes receiving the eluted fractions. The Tris buffer will serve to immediately neutralize the low pH eluent, minimizing exposure of the antigen to low pH conditions.
4. Add 190 μl of the 0.1M glycine, pH 2.8, elution buffer to the immune complex (*see* **Note 16**). Cap tube and invert it several times. Centrifuge at 3,000–5,000g for 1 min into the receiver tubes containing the neutralization buffer. Save the eluted fraction. Place the spin column into a new receiver tube containing neutralization buffer.

5. Repeat step 1 until desired sample is eluted. Sample should elute within the first three fractions. Do not pool fractions. Assess the first three fractions by SDS-PAGE.
6. To preserve activity of the IP resin for reuse, proceed to Section 2.5 immediately following the last elution step.

3.5. Resin Regeneration and Storage Conditions

1. Add 0.5 ml of PBS or TBS to the resin in the spin cup. Cap tube and invert it 10 times. Centrifuge the tube and discard flow-through. Repeat this step once.
2. Add 0.5 ml of PBS or TBS containing 0.02% sodium azide to the spin cup. Place the spin cup into a microcentrifuge tube, cap tube and wrap the cap with laboratory film to prevent gel from drying. Store at 4 °C (*see* Note 17).

3.6. Preparation of Samples for SDS-PAGE

1. Add 20 μl of the IP sample to a microcentrifuge tube.
2. Pipet 5 μl of the 5X Sample Buffer into the microcentrifuge tube.
3. Insert tube into a microcentrifuge tube holder and place it in boiling water. Boil sample for ∼5 min.
4. Allow sample to cool to room temperature and apply to the gel for electrophoresis.
5. Perform SDS-PAGE and analyze results by Coomassie stain, silver stain or western blotting.

4. Notes

1. Materials may be made or purchased individually. Alternatively, kits to perform the described methods are available commercially under the Seize® brand name (Thermo Fisher Scientific, Rockford, IL).
2. In the presence of abundant antigen quantities, a decrease in antigen-binding capacity will be observed with the oriented, crosslinked method as well as the direct antibody immobilization methods (Fig. 4). However, under *limiting* antigen conditions, no significant difference in antigen recovery is observed. Because limiting antigen is usually the norm, the sacrifice made in binding capacity by crosslinking or coupling the antibody to the resin is not apparent under most experimental conditions.
3. For optimal results, use an affinity-purified antibody. Although serum may be used, the antibody that is specific for the antigen of interest may comprise only 2–5% of the total IgG in the serum sample, which results in low antigen yields.
4. Centrifuging at speeds greater than 3,000–5,000g may cause the gel to clump and make resuspending the resin difficult.
5. The amount of antigen needed and the incubation time are dependent upon the antibody-antigen system used and require optimization for each specific system. On average, rabbit serum contains ∼10–12 mg/ml IgG and mouse contains

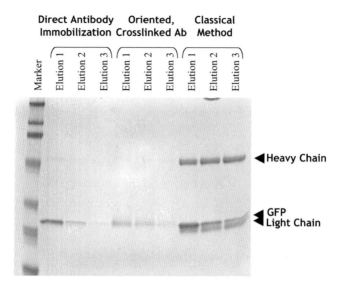

Fig. 4. Comparison of antigen binding efficiency of the three IP methods. Immunoprecipitations were performed on 140 µl of partially purified GFP using 135 µg of affinity-purified goat anti-GFP antibody on 100 µl of gel support (7–8-fold molar excess of antigen). Ten percent of the elution fractions were electrophoresed in 12% polyacrylamide gel and stained with GelCode® Blue Stain Reagent. (Adapted from *(19)*.)

~6–8 mg/ml IgG. Some species may contain up to 30 mg/ml IgG. Ascites fluid contains an average total IgG of ~1–10 mg/ml. Tissue culture supernatant (serum-free medium) contains an average of 0.05 mg/ml IgG. This amount may be too dilute to effectively capture enough antigen for analysis. For best results concentrate cell culture supernatants to at least 500 µg/ml before forming the immune complex.

6. Less immobilized Protein A or G may be used (20–100 µl), however, the amount of DSS should then be scaled proportionately.
7. Alternative buffers may be substituted in place of PBS for the binding and wash steps, provided no primary amines (e.g., Tris, glycine, etc.) are present, the pH is between 7.0 and 8.5 and the salt concentration is not greater than 0.25 M.
8. The flow-through or supernatant can be saved to estimate the amount of antibody bound to the Protein G by comparing its A_{280} reading to that of the antibody sample applied. This calculation can only be performed if the antibody was purified and did not contain a carrier protein such as BSA.
9. If it is known that epitope recognition is mediated by lysine residues in the antibody's antigen recognition site, it may be necessary to titrate the amount of DSS used to minimize antibody inactivation. This is accomplished by performing a 2-fold dilution series of DSS during the antibody immobilization reaction.

Table 2
Amount of AminoLink® Plus coupling gel to use per volume of antibody.*

50% Gel slurry amount	Settled gel amount	Antibody amount	Antibody volume
400 µl	200 µl	100–400 µg	400 µl
200 µl	100 µl	50–200 µg	200 µl
100 µl	50 µl	25–100 µg	200 µl

*Using these guidelines results in a coupling efficiency of ∼85% after 4 h. Coupling overnight is suggested when using more than 400 µg of antibody.

    Because DSS is a hydrophobic molecule, a microprecipitate may form when it is added to an aqueous medium, which results in a cloudy appearance. The reaction will still proceed efficiently and may disappear during conjugation.

10. Glycine at pH 2.8 will elute IgG that is not covalently coupled to the immobilized Protein A or G. Most polyclonal and most monoclonal antibodies can tolerate low pH conditions for short durations. If the antibody immobilized is intolerant of pH conditions 2.5–3.0, wash the resin with a high salt, neutral pH buffer to remove uncoupled antibody instead.

11. Gelatin or carrier protein (e.g., BSA) in the antibody solution will compete for coupling sites. Remove gelatin and carrier proteins by performing a Protein A or Protein G purification and subsequent dialysis into PBS. Alternatively, use the DSS crosslinking procedure to immobilize the antibody to Protein A or G agarose.

12. Amines in the antibody solution (e.g., Tris or glycine) will compete for coupling sites. Remove amines before coupling by dialysis or desalting columns.

13. For optimal results, first-time users should couple 200 µg of affinity-purified antibody to 200 µl of settled gel (Table 2) and perform immunoprecipitations with decreasing amounts of antibody-coupled gel.

14. The importance of running appropriate controls for each method cannot be underestimated *(30)*. Because proteins from the sample may bind to the resin support, it is best if the sample is precleared by incubating the antigen source (protein sample) with Protein G (or A) resin before incubation with the antibody-bound resin. If using the direct immobilization method (Section 2.3), preclear the protein sample by incubating with Tris-quenched AminoLink® resin. Additional controls include running a "no antibody" IP reaction or performing a mock IP reaction using an antibody known to not cross-react with the proteins of interest. If the protein sample source is a transfected lysate, a mock transfected lysate (empty vector) can be used as another negative control.

15. To avoid contamination from residual proteins, before eluting the purified material verify that the gel has been thoroughly washed by performing a protein assay (i.e., A , CoomassiePlus or BCA™ Protein Assay; Thermo Fisher Scientific, Rockford, IL) on the flow-through from the final wash. There should be minimal protein in the final wash fraction. If the material has not been

adequately washed, repeat wash steps before proceeding. Additional washes may be necessary for samples containing high protein concentrations.

16. Alternatively, if the protein or antibody is sensitive to low pH, use a neutral pH system such as ImmunoPure® Gentle Binding and Elution Buffers (Thermo Fisher Scientific, Rockford, IL).
17. Once the antibody has been immobilized, the affinity support may be reused 2–10 times depending upon the stability of the specific antibody.

Acknowledgments

The authors would like to thank Dr. Ling Ren and Daryl Emery for valuable experimental contributions during method development.

References

1. Prinetti, A., Prioni, S., Chigorno, V., Karagogeos, D., Tettamanti, G. and Sonnino. S. (2001) Immunoseparation of sphingolipid-enriched membrane domains enriched in Src family protein tyrosine kinases and in the neuronal adhesion molecule TAG-1 by anti-GD3 ganglioside monoclonal antibody. *J. Neurochem.* **78**,1162–7.
2. Sargsyan, E., Baryshev, M., Szekely, L., Sharipo, A. and Mkrtchian, S. (2002) Identification of ERp29, an endoplasmic reticulum lumenal protein, as a new member of the thyroglobulin folding complex. *J. Biol. Chem.* **277**, 17009–15.
3. Heegaard, N. H., Hansen, M. Z., Sen, J. W., Christiansen, M. and Westermark, P. (2006) Immunoaffinity chromatographic and immunoprecipitation methods combined with mass spectrometry for characterization of circulating transthyretin. *J. Sep. Sci.* **29**, 371–7.
4. Fujita, N., Kaito, M., Tanaka, H., Horiike, S., Urawa, N., Sugimoto, R., Konishi, M., Watanabe, S. and Adachi, Y. (2006) Hepatitis C virus free-virion and immune-complex dynamics during interferon therapy with and without ribavirin in genotype-1b chronic hepatitis C patients. *J. Viral Hepat.* **13**,190–8.
5. Goedert, J.J., Rabkin, C. S. and Ross, S. R. (2006) Prevalence of serologic reactivity against four strains of mouse mammary tumor virus among US women with breast cancer. *Br. J. Cancer* **94**, 548–51.
6. Rascon, A., Lindgren, S., Stavenow, L., Belfrage, P., Andersson, K. E., Manganiello, V.C. and Degerman, E., (1992) Purification and properties of the cGMP-inhibited cAMP phosphodiesterase from bovine aortic smooth muscle. *Biochim. Biophys. Acta.* **1134**,149–56.
7. Stam, H. and Hulsmann, W. C. (1984) Effects of hormones, amino acids and specific inhibitors on rat heart heparin-releasable lipoprotein lipase and tissue neutral lipase activities during long-term perfusion. *Biochim. Biophys. Acta.* **794**, 72–82.

8. Doppler, H., Storz, P., Li, J., Comb, M.J., and Toker, A. (2005) A phosphorylation state-specific antibody recognizes Hsp27, a novel substrate of protein kinase D. *J. Biol. Chem.* **280**, 15013–19.
9. Grønborg, M., Kristiansen, T.Z., Stensballe, A., Andersen, J.S., Ohara, O., Mann, M., Jensen, O.N., and Pandey, A. (2002) A mass spectrometry-based proteomic approach for identification of serine/threonine-phosphorylated proteins by enrichment with phospho-specific antibodies. *Mol. Cell. Proteomics* **1**, 517–27.
10. Schneider, R., Bannister, A.J., Myers, F.A., Thorne, A.W., Crane-Robinson, C., and Kouzarides, T. (2004) Histone H3 lysine 4 methylation patterns in higher eukaryotic genes. *Nature Cell Biol.* **6**, 73–7.
11. Monferran, S., Paupert, J., Dauvillier, S., Salles, B., and Muller, C. (2004) The membrane form of the DNA repair protein Ku interacts at the surface with metalloproteinase 9. *EMBO J.* **23**, 3758–68.
12. Marcora, E., Gowan, K., and Lee, J.E. (2003) Stimulation of NeuroD activity by huntingtin and huntingtin-associated proteins HAP1 and MLK2. *Proc. Natl. Acad. Sci. USA* **100**, 9578–83.
13. Harlow, E. and Lane, D. (ed.) (1999). *Using Antibodies: A Laboratory Manual.* Cold Spring Harbor Laboratory, Cold Spring Harbor, New York.
14. Schilling, B., Murray, J., Yoo, C. B., Row, R. H., Cusack, M. P,, Capaldi, R. A. and Gibson. B. W. (2006) Proteomic analysis of succinate dehydrogenase and ubiquinol-cytochrome c reductase (Complex II and III) isolated by immunoprecipitation from bovine and mouse heart mitochondria. *Biochim. Biophys. Acta.* **1762**, 213–22.
15. Gridley, S., Lane, W. S., Garner, C. W. and Lienhard, G. E. (2005) Novel insulin-elicited phosphoproteins in adipocytes. *Cell Signal.* **17**, 59–66.
16. Barnouin, K. (2004) Two-dimensional gel electrophoresis for analysis of protein complexes. *Methods Mol. Biol.* **261**, 479–98.
17. Faber, E. S., Sedlak, P., Vidovic, M. and Sah, P. (2006) Synaptic activation of transient receptor potential channels by metabotropic glutamate receptors in the lateral amygdala. *Neuroscience* **137**, 781–94.
18. Anzai, N., Miyazaki, H., Noshiro, R., Khamdang, S., Chairoungdua, A., Shin, H. J., Enomoto, A., Sakamoto, S., Hirata, T., Tomita, K., Kanai, Y. and Endou, H. (2004) The multivalent PDZ domain-containing protein PDZK1 regulates transport activity of renal urate-anion exchanger URAT1 via its C terminus. *J. Biol Chem.* **279**, 45942–50.
19. Qoronfleh, M. W., Ren, L., Emery, D., Perr, M. and Kaboord, B. (2003) Use of immunomatrix methods to improve protein—protein interaction detection. *J. Biomed. Biotechnol.* **2003**, 291–8.
20. Sisson, T.H. and Castor, C.W. (1990). An improved method for immobilizing IgG antibodies on protein A-agarose. *J. Immunol. Meth.* **127**, 215–20.
21. Khundmiri, S. J., Rane, M. J., Lederer, E. D. (2003) Parathyroid hormone regulation of type II sodium-phosphate cotransporters is dependent on an A kinase anchoring protein. *J. Biol. Chem.* **278**, 10134–41.

22. Domen, P. L., Nevens, J. R., Mallia, A. K., Hermanson, G. T. and Klenk, D. C. (1990) Site-directed immobilization of proteins. *J. Chromatogr.* **510**, 293–302.
23. Hermanson, G.T., Mallia, A.K., and Smith, P. K. (ed.) (1992). *Immobilized Affinity Ligand Techniques.* Academic Press, Inc., San Diego, CA.
24. Seko, Y., Azmi, H., Fariss, R. and Ragheb, J. A. (2004) Selective cytoplasmic translocation of HuR and site-specific binding to the interleukin-2 mRNA are not sufficient for CD28-mediated stabilization of the mRNA. *J Biol Chem.* **279**, 33359–67.
25. Chen, Y. H. and Lu, Q. (2003) Association of nonreceptor tyrosine kinase c-yes with tight junction protein occludin by coimmunoprecipitation assay. *Methods Mol. Biol.* **218**, 127–32.
26. Ransone, L. J. (1995) Detection of protein-protein interactions by coimmunoprecipitation and dimerization. *Methods Enzymol.* **254**, 491–7.
27. Li, Z., Zou, C. B., Yao, Y., Hoyt, M. A., McDonough, S., Mackey, Z. B., Coffino, P. and Wang, C. C. (2002) An easily dissociated 26 S proteasome catalyzes an essential ubiquitin-mediated protein degradation pathway in Trypanosoma brucei. *J. Biol. Chem.* **277**, 15486–98.
28. Loven, M. A., Muster, N., Yates, J. R. and Nardulli A. M. A novel estrogen receptor alpha-associated protein, template-activating factor Ibeta, inhibits acetylation and transactivation. *Mol. Endocrinol.* **17**, 67–78.
29. Brymora, A., Cousin, M.A., Roufogalis, B.D., and Robinson, P.J. (2001) Enhanced protein recovery and reproducibility from pull-down assays and immunoprecipitations using spin columns. *Anal. Biochem.* **295**, 119–22.
30. Howell, J.M., Winstone, T.L., Coorssen, J.R., and Turner, R.J. (2006) An evaluation of *in vitro* protein-protein interaction techniques; Assessing contaminating background proteins. *Proteomics* **6**, 2050–69.

28

Isolation of Phosphoproteins

Lawrence G. Puente and Lynn A. Megeney

Summary

Phosphorylation is one of the most abundant post-translational modifications on protein and one that frequently has functional biological consequences. For this reason, screening protein samples for phosphorylations has become an important tool in biochemical research. Affinity purification by immunological or chemical reagents can be used to isolate phosphoproteins from other cellular materials.

Key Words: Affinity purification; immunoprecipitation; protein phosphorylation.

1. Introduction

The detection of protein phosphorylation is a rapidly evolving field encompassing many possible approaches. These approaches can be broadly grouped into four categories: labeling, enriching, mass spectrometry, and indirect methods. Labeling approaches use reagents that tag phosphorylated proteins with a detectable marker. Examples include Western blotting, *in vivo* labeling with ^{32}P or ^{33}P, beta-elimination and Michael addition chemistry *(1)*, and the use of phosphoprotein-staining fluorescent dyes *(2,3)*. Affinity methods include immunoprecipitation, immobilized metal affinity chromatography, and proprietary phosphoprotein-binding materials. Mass spectrometry techniques such as precursor ion scan *(4)* or neutral loss scan *(5)* can be used to search for specific mass spectrometric features of phosphoproteins. Finally, indirect measures of protein phosphorylation such as sensitivity to phosphatase treatment can aid in the detection of phosphoproteins.

Iron (Fe^{3+}) and other similar metal ions are known to exhibit affinity for phosphoproteins, phosphopeptides, and phosphoamino acids *(6,7)*. The process of extracting chemicals via metal binding is termed immobilized metal affinity chromatography (IMAC). In IMAC of phosphoproteins and phosphopeptides, the strength of binding generally increases with more phosphorylations that can skew results towards the recovery of highly phosphorylated species. Importantly, negatively charged but nonphosphorylated proteins and peptides can also exhibit metal affinity. In some experimental systems, capping of nonphosphate negative charges by methylation has been introduced to render IMAC more selective for phosphorylation *(8)*. In the phosphoprotein field most interest in IMAC has focused on the selective enrichment of phosphopeptides from peptide mixtures *(8,9)* and the use of IMAC for the selective isolation of intact phosphoproteins for gel-based workflows has been limited. Recently however an IMAC method was introduced that is reported to be compatible with one-dimensional (1D) and two-dimensional (2D) gel electrophoresis and is proposed to provide efficient enrichment of phosphoproteins *(10)*.

Immunoprecipitation utilises the exquisite specificity of monoclonal antibody binding to recognize and isolate targets of interest. Antibodies against phosphorylated forms of many different proteins are available commercially. Antibodies raised against serine or threonine phosphorylated proteins are usually specific to one particular phosphoprotein. Despite efforts to create pan antiserine or pan antithreonine antibodies via inoculation with several peptides or via mixing several antibodies, broad-spectrum antiserine and antithreonine antibodies have so far found relatively limited application. The case is different however for tyrosine phosphorylation. Some monoclonal antibodies against phosphorylated tyrosine are highly effective at recognizing and precipitating a wide variety of tyrosine-phosphorylated proteins.

Some commercially available systems for purification of phosphoproteins use proprietary phosphoprotein-affinity materials, the identities of which have not been publicly disclosed. The Qiagen Phosphoprotein Purifcation kit is one such system. Published studies suggest that this system is highly effective at enriching for phosphoproteins *(11–13)*.

Here we present two affinity-based methods for selective isolation of phosphorylated proteins that are compatible with gel-based proteomics and are readily accessible for most molecular biology laboratories. Immunoprecipitation is an effective technique for the isolation of tyrosine-phosphorylated proteins. For serine and threonine-phosphorylated proteins, the use of commercially prepared phosphoprotein-affinity matrices provides a rapid and simple means of phosphoprotein isolation (*see* Fig. 1).

Isolation of Phosphoproteins

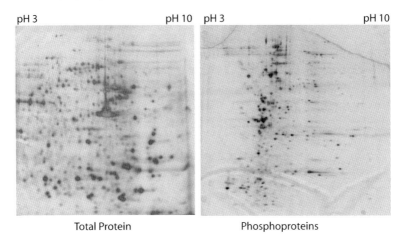

Fig. 1. Myoblast proteins before and after phosphoprotein purification. Protein from C2C12 myoblasts was separated by two-dimensional gel electrophoresis and detected by silver stain. At left, total protein was recovered by lysing cells in a 7 M urea buffer. At right C2C12 proteins recovered from phosphoprotein purification as described in this chapter. The first dimension of separation was isoelectric focusing on a pH 3–10 gradient. Note that phosphorylated proteins tend to have an acidic nature and therefore migrate towards the left side of the gel.

2. Materials

2.1. Immunoprecipitation of Tyrosine-Phosphorylated Proteins

1. Phosphate-buffered saline (PBS): 150 mM NaCl, 1.5 mM KH_2PO_4, 5 mM Na_2HPO_4. Adjust pH to 7.4 using HCl. Sterile-filter and store at 4 °C.
2. PLC-γ buffer: 1% Triton X-100, 50 mM HEPES, 150 mM NaCl, 10% glycerol, 1 mM EGTA, 1.5 mM $MgCl_2$. Alternatively, Modified RIPA buffer: 1% NP-40, 50 mM Tris-HCl pH 7.4, 150 mM NaCl, 1 mM EDTA may be used (*see* Note 1). Protease and phosphatase inhibitors should be added to the buffer just before use: 10 mM NaF, 1mM PMSF (*see* Note 2), 1 µg/ml aprotinin, 1 µg/ml pepstatin, 1 µg/ml leupeptin, 1 mM activated sodium orthovanadate. This protease-inhibitor supplemented buffer is used as the lysis buffer. Addition of 10 mM sodium pyrophosphate and 1 mM beta-glycerophosphate to the lysis buffer may provide improved results. If necessary, a more stringent lysis buffer can be made by adding 1% sodium deoxycholate.
3. Antiphosphotyrosine antibody (*see* Note 3). A variety of pan antiphospho-tyrosine antibodies are commercially available; some common choices are PY100, 4G10, and PY20 (*see* Note 4).
4. Protein A sepharose beads (*see* Note 5). For certain antibodies protein G beads may be preferred. Some antibodies can be purchased already bound to agarose or sepharose beads.

5. The choice of elution buffer depends on the downstream processing to be performed on the recovered proteins. If polyacrylamide gel electrophoresis followed by Western blotting or silver staining is to be performed, a standard SDS-PAGE sample buffer (*e.g.*, Laemmli) can be used as elution buffer.
6. A centrifuge with the capability to exert 15,000–20,000*g*. A benchtop microfuge style unit is the most convenient type for this procedure. A refrigerated centrifuge is preferable.

2.2. Phosphoprotein-Affinity Purification

1. Cell sample (approximately 10 million cells will be required) (*see* Note 18)
2. Cell wash buffer: use a serum-free and phosphate-free cell culture medium, *e.g.*, DMEM.
3. Qiagen Phosphoprotein Purification Kit (#37101).
4. Retort stand and clamps to hold the enrichment columns.
5. Centrifuge capable of subjecting a 5-ml sample to 10,000*g*.
6. Centrifuge capable of subjecting a 1.5-ml tube to 10,000*g*.
7. Ultrafiltration spin columns (included in Qiagen Kit). One per sample to be analyzed.
8. Diafiltration buffer: 10 mM Tris-HCl pH 7.0
9. Isoelectric focusing buffer: 7 M urea, 2 M thiourea, 4% CHAPS, 1% DTT.

3. Methods

3.1. Immunoprecipitation of Tyrosine Phosphorylated Proteins

1. Wash cells (*see* Notes 6, 7, and 8) 2–3 times in cold PBS then remove excess liquid.
2. Add lysis buffer to cells (0.5 ml per 10-cm plate) and incubate on ice for 20 min. Periodically agitate the cells.
3. Scrape cells from the plate(s) and collect all material in a microfuge tube.
4. Centrifuge the tube at high speed (15,000–20,000*g*) for 10 min at 4 °C to remove nuclei and insoluble materials (*see* Note 9). Transfer the supernatant to a fresh 1.5-ml microfuge tube. This solution is the lysate.
5. Add antibody to the lysate at the recommended dilution ratio (*see* Note 10).
6. If antibody is not prebound to beads, incubate lysate and antibody for at least 15 min on ice. For antibodies prebound to beads, or where recommended by the antibody supplier, incubate the sample/antibody mixture overnight at 4 °C. Gently rotate or agitate tubes during the incubation.
7. If antibody is not prebound to beads, add 20 µl of a 50% slurry of protein A beads and incubate for 1–3 h at 4 °C with rotation of the tube(s) (*see* Note 11).
8. Wash beads 3–5 times in ice cold lysis buffer. To do this, pellet the beads (with bound antibody and protein) by a brief 10–30 sec centrifugation, aspirate the liquid portion down to the top of the pellet, then resuspend the bead pellet in cold lysis buffer (*see* Note 12).

Isolation of Phosphoproteins

9. After removing all supernatant, elute bound proteins from the beads by adding elution buffer and boiling for 2–3 min (*see* Note 13).
10. The sample is ready to be further analyzed by polyacrylamide gel electrophoresis or other methods.

3.2. Phosphoprotein-Affinity Purification

1. Prepare the lysis and elution buffers from material supplied in the kit (*see* Note 17).
2. Wash cells 2–3 times in cell wash buffer (*see* Notes 18, 19).
3. For nonadherent cultured cells, pellet the cells using a clinical centrifuge (150g for 3–5 min) and resuspend the cell pellet in lysis buffer. For adherent cells, remove the wash buffer and add lysis buffer directly to the cell culture plates. Alternatively, adherent cells may be de-adhered using trypsin treatment (*see* Note 20) before washing and lysis.
4. Incubate the cells at 4°C or on ice for 20–30 min with occasional agitation.
5. Clarify the lysate by centrifugation at 10,000g for 30 min at 4 °C (*see* Note 21).
6. During the centrifugation step, prepare a phosphoprotein-purification column for use by mounting it vertically in a clamp and allowing the storage solution to flow out.
7. Wash the column by adding 4 ml of lysis buffer. The column can be kept at room temperature throughout the remainder of the procedure.
8. After centrifugation, reserve the supernatant and discard the pellet.
9. Determine protein concentration in the supernatant by any suitable method (*e.g.*, Bradford assay).
10. To a volume of lysate containing 2.5 mg of protein, add additional lysis buffer to bring the total volume of the sample to 25 ml (*see* Note 22).
11. Apply the diluted sample to the column and allow all material to flow through.
12. Wash the column with 6 ml of lysis buffer.
13. Elute phosphoproteins from the column in four fractions by adding four 500-μl aliquots of elution buffer. The phosphoproteins should be concentrated in elution fractions 3 and 4 (*see* Note 23). A protein concentration assay may be used to determine which fractions contain the most eluting protein.
14. Apply elution fraction 3 to the upper sample reservoir of a Nanosep ultrafiltration column (*see* Note 24).
15. Centrifuge the ultrafiltration column at 10,000g until approximately 50 μl remains in the upper reservoir. The flow-through can be discarded.
16. Add elution fraction 4 to the upper reservoir of the ultrafiltration column.
17. Centrifuge repeatedly for up to 10 min at a time until the sample has been concentrated to approximately 50 μl.
18. Add 450 μl of 10 mM Tris-HCl pH 7.0.
19. Centrifuge until sample has been concentrated to approximately 50 μl.
20. Add a second 450-μl aliquot of 10 mM Tris-HCl pH 7.0 and centrifuge again until the sample is concentrated to approximately 50 μl.

21. Withdraw the retained sample from the upper reservoir and place in a microfuge tube.
22. If 2D gel electrophoresis is to be performed, add 250 µl of isoelectric focusing buffer to the sample (*see* Note 23). Otherwise, store or dilute the sample in an appropriate buffer for further processing.

4. Notes

1. The choice of buffer is cell and antibody dependent. For immunoprecipitation of phosphoproteins PLC-γ buffer including sodium pyrophosphate is typically a good choice.
2. PMSF is unstable but stock solutions in isopropanol can be stored safely at room temperature and added to lysis buffers just before use.
3. The protocol can be applied using other types of antibodies however the effectiveness of any given antibody as an immunoprecipitating agent and the optimum immunoprecipitation conditions must be determined empirically.
4. Although pan antiphosphotyrosine antibodies recognize a wide variety of tyrosine-phosphorylated proteins, not every such protein can be recognized by a given antibody. The binding range of any given antibody must be determined empirically.
5. Protein A or G beads are shipped in solutions containing preservatives to prevent bacterial growth. Beads should be washed before use.
6. This procedure can be performed on cells extracted from tissue as well as from cultured cells.
7. Because tyrosine-phosphorylated proteins are relatively rare (typically 2–3%), prepare enough cells to obtain at least 1 mg of protein. Depending on the detection method to be used, even more protein may be required.
8. To generate a positive control sample, treat cells with 1 mM pervanadate for 30 min to inhibit tyrosine phosphatases and increase intracellular tyrosine phosphorylation.
9. Some proteins translocate to the nucleus when phosphorylated; these proteins will not be detected efficiently under the immunoprecipitation conditions given. If desired, the nuclear pellet can be solubilized in 2% sodium dodecyl sulfate (SDS) with physical agitation (*e.g.*, repeated pipeting) then diluted in lysis buffer. Preclearing this lysate before performing immunoprecipitation is recommended (*see* Note 16).
10. Check with the antibody supplier for a recommended dilution for immunoprecipitation. If a recommended dilution for the antibody is not known, 1:100 (*i.e.*, 10 µl/ml) is usually a reasonable starting point. For antibody that is prebound to beads, using 20 µl of a 50% bead slurry is typical.
11. If the antibody you choose does not bind effectively to protein A or G beads, a rabbit anti-isotype secondary antibody (*e.g.*, rabbit anti-rat) can be added between the primary antibody and protein A incubations.

12. Agarose beads can be damaged by excessive centrifugation. This will result in a bead pellet that is difficult to resuspend and will increase nonspecific protein retention.
13. The heavy and light chains of the antibody will also be recovered in the final sample.
14. Samples should be kept cold (0–4 °C) throughout the immunoprecipitation procedure.
15. It is strongly recommended to use fresh (never frozen) protein samples for all phosphoprotein enrichment procedures.
16. If background protein precipitation is excessive, preclear the lysate by adding beads with no antibody and incubating for 30–60 min at 4 °C. Pellet beads by pulse centrifugation and use the precleared supernatant for immunoprecipitation.
17. The portion of lysis buffer used for cell lysis should contain the supplied protease inhibitors and nuclease. The remaining lysis buffer used for column equilibration and washing does not require additional protease inhibitors or nucleases.
18. The columns are designed to be loaded with 2.5 mg of protein. Typically 2–3 10-cm plates of cultured cells will yield enough protein. Expected yield is 175–375 µg of phosphoprotein. The maximum binding capacity of the columns is estimated to be 500 µg of phosphoprotein.
19. Do not use phosphate-buffered saline (PBS) as the wash buffer. The wash buffer must not contain any free phosphates. If detaching adherent cells by trypsin-treatment, perform trypsinization *before* the phosphate-free, serum-free wash.
20. If trypsinizing adherent cells first remove all culture media and wash cells with PBS, then incubate cells in trypsin solution: 0.25% trypsin, 1mM EDTA in Hank's Balanced Salt Solution or phosphate-buffered saline, for 2–4 min at 37 °C. Prepared trypsin solution can be purchased from commercial sources. Wash the detached cells once with serum-containing media to deactivate trypsin then proceed to serum-free, phosphate-free washes (*see* Note 19).
21. Some lysates, especially those containing higher concentrations of protein, develop a yellow color during cell lysis or during the postlysis centrifugation. This change does not seem to affect the procedure.
22. This adjusts protein concentration of the supernatant to 0.1 mg/ml.
23. Samples may be frozen for storage at this step if desired.
24. The supplied ultrafiltration devices do not efficiently retain proteins of less than 10 kDa in size. Similar devices with lower molecular mass cut-offs can be purchased.

References

1. Adamczyk, M., Gebler, J. C., and Wu, J. (2001) Selective analysis of phosphopeptides within a protein mixture by chemical modification, reversible biotinylation and mass spectrometry *Rapid Commun Mass Spectrom* **15,** 1481–1488.

2. Goodman, T., Schulenberg, B., Steinberg, T. H., and Patton, W. F. (2004) Detection of phosphoproteins on electroblot membranes using a small-molecule organic fluorophore *Electrophoresis* **25,** 2533–2538.
3. Schulenberg, B., Goodman, T. N., Aggeler, R., Capaldi, R. A., and Patton, W. F. (2004) Characterization of dynamic and steady-state protein phosphorylation using a fluorescent phosphoprotein gel stain and mass spectrometry *Electrophoresis* **25,** 2526–2532.
4. Kocher, T., Allmaier, G., and Wilm, M. (2003) Nanoelectrospray-based detection and sequencing of substoichiometric amounts of phosphopeptides in complex mixtures *J Mass Spectrom* **38,** 131–137.
5. Bateman, R. H., Carruthers, R., Hoyes, J. B., Jones, C., Langridge, J. I., Millar, A., and Vissers, J. P. (2002) A novel precursor ion discovery method on a hybrid quadrupole orthogonal acceleration time-of-flight (Q-TOF) mass spectrometer for studying protein phosphorylation *J Am Soc Mass Spectrom* **13,** 792–803.
6. Andersson, L., and Porath, J. (1986) Isolation of phosphoproteins by immobilized metal (Fe^{3+}) affinity chromatography *Anal Biochem* **154,** 250–254.
7. Muszynska, G., Andersson, L., and Porath, J. (1986) Selective adsorption of phosphoproteins on gel-immobilized ferric chelate *Biochemistry* **25,** 6850–6853.
8. Ficarro, S. B., McCleland, M. L., Stukenberg, P. T., Burke, D. J., Ross, M. M., Shabanowitz, J., Hunt, D. F., and White, F. M. (2002) Phosphoproteome analysis by mass spectrometry and its application to Saccharomyces cerevisiae *Nat Biotechnol* **20,** 301–305.
9. Posewitz, M. C., and Tempst, P. (1999) Immobilized gallium(III) affinity chromatography of phosphopeptides *Anal Chem* **71,** 2883–2892.
10. Dubrovska, A., and Souchelnytskyi, S. (2005) Efficient enrichment of intact phosphorylated proteins by modified immobilized metal-affinity chromatography *Proteomics* **5,** 4678–4683.
11. Metodiev, M., Timanova, A., and Stone, D. E. (2004) Differential phosphoproteome profiling by affinity capture and tandem matrix-assisted laser desorption/ionization mass spectrometry *Proteomics* **4,** 1433–1438.
12. Puente, L. G., Borris, D. J., Carriere, J. F., Kelly, J. F., and Megeney, L. A. (2006) Identification of candidate regulators of embryonic stem cell differentiation by comparative phosphoprotein affinity profiling *Mol Cell Proteomics* **5,** 57–67.
13. Puente, L. G., Carriere, J. F., Kelly, J. F., and Megeney, L. A. (2004) Comparative analysis of phosphoprotein-enriched myocyte proteomes reveals widespread alterations during differentiation *FEBS Lett* **574,** 138–144.

29

Glycoprotein Enrichment Through Lectin Affinity Techniques

Yehia Mechref, Milan Madera, and Milos V. Novotny

Summary

Posttranslational modifications (PTM) of proteins are among the key biological regulators of function, activity, localization, and interaction. The fact that no more than 30,000–50,000 proteins are encoded by the human genome underlines the importance of posttranslational modifications in modulating the activities and functions of proteins in health and disease. With approximately 50% of all proteins now considered to be glycosylated, its physiological importance in mammalian systems is imperative. Aberrant glycosylation has now been recognized as an attribute of many mammalian diseases, including hereditary disorders, immune deficiencies, neurodegenerative diseases, cardiovascular conditions, and cancer. As many potential disease biomarkers may be glycoproteins present in only minute quantities in tissue extracts and physiological fluids, glycoprotein isolation and enrichment may be critical in a search for such biomarkers. For decades, efforts have been focused on the development of glycoprotein enrichment from complex biological samples. Logically, the great majority of these enrichment methodologies rely on the use of immobilized lectins, which permit selective enrichment of the pools of glycoproteins for proteomic/glycomic studies. In this chapter, lectin affinity chromatography in different formats are described, including tubes; packed columns, and microfluidic channels.

Key Words: Agarose lectin material; glycoprotein enrichment; human blood serum glycoproteins; lectin affinity chromatography; silica-based lectin material.

1. Introduction

Glycosylation occurs with proteins small and large. With an increase of molecular size, the occurrence of several glycosylation sites typically increases, adding to the complexity of glycosylation patterns and the usual difficulty of resolving fine structural differences in a large biopolymer. The successful attempts to resolve partially or completely different glycoforms of native glycoproteins are thus generally confined to relatively small proteins or glycopeptides. In biological materials such as cellular extracts and physiological fluids, glycoproteins are often encountered in minute quantities, placing high demands on both the measurement sensitivity and proper isolation procedures. A combination of orthogonal separation techniques and the use of affinity principles are the most commonly practiced isolation/enrichment strategies. Miniaturization of these separation methodologies represents a general trend in bioanalysis and is thus applicable to all stages of modern glycoanalysis.

Glycoproteins of interest may be encountered as either soluble or membrane-bound molecules. The isolation strategies will vary, depending on whether such glycoproteins occur in cytosolic space, nucleus, extracellular space, cellular or subcellular membranes, etc. Also, any isolation protocol has to be adjusted to take into account the physicochemical properties of a given glycoprotein or a glycoprotein class. For example, detergents must be included in the extraction buffer to yield membrane-bound glycoproteins. For the extraction and fractionation of soluble glycoproteins, detergents are usually not needed. However, the addition of certain detergents can occasionally increase extraction yields and reduce contamination during purification.

The methods used in any glycoprotein isolation and enrichment protocols are very dependent on whether or not a biological activity must be retained for further studies. Many glycoproteins, including several membrane-bound molecules, can be easily denatured on contact with the surface of glassware or a chromatographic packing. If the sample is to be subjected to a protease cleavage or a release of glycans, isolation or purification under harsher conditions is unlikely to affect adversely the final analysis. Generally, glycoproteins can be purified by most conventional protein separation methodologies, including gel electrophoresis and various chromatographic forms such as ion-exchange, size exclusion, reversed phase using C_{18}, C_8, or C_4 columns, hydrophobic interaction, and affinity *(1,2)*. A most useful, specific isolation principle involves the use of lectins that are immobilized on chromatographic resins.

Electrophoretic separations in gels are now widely applied to a variety of problems in the isolation and analysis of proteins. Simplicity of the apparatus and robustness are the chief selling points of both sodium dodecyl-sulfate polyacrylamide gel electrophoresis (SDS-PAGE) and two-dimensional electrophoresis (2-DE) in gels that combines the separation mechanisms based on molecular size and isoelectric points. Both techniques have found their wide application in glycoprotein isolation/analysis. Because of the recently advanced sensitivity of MS measurements, it has become feasible to analyze the contents of individual spots on a gel.

SDS-PAGE has frequently adequate resolving power to migrate proteins according to their size and to provide a rough estimate of their molecular weights. It is particularly informative to compare the "profiles" of proteins in a complex sample with those processed through a lectin chromatographic column, indicating which components of a mixture are glycosylated. Typical gel/buffer systems used in this procedure are essentially those described in the pioneering work by Laemmli *(3)*. However, SDS-PAGE of glycoproteins suffers from certain pitfalls. A positive bias in the estimation of molecular weights can be observed because of a lower SDS binding in presence of carbohydrate structures. Conversely, negative deviations are experienced because of the presence of charged sialic acids that contribute to the overall electrophoretic mobilities of glycoproteins. The glycoprotein bands observed on gels are often broad because of the tendency to resolve microheterogeneities.

Lectins are specialized proteins that have been isolated from various plants and animal sources. For number of years, they have been used to study interactions with glycoconjugates *(4,5)*. Lectins have also been widely used to isolate, purify and characterize glycoproteins and glycolipids in various modes of affinity chromatography *(4)*. The lectin affinity separations in conventional columns have now been developed into a powerful means for the purification of glycoproteins before structural studies *(6)*. These techniques are based on a reversible biospecific interaction of certain glycoproteins with the lectins immobilized to a solid support. Over the years, various lectins with a high specificity toward oligosaccharides have been immobilized to agarose and other separation matrices. Some of these materials have been commercially available to isolate glycoproteins on the basis of their different glycan structures.

The choice of affinity matrix is often critical and may depend on a particular application. Generally, a medium has to be insoluble in the commonly used aqueous buffers. It must either possess the reactive groups allowing a direct attachment of a lectin protein, or at least there must be a possibility of a facile derivatization of the surface to introduce such reactive groups into the matrix. This process is called the activation of a matrix. Additionally, the appropriate affinity support should be of a hydrophilic nature, causing minimum nonspecific binding. Particle matrixes are additionally required to be macroporous to allow the interaction of high-molecular-weight proteins with the immobilized lectins that are located predominantly inside the pores. However, in certain cases (when the capacity of affinity media is not an issue), lectin can be immobilized onto totally nonporous particles or solid surfaces to achieve fast, small-scale interactions.

2. Materials
2.1. Lectin Materials and Buffers

1. Sepharose-immobilized lectins (Vector Laboratories, Burlingame, CA).
2. Binding buffer: 20 mM Tris-HCl buffer, pH 7.8, 1 mM $MnCl_2$, 1 mM $CaCl_2$, and 0.5-M sodium chloride.
3. Elution buffer: 20 mM Tris-HCl buffer, pH 7.8, 0.5-M sodium chloride, and 0.1–0.5 M methyl-α-D-glucopyranoside (methyl-α-D-glucoside) or methyl-α-D-mannopyranoside for Con A lectin.
4. Other appropriate hapten sugars are used with other lectin materials.

2.2. Immobilization of Lectin to Porous Silica

1. Lectins (EY Laboratories, Inc, San Mateo, CA) prepared in 0.1 M sodium bicarbonate buffer, pH 8.5, containing 0.5 M sodium chloride.
2. Molecular-sieve-dried toluene.
3. 3-glycidoxypropyltrimethoxysilane (epoxysilane), triethylamine.
4. 20 mM sulfuric acid aqueous solution.
5. Sodium periodate.
6. Sodium borohydride.
7. Sodium cyanoborhydride prepared in 0.1 M sodium bicarbonate buffer, pH 8.5, containing 0.5 M sodium chloride.

2.3. 2-Dimensional Gel Electrophoresis (2-DE)

1. IPG strip (18 cm, 3–10 NL) and IPGphor focusing chamber (Amersham Pharmacia Biotech).
2. IPG strip rehydration buffer: 8 M urea, 2% CHAPS, 50 mM dithiothreitol (DTT), 0.2% Bio-Lyte 3/10 ampholyte, 0.001% bromophenol blue.
3. Equilibration solution I (with DTT): 6 M urea, 0.375 M Tris-HCl, pH 8.8, 2% SDS, 20% glycerol, 2% (w/v) DTT.
4. Equilibrating solution II (with iodoacetamide): 6 M urea, 0.375 M Tris-HCl, pH 8.8, 2% SDS, 20% glycerol, 2.5% (w/v) iodoacetamide.

2.4. SDS-Polyacrylamide Gel Electrophoresis (SDS-PAGE)

1. Separation buffer: 1.5 M Tris-HCl, pH 8.8, 0.5% SDS, store at 4 °C. Warm to room temperature before use if precipitation is noticed.
2. Stacking buffer: 0.5 M Tris-HCl, 6.8, 0.5% SDS, store at room temperature.
3. Sample buffer: prepared by mixing 1 ml of 0.5 M Tris-HCl, pH 6.8, 3.8 ml deionized water, 0.8 ml glycerol, 10% (w/v) sodium dodecyl sulfate (SDS), 0.4 ml 2-mercaptoethanol and 1% (w/v) bromophenol blue. This SDS reducing buffer should be stored at room temperature.
4. Acrylamide/bis solution (30% T, 2.67% C), N,N,N',N'-tetramethylethylenediamine (TEMED, Bio-Rad, Hercules, CA).

5. 10% Ammonium persulfate (APS): prepared by dissolving 10 mg ammonium persulfate in 100 μl deionized water. This should be prepared fresh.
6. Water-saturated isobutanol: prepared by shaking equal volumes of water and isobutanol in a glass bottle and allowed to separate. The top layer is used. This solution should be stored at room temperature.
7. Staining solution 0.1% Coomassie blue R-250 (Bio-Rad, Hercules, CA) prepared in 40% methanol, 10% acetic acid.
8. Protein molecular weight markers (Bio-Rad, Hercules, CA).

2.5. MALDI/TOF MS Matrix Solution

1. Sinapinic Acid Matrix solution: prepared at 10 mg/ml concentration in 70:30 water/acetonitrile (0.1 TFA) (*see* Note 9–10).
2. Sample Solution: 3% acetonitrile aqueous solution containing 0.1% TFA.

2.6. Immobilization of Lectin to monoliths

1. Ethylene dimethyl acrylate (EDMA) or Trimethylpropane trimethylacrylate as the cross-linking monomer and glycidyl methacrylate (GMA) as the active monomer.
2. Free Con A and WGA.
3. Cyclohexanol.
4. Dodecanol.
5. 2,2'-azobisisobutyronitrile (used as polymerization initiator).
6. Sodium periodate.

3. Methods

3.1. Lectin Immobilized to Agarose Gels

The first attempts to use agarose as a chromatographic support for low-pressure lectin affinity chromatography in enrichment of glycoproteins date back to the late 1980s, with mainly *Concanavalin A* (Con A), *wheat germ agglutinin* (WGA), or *Len culinaris* (lentil lectin) used for the affinity purification of membrane glycoproteins *(7)* or certain types of cell lines *(8)*. Agarose, as a linear polysaccharide consisting of alternating D-galactose and 3,6-anhydro-L-galactose units, connected by β(1,4) *O*-glycosidic bond, offers several properties meeting the critical requirements for being used as an affinity matrix: hydrophilicity; low tendency to nonspecific binding; high permeability; and exclusion limit. Because of the polysaccharide backbone, the agarose surface features a substantial number of free hydroxyl groups that do not usually permit a direct attachment of lectins as affinity ligands, so they have to be modified before the immobilization of lectins.

3.1.1. Isolation of Glycoproteins from VNO.

As a typical example for lectin affinity chromatography using agarose gels, vomeromodulin, a putative pheromone transporter of the rat vomeronasal organ (VNO), was isolated by lectin chromatography, purified, and subjected to MS analysis *(9)*.

1. The vomeronasal organs of adult female and male rats are dissected and broken into pieces on dry ice. Briefly, the equivalents of approximately 1 mg of tissues are dissected in dry ice and homogenized in a polytron tissue disruptor with 9 volumes of homogenizing buffer (50 mM Tris–HCl, pH 7.5) containing 50 µl of protease inhibitor cocktail.
2. The homogenizing buffer with the extract is centrifuged at 10,000*g* for 1 h.
3. The supernatant is lyophilized and reconstituted in 2 ml of binding buffer before lectin chromatography.
4. Lectin chromatography is performed using a column packed with 1 ml Con A–Sepharose prepared according to the following procedure.
5. Lectin materials and all buffers are first equilibrated to the temperature at which the purification will be performed (4 °C is recommended). The column end pieces are flushed with the binding buffer to eliminate air. Next, column outlet is closed leaving 1–2 cm of binding buffer in the column. Gently, 1 ml of lectin media is transferred to the column. The column is then immediately filled with binding buffer. The column top is then mounted and connected to a pump flowing at 200 µl/min (*see* Note 1). The column must be washed with at least 10 bead volumes (*see* Note 2).
6. The sample resuspended in the binding buffer is then pumped through the column at a flow rate of 200 µl/min and at 4°C. During sample application, flow rate is the most significant factor to obtain maximum binding.
7. The column is then washed with 5–10 column volumes of binding buffer or until no material appears in the eluent (monitored by UV absorption at $A_{280\ nm}$).
8. Bound glycoproteins are then eluted with 5 column volumes of elution buffer (*see* Note 3–5), and dialyzed over night using 3,000 Da cut-off dialysis membrane to eliminate the hapten sugars. This elution procedure is commonly used; however, other procedures are utilized to elute tightly bound substances (*see* Note 6–8). The dialyzed sample is then lyophilized.
9. The lectin-enriched sample was then subjected to SDS-PAGE.
10. The SDS-PAGE analysis is performed using Mini-PROTEAN II Cell (Bio-Rad, Hercules, CA).
11. A 1.5-mm thick, 12% gel is prepared by mixing 3.35 ml deionized water, 2.5 ml 1.5 M Tris-HCl, pH 8.8, 100 µl 10% SDS solution, 4 ml acrylamide/bis solution (degassed for at least 15 min), 50 µl 10% APS, and 5 µl TEMED. TEMED and APS are added immediately before gel casting. The mixture is then poured between two glass plates, leaving enough space for a stacking gel. Water-saturated isobutanol is then overlaid to prevent drying of the gel. The gel commonly polymerizes in 45 min.
12. The water-saturated isobutanol is then removed and the top of the gel is washed with water to eliminate any traces of water-saturated isobutanol.

13. The stacking gel (4% gel, 125 mM Tris-HCL, pH 6.8) is then prepared by mixing 6.1 ml deionized water, 2.5 ml 0.5 M Tris-HCl, pH 6.8, 100 µl 10% SDS, 1.33 ml acrylamide-bis (30% stock solution) 50 µl 10% APS, and 10 µl TEMED. Again, TEMED and APS are added immediately before gel casting. This mixture is then poured and sample comb is inserted. The stacking gel commonly polymerizes in 45 min.
14. The running buffer is prepared by diluting the separation buffer five times with deionized water.
15. Once the stacking gel is polymerized, the sample comb is carefully removed and the wells are washed with separation buffer using a syringe or a pipette.
16. The unit is then assembled and the diluted separation buffer is added to the upper and lower chambers. The sample is then loaded into the wells sandwiched between two molecular weight markers. Samples and molecular weight markers are prepared by resuspending the lyophilized sample and the molecular weight markers in the sample buffer, and heated at 95 °C for 4 min.
17. The Mini-PROTEAN II Cell recommended power condition for optimal resolution with minimal distortion is 200 volt, constant voltage setting. The usual run time is 45 min or until the bromophenol blue band reaches the end of the gel.
18. The cell is then disassembled and the gel is carefully washed with water and stained for ½ hour in the staining solution. Next, the gel is destained with several changes of 40% methanol, 10% acetic acid to remove background staining (usually 1–3 h).
19. The SDS–PAGE run of the Con A-enriched fraction was shown to correspond to three bands, which included a 70-kDa protein corresponding to vomeromodulin (Fig. 1).

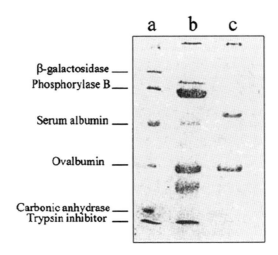

Fig. 1. SDS–polyacrylamide gel electrophoresis of the vomeronasal tissue extract after lectin chromatography: (a) protein M_r markers; (b) proteins unretained by Concanavalin A; (c) retained fraction.
Source: Reproduced from (9), with permission.

3.1.2. Isolation of MUP Glycoproteins from Mouse Urine

In another application, lectin chromatography was used to enrich a minor component of the major urinary protein (MUP) complex of the house mouse, which was then ascertained to be a previously suspected glycoprotein *(10)*. The minor glycosylated protein from the MUP complex is isolated and enriched using three chromatographic steps: (i) fractionation on a Superdex 75 column (prompting fractionation on the basis of molecular size) and (ii) lectin chromatography on Con A column (enriching glycoproteins).

1. The glycosylated protein is at trace levels compared to the other MUPS. Therefore, its isolation and enrichment is only possible using eleven 50 μl male urine aliquots.
2. Each aliquot is subjected to size exclusion chromatography on a Pharmacia Smart System (Piscataway, NJ), using a Pharamcia HR 5/20 column packed with Superdex 75 media. This step is used to remove salts and other low molecular weight components associated with urine.
3. The column is eluted with a buffer consisting of 10 mM Tris-HCl, pH 7.8, and 0.15 M sodium chloride, at a flow rate of 30 μl/min.
4. Lectin affinity chromatography is performed as described above.
5. The glycosylated protein is more acidic then the other MUPs, as indicated from its isoelectric focusing point (pI = 4.0). This agrees with the fact that sialic acid residues were associated with this protein. Both the ion-exchange (upper trace) and lectin affinity chromatograms (lower trace) are depicted in Fig. 2.
6. MALDI/TOF mass spectrometry analysis of the enriched glycosylated MUP is performed on Voyager-DE RP Biospectrometry Workstation (Applied Biosystems, Framingham, MA) equipped with a pulsed nitrogen laser (337 nm). A 0.5-μl aliquot of the glycosylated MUP and 0.5 μl aliquot of sinapinic acid matrix solution are spotted and mixed on stainless steel MALDI plate and allowed to dry at room temperature. A MALDI mass spectrum of the enriched glycoprotein is shown in Fig. 2 inset in which the observed mass-to-charge value is higher than the mass predicted from cDNA sequence, suggesting the presence of posttranslational modification (glycosylation), which was eventually established as tri- and tetraantennary complex type with a wide glycosylation heterogeneity, partially contributed to both the degree of sialylation and the linkages of galactose residues.

3.1.3. Enrichment of Glycoproteins from Snake Venom using Con A-Sepharose

Recently, agarose-based lectin affinity chromatography was used to enrich glycoproteins from ten samples of venom from eight snake species including the *Naja naja kaonthia, Ophiophagus hannah, Bungarus fasciatus, Vipera russelli sigmensis, Naja naja atra, Deinagkistrodon acutus, Bungarus multicinctus, Vipera russelli formosensis, Tremeresurus mucrosquamatus, Trimeresurus*

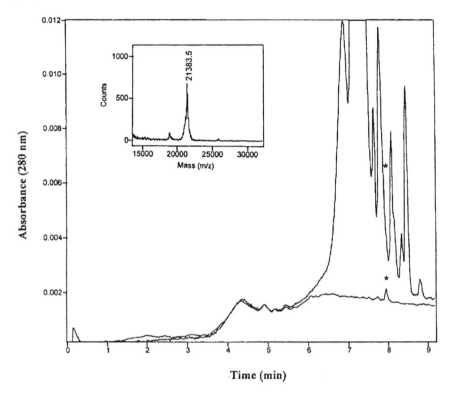

Fig. 2. Anion-exchange chromatogram of the isolated MUP components (upper trace) and the Concanavalin A bound fraction (lower trace); inset is the mass spectrum of isolated glycoprotein, as indicated with asterisk.
Source: Reproduced from *(10)*, with permission.

Steinegeri (11). Glycoprotein enrichment in this application was performed using the batch approach in which lectin and samples are mixed in a tube overnight

1. All venom samples were lyophilized before use. The lyophilized samples are resuspended in lectin-binding buffer to the concentration of 10 mg/ml.
2. A 1-ml aliquot of the solution is added to 0.1-ml slurry of Con A-Sepharose gel placed in an Eppendorf microtube.
3. The mixture is incubated at room-temperature overnight with continuous shaking.
4. Unbound proteins are washed five times with 2 ml of binding buffer.
5. This is followed by washing twice with double distilled water (2 ml).
6. The bound glycoproteins are extracted from the Sepharose-Con A gel with 7 M urea, 4% CHAPS, 4 mM Tris base, 2 M Thiourea, 2% IPG buffer, 65 mM DTT, 0.4 ml.

7. The negative control is a lysis buffer extract of Sepharose-Con A gel. In an additional control experiment, the bound glycoproteins are eluted from Sepharose-Con A gel with 1 ml of 0.2 M R-methylmannose. The eluted glycoproteins mixture is lyophilized and dissolved in a lysis buffer for 2-DE analysis.
8. Both Con A-binding and nonbinding fractions are subjected to 2-DE analysis.
9. A 0.50-mg aliquot of each of the lectin-binding fraction, the nonbinding fraction (Con A-agarose eluent), the control and *N. naja kaouthia* venom are separately solubilized in a lysis buffer and were separated between pH 3 and 10 in the first dimension of 2-DE. The following procedure is utilized with PROTEAN IEF Cell and Mini-PROTEAN II Cell (Bio-Rad, Hercules, CA).
10. Each sample is sonicated, centrifuged, and applied to an IPG strip (Amersham Pharmacia Biotech, 18 cm, 3–10 NL) by an IPG strip rehydration application method. The IPG strip is rehydrated for 12 h on an IPGphor machine (Amersham Pharmacia Biotech). An electric current is applied at 30 V during the rehydration time. The isoelectric focusing (IEF) running conditions are: 100 V for 1 h, 250 V for 1 h, 500 V for 1 h, 1000 V for 1 h, and 6000 V for up to 30 kVh. SDS-PAGE. After IEF, IPG strips are equilibrated for 15 min in an equilibration solution I, followed by another 15 min in the equilibration solution II.
11. IPG gel strips were embedded on top of the second dimension gels and covered with 0.5% agarose.
12. SDS-PAGE was carried out on 10–20% acrylamide gradient gel (18 × 18 cm) at 45 mA/gel until the bromophenol blue dye front reached the bottom of the gel.
13. The 2-DE images revealed that Con A-binding glycoproteins, eluted from agarose gel by using a lysis buffer, have moderate to high molecular weight and separated in the upper part of the 2-DE gel (Fig. 3B). The protein pattern (Fig. 3B) is similar to that for *N. naja kaouthia* total proteins at the upper part of the 2-DE gel (Fig. 3A). The unbound proteins from Con A-affinity contain low molecular weight proteins (Fig. 3D). The Con A-agarose extracted with a lysis buffer in the control experiment showed about 20 spots (Fig. 3C).

3.2. Lectin Immobilized to Porous Silica

Because of its mechanical and chemical stability, hydrophilicity, and porosity combined with a well-defined spherical shape of particles, porous silica has become a popular chromatographic matrix used in many high-resolution separation techniques. These distinctive properties also permit porous silica to be used as a stationary phase for high-performance affinity chromatography, as was demonstrated already in the late 1980s as an interesting alternative to low-pressure fractionation techniques. In comparison to its low-pressure counterpart, the silica-based high-performance lectin affinity chromatography (HPLAC) offers many advantages, such as sharper elution profiles, faster analyses, and the possibility of interfacing to various high-pressure, valve-based systems that may provide high throughput and automation. However, despite these evident

Glycoprotein Enrichment Through Lectin Affinity Techniques 383

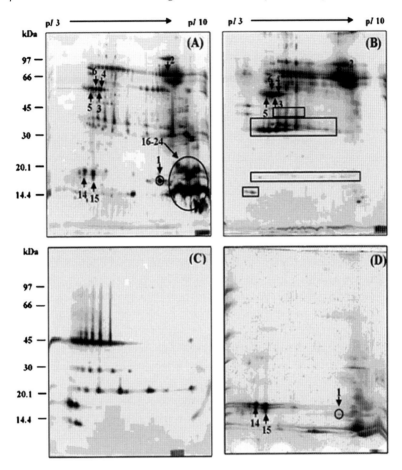

Fig. 3. 2-DE gel images of *N. naja kaouthia* venom: (A) whole venom; (B) venom from Con A-agarose binding fraction; (C) control, Con A-agarose extracted with lysis buffer; and (D) venom from Con A-agarose nonbinding fraction. All samples were dissolved in the same lysis buffer and use the same conditions for 2-DE. Squared spots are Con A-agarose proteins. Numbers are in order of identified proteins.
Source: Reproduced from *(11)*, with permission.

merits, silica-based affinity matrixes suffer from several disadvantages that may limit their utility in certain applications. First, although the stability of porous silica under acidic conditions is very good, shifting the pH towards basic values (above 8.0) significantly increases its solubility. It is thus highly recommended to perform coupling and fractionation procedures in buffers with the pH-values maintained below pH 8.0 to prevent a ligand leakage and protect the stability of siliceous matrixes. Second, the presence of surface silanol groups introduces a significant source of nonspecific binding, as based mainly on their ionic interactions. For this reason, and also because of the low reactivity, silanol groups have to be modified before the immobilization of lectins, which usually requires harsher conditions accompanied by longer reaction times.

3.2.1. Immobilization of Lectins to Porous Silica.

The activation of silica surface is done exclusively through the silanization process, meaning that the reactive groups capable of forming a covalent bond with the lectin are introduced by the treatment of silica with functionalized silanes in either organic *(12)* or aqueous *(13)* solvents. A derivatized silica can then be subsequently modified to increase the number of possibilities for lectin attachment, as each particular way offers different coupling yields. Such a derivatization scheme, comprising several subsequent or parallel reaction steps, has been proposed by Larson and coworkers *(12)*, as is depicted in Figure 4. Bulk macroporous silica is first treated with 3-glycidoxypropyltrimethoxysilane to introduce active epoxy groups onto the surface. This reaction is usually done in dry toluene, in presence of a trace of triethylamine as a catalyst, and at elevated temperatures under reflux. A stream of nitrogen is continuously provided to maintain the anhydrous conditions, because epoxysilane is highly susceptible to hydrolysis in aqueous media. The resulting epoxy-silica can either react with the lectin directly, permitting a covalent attachment via free amino, hydroxyl and thiol groups, or it can be further converted to aldehyde- and tresyl-silica, which provide an alternative lectin attachment. As shown in Figure 4, several ways of derivatization of silica surface, starting with introduction of the active epoxy groups through treatment with 3-glycidoxypropyltrimethoxysilane, offer different coupling yields. A direct attachment of lectins to epoxy-activated silica is not very popular primarily because of its low reaction efficiency and also because of the fact that a lectin can be conjugated via more than one functional group, a possibility that may sterically hinder the binding site. Conversely, an immobilization of lectins to aldehyde- and tresyl-activated silicas has been demonstrated to result in coupling yields up to 97 % and used successfully for a large-scale fractionation of human blood serum glycoproteins *(14)* or structurally different oligosaccharides *(15)*. The most effective immobilization approach utilized with silica is described next.

1. A 1-g aliquot of macro porous silica (10 μm, 1000 Å) is sequentially washed with water, 6 M HCl and water again. It is then dried at 150 °C overnight.
2. The dry silica is then resuspended in 15 ml of molecular-sieve-dried toluene. Next, a 200-μl aliquot of 3-glycidoxypropyltrimethoxysilane (epoxysilane) is added followed by the addition of a 5-μl aliquot of triethylamine.
3. The suspension is then gently stirred under a reflux at 105 °C for 16 h.

Glycoprotein Enrichment Through Lectin Affinity Techniques

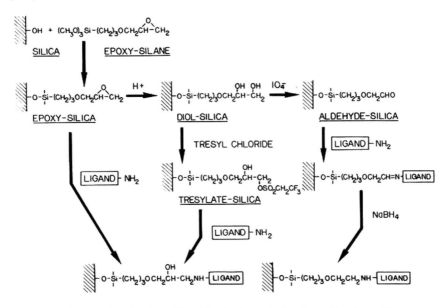

Fig. 4. Different chemistries utilized for the immobilization of lectin to silica material. *Source*: Reproduced from *(12)*, with permission.

4. The resulting epoxy-silica is consecutively washed with toluene, acetone, ether on a sintered glass frit and dried under reduced pressure.
5. A 1-g aliquot of freshly prepared epoxy silica is then resuspended in 100 ml of 20 mM sulfuric acid and hydrolyzed with a gentle stirring at 90 °C for 3 h.
6. The resulting diol-silica is washed with water, ethanol, ether and dried under reduced pressure. Next, a 1-g aliquot of the freshly prepared diol-silica is mixed with 20 ml of acetic acid/water (90/10) before the addition of 1 g of sodium periodate.
7. Diol functional groups are then oxidized to aldehydes through stirring the suspension at room temperature for 2 h.
8. The resulting aldehyde-silica is washed with water, ethanol, ether and dried under reduced pressure.
9. Lectin is immobilized to aldehyde silica by mixing a 125-mg aliquot of freshly prepared activated-silica with a 1 ml of solution containing 10 mg Con A, 10 mg sodium cyanoborohydride prepared in 0.1 M sodium bicarbonate with 0.5 M sodium chloride. The suspension is then slowly stirred at room temperature for 3 h.
10. A 5-ml aliquot of sodium borohydride is added in portions over 30 min and the reaction is allowed to proceed for 1 h with stirring to reduce excessive aldehyde groups to diols.
11. The resulting Con A-silica is washed sequentially with 0.1 M sodium bicarbonate containing 0.5 M sodium chloride, and water. Finally the lectin-silica is washed and stored in the binding buffer at 4 °C before use.

12. Lectin microcolumns were packed in 5 cm of 0.25 mm inner diameter Polyether Ether Ketone (PEEK) tubing that is initially washed with acetone to remove particles and contaminants and it is allowed to dry at room temperature.
13. A 0.5-μm stainless steel frit with a PEEK ring is placed in a PEEK union that is fitted to one end of the PEEK tubing. The other end is connected to a Capillary Perfusion Toolkit packing apparatus (Perseptive Biosystems), which is filled with 2 ml of lectin-silica slurry, while the rest of the volume is filled with lectin binding buffer.
14. The packing apparatus is then sealed and connected to HPLC pump previously washed with lectin binding buffer. Packing should proceed until the pressure reaches about 2,500 psi (175 bar), then continue pumping for 15 min at the end of which the flow is stopped and system is allowed to depressurize.
15. The HPLC pump is carefully disconnected from the packing apparatus.
16. The second union with a stainless steel frit is used to cap the upper end of lectin microcolumn.
17. The lectin microcolumn is then washed with the binding buffer at 2 μl/min for 20 min and stored at 4 °C.
18. Figure 5 depicts a microscopic picture of the derivatized silica (a) under white light, (b) under the light of a wavelength suitable for the excitation of FTIC, and (c) under the light of a wavelength suitable for the excitation of Texas red. The even distribution of the emission observed in Figure 5b demonstrates an efficient lectin coupling, whereas the light observed in Figure 5c illustrates the efficiency of glycoprotein trapping. The performance of the silica-based lectin materials was also evaluated against their Sepharose counterpart. The trapping capabilities of both the Sepharose-based Con A material (Fig. 5d) and silica-based Con A material (Fig. 5e) with ribonuclease B as the substrate were compared. The trapping efficiencies of both materials are very comparable, as was deduced from the similar peak areas resulting from these analyses. However, a shorter analysis time observed for the silica-based lectin can be attributed to the ability of these materials to withstand high back-pressures without adversely affecting their trapping efficiencies. This feature cannot be offered by the Sepharose-based material that endures gel shrinkage at high backpressures, thus lengthening the analysis time.

3.2.2. Enrichment of Human Blood Serum Glycoproteins using Silica-Based Microcolumn Lectin Affinity Chromatography Coupled On-Line with High-Temperature Reversed-Phase Chromatography.

1. Recently, the lectin microcolumns have been used in conjunction with high-temperature reversed-phase chromatography to analyze human blood serum enriched glycoproteins (16). This system consists of a sample injector, lectin microcolumn, 10-port valve, isocratic and gradient pump, and C4 trap with Poroshell C8 reversed-phase microcolumn (0.5 x 75 mm). The setup scheme of this system is depicted in Figure 6.

Glycoprotein Enrichment Through Lectin Affinity Techniques 387

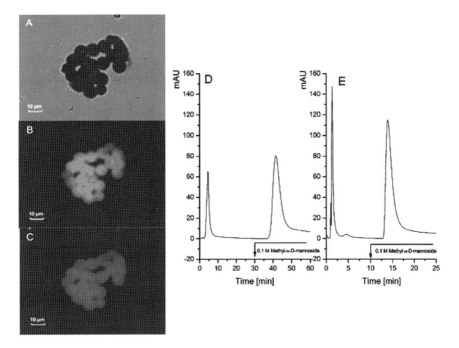

Fig. 5. Microscopic photographs of the lectin beads coupled to FITC-Con A and treated with Texas red-ovalbumin. Pictures were acquired under (a) white light, (b) light with excitation wavelength of 480 nm suitable for FITC and (c) light with excitation wavelength of 545 nm, suitable for Texas red. Binding capacity of Con A Sepharose lectin (A) versus Con A silica-based lectin (B). Conditions: Columns, Con A Sepharose (1 × 50 mm, 10 µm, 1000 Å) (A) and silica-based Con A lectin with 60 mg Con A/g silica (1 × 50 mm, 10 µm, 1000 Å) (B); buffers: loading buffer, 10 mM Tris/HCl (pH 7.4), 0.5 M sodium chloride, 1 mM calcium chloride, 1 mM manganese chloride, 1 mM magnesium chloride, 0.02% sodium azide; elution buffer, the same as the loading buffer containing 0.1 M methyl-α-D-mannoside; flow-rate, 50 µl/min; 50 µg injected; UV detection at 280 nm.

Source: Reproduced from *(31)*, with permission.

2. The system is initially configured so that the C4 trap is connected to the Poroshell microcolumn (maintained at 70 °C, Agilent Technologies, Palo Alto, CA) while the sample injector with lectin microcolumn are connected to the waste. This configuration will permit the trapping of glycoprotein in the lectin microcolumn. A 10-µl aliquot of human blood serum is injected, and passed through the lectin microcolum at a flow rate of 2 µl/min for 20 min during which glycoproteins are bound to the lectin media.
3. Next, the 10-port valve is switched back to the first position where C4 trap is connected to the injector with the lectin microcolumn. A 20-µl aliquot of the

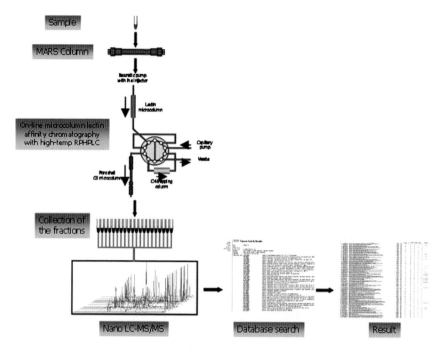

Fig. 6. Experimental workflow chart. *Source*: Reproduced from *(16)*, with permission.

elution buffer is injected to displace the bound glycoproteins from the lectin media. After 20 min, the 10-port valve configuration is switched back to the second position and the trapped glycoproteins are fractionate on a Poroshell microcolumn with a linear gradient from 3 to 100% solvent B (3% acetonitrile, 0.1% formic acid) over 20 min. Finally, 20 fractions are collected into a 96-well plate at the rate of 1 fraction/min.

3.3. Lectins Immobilized to Monolithic Capillaries

The versatile features of monolithic stationary-phases have prompted their wide use as chromatographic media in many capillary electrochromatography (CEC) and nano-LC applications *(17–26)*. These desirable features include: (i) nearly limitless avenues to explore in polymer design and modification; (ii) the ease of fabrication, allowing a monolith confinement in large-diameter columns, capillaries and microfluidic channels; and (iii) improved solute mass transfer because of the formation of macroporous structures, allowing the use of high flow rates without a penalty in resolution. The monolithic media are commonly designed in such a way that the mobile phase flows freely through macropores, while the surface area necessary for the separation of analytes is provided by the mesopores. Thus far, there have been only a limited number of applications demonstrating the utility of monolithic columns with immobilized affinity ligands for performing affinity-based analysis. Pen et al. *(27)* and Bedair and El Rassi *(28,29)* have recently demonstrated some potential of affinity monoliths in nano-LC and nano-LC and CEC, respectively. Bedair and El Rassi addressed the possibility of using monoliths for the preparation of lectin affinity-based media for nano-LC glycoprotein enrichment *(29)*. The application of lectin affinity monoliths was further expanded by Mao et al. *(30)* to include affinity-based separations in microfluidic chips.

1. Typically, monolithic stationary phases for affinity separations are prepared using ethylene dimethyl acrylate (EDMA) or trimethylpropane trimethylacrylate as the cross-linking monomer and glycidyl methacrylate (GMA) as the active monomer for a subsequent attachment. Monolithic capillary columns (50 × 4.6 mm i.d.) for affinity chromatography can be prepared by an in situ polymerization procedure using GMA as a monomer and trimethylopropane trimethylacrylate (TRIM) and EDMA as cross-linkers *(27)*.
2. Con A and WGA are immobilized to the monolithic capillary columns prepared through polymerization of GMA with EDMA or [2-(methacryloyloxy)ethyl]trimethylammonium chloride (MAETA) to yield neutral and cationic macroporous polymers, respectively *(29)*. Wider macropores are observed for the neutral monoliths with an immobilized lectin. For the fabrication of neutral monoliths, the polymerization solutions weighing 2 g each are prepared from GMA 18% (w/w), EDMA 12% (w/w), cyclohexanol 58.8% (w/w) and dodecanol 11.2% (w/w). For the preparation of positively charged monoliths, MAETA 1% (w/w) is added to the polymerization solution (2 g) consisting of GMA 17% (w/w), EDMA 12% (w/w), cyclohexanol 59.5% (w/w) and dodecanol 10.5% (w/w). In all cases, 2,2'-azobisisobutyronitrile (1.0%, w/w, with respect to monomers) is added to the polymerization solution as the initiator.
3. The solution is then degassed by purging with nitrogen for 5 min. A 40-cm long pretreated capillary is filled with the polymerization solution, up to 30 cm, by immersing the inlet of the capillary in the solution vial and applying vacuum to the outlet.
4. The capillary ends are then sealed with a rubber septum, and the capillary is submerged in a 50 °C water bath for 24 h. The resulting monolithic column is washed with acetonitrile and then with water using an HPLC pump.
5. The immobilization of lectin is achieved through the reaction between the epoxy groups found in the polymer structure and the ε-amino groups of the lectin. The glycidyl methacrylate monolithic column is first rinsed thoroughly with water and then filled with 0.1 M HCl solution and heated at 50 °C for 12 h to hydrolyze the epoxy groups. The column is then rinsed with water, followed by a freshly prepared solution of 0.1 M sodium periodate for 1 h to oxidize the ethylene glycol on the surface of the column to an aldehyde. Lectin is then immobilized on the monolithic material by pumping a solution of lectin at 5 mg/ml in 0.1M sodium acetate, pH 6.4, containing 1 mM of $CaCl_2$, 1 mM $MgCl_2$, 1 mM $MnCl_2$,

0.1 M and 50 mM sodium cyanoborohydride through the column overnight at room temperature.

6. The resulting column is then rinsed with water and cut to an effective length of 25 cm and a total length of 33.5 cm.

7. The monolithic columns are then utilized in conjunction with nano-LC in a two-dimensional format to isolate (enrich) and separate glycoproteins *(29)*. This is demonstrated for a mixture of five model proteins, including glucose oxidase, human transferrin, conalbumin (an ovotransferrin), trypsinogen, and lactalbumin. The first three are glycoproteins.

8. The mixture is introduced into the 2D system containing a short Con A column connected to a C17 monolith. While glucose oxidase and human transferrin have affinity to Con A, ovotransferrin does not. Therefore, glucose oxidase and human transferrin are captured by the Con A material, while the other proteins passed through and accumulated on the top of the C17 monolith. After this sample introduction step, the two columns are disconnected, and the C17 monolithic column is kept in the cartridge holder, while the lectin column with the adsorbed glycoproteins is taken out and let aside for further use.

9. Elution from the C17 monolith by step gradients of an increased percentage of acetonitrile in the mobile phase allows the separation of ovotransferrin, trypsinogen, and lactalbumin (Fig. 7a), while the two glycoproteins are "parked" on the Con A monolith.

10. In a second step, the Con A column is reconnected to the C17 monolithic column and subsequently eluted with the hapten sugar Me-α-D-Man, under which condition the enriched glycoproteins are moved to the C17 monolith. Thereafter, the Con A column is disconnected again and taken out from the capillary cartridge holder, and the C17 monolith is eluted by step gradients, which resulted in separating the two glycoproteins (Fig. 7b). This approach is an example of peak-parking in glycoproteomic analyses.

11. While the ability of interfacing such a scheme to mass spectrometry was not demonstrated, the potential of this approach is undoubtedly clear. Configuring the monolith-based lectin and monolith-based reversed-phase columns into a valving system will allow automation and high-throughput analyses.

3.4. Lectins Immobilized to Monolithic Microfluidics

Lectin affinity chromatography could be miniaturized into a microfluidic device using monoliths fabricated by polymerization of GMA and EDMA *(30)*. The monolith preparation and lectin immobilization is performed as described for the capillaries.

1. The glass chips are made by the standard photolithography and wet chemical etching techniques. The cross-section of the microchannel is 70 × 20 μm. The layout of this chip is shown in Figure 8. In addition to the required primary reservoirs, the reservoirs 5 and 6 are added for the microchannel washing to achieve a stable and reproducible electroosmotic flow (EOF) (Fig. 8A).

Glycoprotein Enrichment Through Lectin Affinity Techniques

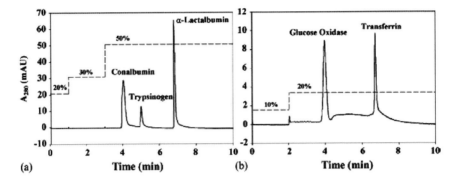

Fig. 7. Two dimensional separations of a mixture of some proteins using a Con A column (12 ×100 um i.d.) in the first dimension followed by reversed-phase LC on a neutral C17 capillary column (25 cm effective length, 33.5 cm total length×100 μm I.D.) in the second dimension. (a) Chromatogram obtained on the C17 column for the unretained proteins on the Con A column (passed through fraction); mobile phase used was 20 mM BisTris, pH 6.0, at 20% (v/v) acetonitrile for 1 min and then the acetonitrile content was increased stepwise to 30% at 1 min, and to 50% at 3.0 min. (b) Chromatogram obtained on the C17 column for the two glycoproteins transferred from the Con A column (i.e., previously captured by Con A column); mobile phase used was 20 mM BisTris, pH 6.0, at 10 % (v/v) for 2 min and then the acetonitrile content was increased to 20% at 2.0 min. Con A column conditions are as follows. Binding mobile phase: 20 mM BisTris, pH 6.0, containing 100 mM NaCl, 1 mM $CaCl_2$, 1 mM $MnCl_2$, 1 mM $MgCl_2$; eluting mobile phase: 0.2 M Me-α-D-Man in the binding mobile phase. Pressure drop: 1.0 MPa for both running mobile phase and sample injection. Sample injection: 12 sec.
Source: Reproduced from *(29)*, with permission.

2. A two-step procedure is utilized to immobilize lectin in the microfluidic channels. A porous monolith matrix is synthesized by UV-initiated polymerization, and then lectin is immobilized on the monolith matrix.
3. Before the fabrication of the monolith, the wall of the microchannels is pretreated with 3-(trimethoxysilyl) proxyl methacrylate (binding silane). The microchannels are flushed with 0.2 mol/L NaOH for 30 min; washed with water, followed by 0.2 mol/L HCl for 30 min; then washed with water and acetone and dried at 150 °C for 12 h.
4. The microchannels are then filled with a 30% (v/v) 3-(trimethoxysilyl) proxyl methacrylate solution in acetone. The device was sealed and kept overnight at room temperature and then rinsed with acetone and dried. The microchannels are filled with the monolith material mixture consisting of 24% glycidyl methacrylate (v/v %), 15 % ethylene dimethacrylate (v/v %), 60% dodecanol (v/v %), and 1% 2-dimethoxy-2-phenylacetophenone (w/v %). The sections of the chip that

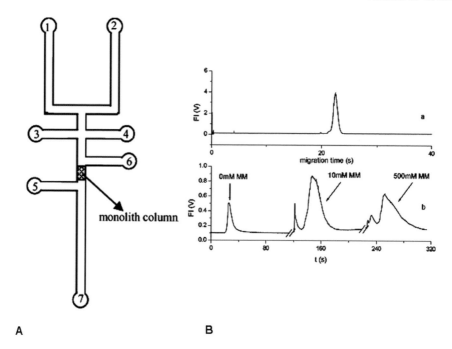

Fig. 8. (A) Layout of the microfluidic chip: i, running buffer reservoir (1); ii, eluent buffer reservoir (2); iii, sample reservoir (3); iv, sample waste reservoir (4); v, vi, washing reservoir; vii, waste reservoir (7). (B) Electropherograms of ovalbumin by CCZE and the PSA-affinity microfluidic chip-based method. (a) Electropherogram of ovalbumin by CCZE. (C) Electropherogram of ovalbumin by PSA- affinity microfluidic chip.
Source: Reproduced from *(30)*, with permission.

should not contain a monolith are covered with an opaque mask. The length of the unmasked area is kept at 500 μm.

5. The polymerization is initiated by exposing the masked chip to UV light (500 W) for 6 min. After the completion of the reaction, methanol is passed through the monolith column to remove the porogenic solvents and other unreacted soluble compounds by a vacuum device. The washing process is continued for 2 h.
6. Immobilization of lectins onto the monolith matrix took place via the epoxy groups found naturally in the polymer structure of the monolith material and the ε-amino groups of the lectins. The microchannel containing monolith is washed with increasing polar solvents (methanol, methanol/water, 1:1 (v/v)), then filled with a 100 mM carbonate buffer (pH 9.5) and equilibrated for 2 h before a 5.0 mg/ml solution of PSA in the same carbonate buffer was introduced. The immobilization reaction was allowed to proceed for 16 h at 30 °C. The microchannels were washed with the same carbonate buffer for 1 h. The extra epoxy groups were subsequently blocked by reacting with 1 M ethanolamine for 1 h at room temperature. The

microchannels were thoroughly washed with running buffer. The microchannels beside the PSA monolith column were washed with 1 M NaOH.
7. Using EOF as the driving force, the enrichment of three glycoproteins (turkey ovalbumin, chicken ovalbumin, and ovomucoid) is achieved with this microfluidic system.
8. All the glycoproteins are successfully separated into several fractions, with different affinities toward the immobilized *Pisum sativum agglutinin* lectin. The integrated system reduces the time conventionally required for the lectin affinity chromatography by ∼ 30-fold, thus allowing a complete analysis to be performed in 400 s. Also, lectin affinity chromatography is substantially simplified through integration of the different components into one microfluidic device, so that minimizing sample handling has allowed the enrichment of glycoproteins at 300 pg level (30). Further improvements in sensitivity are expected upon coupling of this microfluidic chip to mass spectrometry.

4. Notes

1. In the case of agarose-lectin, it is not recommended to exceed the maximum operating pressure of the medium or column, because medium is easily damaged at high flow rates.
2. The medium must be thoroughly washed to remove the storage solution, usually 20% ethanol, because residual ethanol may interfere with subsequent procedures.
3. Recovery from Sepharose-lectin is decreased in the presence of detergents. If the glycoprotein of interest needs the presence of detergent and has affinity for either lentil lectin or wheat germ lectin, such lectins may provide a suitable alternative to improve recovery.
4. For complex samples containing glycoproteins with different affinities for the lectin, a continuous gradient or step elution may improve resolution.
5. Recovery from lectin media can sometimes be improved by pausing the flow for some minutes during elution.
6. Tightly bound substances may be eluted by lowering the pH. Note that elution below pH 4.0 is not recommended and that below pH 5.0 Mn^{2+} will begin to dissociate from the Con A and the column will need to be reloaded with Mn^{2+} before reuse.
7. Lectin columns can be cleaned by washing with 10 column volumes of 0.5 M NaCl, 20 mM Tris-HCl, pH 8.5, followed by 0.5 M NaCl, 20 mM acetate, pH 4.5. Repeat 3 times before re-equilibrating with binding buffer.
8. Remove strongly bound substances by washing with 0.1 M borate, pH 6.5 at a low flow rate, or washing with 20% ethanol or up to 50% ethylene glycol, or washing with 0.1% Triton X-100 at 37 °C for one minute. Re-equilibrate immediately with 5 column volumes of binding buffer after any of these wash steps.

9. In the case of a contaminated sample, an ideal sinapinic acid matrix solution is prepared at 10 mg/ml concentration in 50:50 water/acetonitrile acqueous solution containing 0.1% TFA.
10. Sinapinic acid powder may also contain orange crystals that should not be used when preparing the matrix solution.

Acknowledgements

This work was supported by Grant No. GM24349 from the National Institute of General Medical Sciences, U.S. Department of Health and Human Services and a center grant from the Indiana 21st Century Research and Technology Fund. This work was also supported by the National Institute of Health through the National Center for Research Resources (NCRR) by grant number RR018942 for the National Center for Glycomics and Glycoproteomics.

References

1. Kellner, R., Lottspeich, F., and Meyer, H. E. (1999) Microcharacterization of Proteins, Wiley-Vch, Weinheim.
2. Walker, J. M. (Ed.) (1996) The Protein Protocols Handbook, Humana Press, Humana Press, Totowa.
3. Laemmli, U. K. (1970) Cleavage of structural proteins during the assembly of the head of bacteriophage T4. *Nature* **227**, 680–85.
4. Cummings, R. D. (1994) Use of Lectins in Analysis of Glycoconjugates. *Methods Enzymol.* **230**, 66–86.
5. Gravel, P., Walzer, C., Aubry, C., Balant, L. P., Yersin, B., Hochstrasser, D. F., and Guimon, J. (1996) New alterations of serum glycoproteins in alcoholic and cirrhotic patients revealed by high resolution two-dimensional gel electrophoresis. *Biochem. Biophys. Res. Commun.* **220**, 78–85.
6. Carlsson, J., Janson, J.-C., and Sparrman, M. (1989) Affinity chromatography, in Protein Purification (Janson, J.-C., and Rydén, L., eds.), VCH Publishers, Inc., Uppsala, pp. 275–329.
7. Lotan, R., Beattie, G., Hubbell, W., and Nicolson, G. L. (1977) Activities of lectins and their immobilized derivatives in detergent solutions – implications on use of lectin affinity chromatography for purification of membrane glycoproteins. *Biochemistry* **16**, 1787–94.
8. Kinzel, V., Kubler, D., Richards, J., and Stohr, M. (1976) Lens Culinaris Lectin Immobilized on Sepharose – Binding and Sugar-Specific Release of Intact Tissue-Culture Cells. *Science* **192**, 487–89.
9. Mechref, Y., Ma, W., Hao, G., and Novotny, M. V. (1999) N-Linked oligosaccharides of vomeromodulin, a putative pheromone transporter in rat. *Biochem. Biophys. Res. Commun.* **255**, 451–55.
10. Mechref, Y., Zidek, L., Ma, W., and Novotny, M. V. (2000) Glycosylated major urinary protein of the mouse: characterization of its N-linked oligosaccharides. *Glycobiology* **10**, 231–35.

11. Nawarak, J., Phutrakul, S., and Chen, S.-T. (2004) Analysis of lectin-bound glycoproteins in snake venom from the elapidae and viperidae families. *J. Proteome Res.* **3**, 383–92.
12. Larsson, P.-O. (1984) High-performance liquid affinity chromatography. *Methods Enzymol.* **104**, 212–23.
13. Voyksner, R. D., Chen, D. C., and Swaisgood, H. E. (1990) Optimization of immobilized enzyme hydrolysis combined with high-performance liquid-chromatography thermospray mass-spectrometry for the determination of neuropeptides. *Anal. Biochem.* **188**, 72–81.
14. Borrebaeck, C. A. K., Soares, J., and Mattiasson, B. (1984) Fractionation of glycoproteins according to lectin affinity and molecular-size using a high-performance liquid-chromatography system with sequentially coupled columns. *J. Chromatogr.* **284**, 187–92.
15. Green, E. D., Brodbeck, R. M., and Baenziger, J. U. (1987) Lectin affinity high-performance liquid-chromatography – interactions of N-glycanase-released oligosaccharides with leukoagglutinating phytohemagglutinin, concanavalin-A, datura-stramonium agglutinin, and vicia-villosa agglutinin. *Anal. Biochem.* **167**, 62–75.
16. Madera, M., Mechref, Y., Klouckova, I., and Novotny, M. V. (2006) Semiautomated high-sensitivity profiling of human blood serum glycoproteins through lectin preconcentration and multidimensional chromatography/tandem mass spectrometry. *J. Proteome Res.*, **5**, 2348–63.
17. Palm, A., and Novotny, M. V. (1997) Macroporous Polyacrylamide/Poly(ethylene glycol) Matrixes as Stationary Phases in Capillary Electrochromatography *Anal. Chem.* **69**, 4499–507.
18. Palm, A. K., and Novotny, M. V. (2004) Analytical characterization of a facile porous polymer monolithic trypsin microreactor enabling peptide mass mapping using mass spectrometry. *Rapid Commun. Mass Spectrom.* **18**, 1374–82.
19. Palm, A. K., and Novotny, M. V. (2005) A monolithic PNGase F enzyme microreactor enabling glycan mass mapping of glycoproteins by mass spectrometry. *Rapid Commun. Mass Spectrom.* **19**, 1730–38.
20. Williams, S. L., Eccleston, M. E., and Slater, N. K. H. (2005) Affinity capture of a biotinylated retrovirus on macroporous monolithic adsorbents: Towards a rapid single-step purification process. *Biotechnol. Bioeng.* **89**, 783–87.
21. Que, A. H., Mechref, Y., Huang, Y., Taraszka, J. A., Clemmer, D. E., and Novotny, M. V. (2003) Coupling capillary electrochromatography with electrospray Fourier transform mass spectrometry for characterizing complex oligosaccharide pools. *Anal. Chem.* **75**, 1684–90.
22. Guo, L., Eisenman, J. R., Mahimkar, R. M., Peschon, J. J., Paxton, R. J., Black, R. A., and Johnson, R. S. (2002) A proteomic approach for the identification of cell-surface proteins shed by metalloproteases. *Mol. Cell. Proteomics* **1**, 30–36.
23. Josic, D., and Buchacher, A. (2001) Application of monoliths as supports for affinity chromatography and fast enzymatic conversion. *J. Biochem. Bioph. Methods* **49**, 153–74.

24. Josic, D., Buchacher, A., and Jungbauer, A. (2001) Monoliths as stationary phases for separation of proteins and polynucleotides and enzymatic conversion. *J. Chromatogr. B* **752**, 191–205.
25. Svec, F. (2001) Capillary column technology: continuous polymer monoliths, in Capillary Electrochromatography (Deyl, Z., and Svec, F., Eds.), Elsevier, Amesterdam, pp. 183–240.
26. Svec, F., Peters, E. C., Sykora, D., and Frechet, J. M. J. (2000) Design of the monolithic polymers used in capillary electrochromatography columns. *J. Chromatogr. A* **887**, 3–29.
27. Pan, Z., Zou, H., Mo, W., Huang, X., and Wu, R. (2002) Protein A immobilized monolithic capillary column for affinity chromatography. *Anal. Chim. Acta* **466**, 141–50.
28. Bedair, M., and Rassi, Z. E. (2004) Affinity chromatography with monolithic capillary columns I. Polymethacrylate monoliths with immobilized mannan for thye separation of mannose-binding proteins by capillary electrochromatography and nano-scale liquid chromatography. *J. Chromatogr. A* **1044**, 177–86.
29. Bedair, M., and Rassi, Z. E. (2005) Affinity chromatography with monolithic capillary columns: II. Polymethacrylate monoliths with immobilized lectins for the separation of glycoconjugates by nano-liquid affinity chromatography. *J. Chromatogr. A* **1079**, 236–45.
30. Mao, X., Luo, Y., Dai, Z., Wang, K., Du, Y., and Lin, B. (2004) Integrated lectin affinity microfluidic chip for glycoform separation. *Anal. Chem.* **76**, 6941–47.
31. Madera, M., Mechref, Y., and Novotny, M. V. (2005) Combining lectin microcolumns with high-resolution separation techniques for enrichment of glycoproteins and glycopeptides. *Anal. Chem.* **77**, 4081–90.

30

Isolation of Bacterial Cell Membranes Proteins Using Carbonate Extraction

Mark P. Molloy

Summary

Producing high quality two-dimensional electrophoretic gels of bacterial cell membrane proteins is challenging because of non-membrane protein impurities within cell membrane preparations and the intractability of membrane proteins for solubilisation. Incubation of cell membrane preparations with sodium carbonate solution enriches for cell membrane proteins by stripping away other loosely associated protein contaminants from membranes (e.g., ribosomes, elongation factors). With the aid of strong sample solubilising reagents compatible with two-dimensional electrophoresis (2-DE), many of the carbonate-enriched membrane proteins can be separated using standard 2-DE techniques.

Key Words: Carbonate extraction; hydrophobicity; membrane proteins; sample solubilisation; two-dimensional gel electrophoresis.

1. Introduction

Carbonate extraction is a facile approach for purification and enrichment of membrane proteins from bacteria *(1)*. The method is based on the high alkalinity of sodium carbonate solution (pH 11) to solubilise and strip loosely associated proteins from cell membranes *(2)*. Cell membranes containing integral membrane proteins and some *N*-acyl diglyceride attached lipoproteins remain insoluble in the carbonate solution and can be recovered by centrifugation. Upon solubilisation of this pellet using reagents compatible with two-dimensional gel electrophoresis (2-DE), the membrane proteins can be displayed and analysed using standard proteomic techniques. One of

From: *Methods in Molecular Biology, vol. 424: 2D PAGE: Sample Preparation and Fractionation, Volume 1*
Edited by: A. Posch © Humana Press, Totowa, NJ

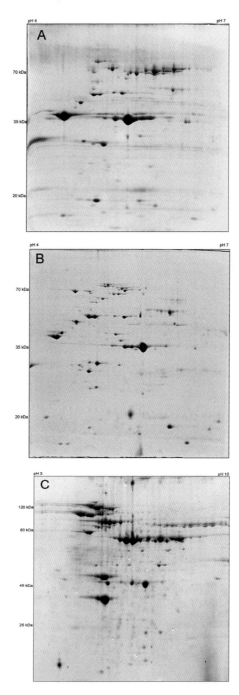

Fig. 1. Coomassie Blue stained two-dimensional electrophoresis gel of carbonate extracted bacterial membrane proteins from (**A**) *S. typhimurium*, (**B**) *K. aerogens*, (**C**) *C. crescentus*. (FIGSRC)Reproduced with permission from *(3)*.

the key benefits of this enrichment approach compared to density centrifugation isolation methods is that most of the "contaminating" highly abundant cytosolic proteins (e.g., ribosomal proteins, elongation factors) are removed from carbonate enriched membrane preparations.

Carbonate enriched cell membrane pellets can be solubilised for 2-DE with the aid of strong solubilisation reagents including thiourea and sulfobetaine zwitterionic surfactants. Using 2-DE gels, this approach has been successful in recovering outer membrane proteins from Gram-negative bacteria including, but not limited to, *Escherichia coli*, *Salmonella typhimurium*, *Klebsiella aerogens*, *Pseudomonas aeruginosa,* and *Caulobacter crescentus* (*1,3,4*). Example 2-DE gels of carbonate extracted membrane proteins from three Gram-negative bacteria are shown in Figure 1.

2. Materials

1. Wash solution: 50 mM Tris-HCl, pH 7.5.
2. DNase I.
3. Carbonate extraction solution: ice-cold 100 mM sodium carbonate (*see* **Note 1**).
4. Sample solubilisation solution: 7 M urea, 2 M thiourea, 1% (w/v) amidosulfobetaine 14 (ASB14), 0.5% (v/v) Triton X-100, 30 mM dithiotheritol, 40 mM Tris-base, 0.5% Biolytes pH 3–10 (*see* Note 2).

3. Methods

3.1. Cell Harvesting

1. Collect bacterial cells from 200–400 ml of liquid culture by centrifugation at 2,500g for 8 min.
2. Discard the media and resuspend cells in 20 ml wash solution with gentle pipeting.
3. Collect the cells by centrifugation at 2,500g for 8 min.
4. Discard the wash solution, then resuspend the equivalent of 20 mg cellular protein (*see* Note 3) in 5 ml of wash solution containing 0.7 mg DNase I.
5. Rupture the cells by 2× passage through a cold Aminco French press at 9.65×10^7 Pa (14,000 psi).
6. Discard any unbroken cells by centrifugation at 2,500g for 8 min (*see* Note 4).
7. Retain the supernatant for carbonate extraction.

3.2. Carbonate Extraction

1. Add the supernatant directly to 50 ml of carbonate extraction solution. Slowly stir the solution for 1 h in an ice bath.
2. The cell membranes are collected by ultracentrifugation of the carbonate extraction solution using a Beckman Type 55.2 Ti rotor or equivalent at 115,000g for 1 h at 4°C (*see* Note 5).

3. Discard the supernatant and wash the membrane pellet be resuspending in 2 ml wash solution by pipet mixing (*see* Note 6).
4. Collect the membrane pellet by ultracentrifugation as described in Step 2 above.

3.3. Two-Dimensional Gel Electrophoresis

1. Solubilise the membrane pellet for 2-DE in 1.2 ml sample solubilisation solution to give approximately 1 mg/ml (*see* Note 7).
2. Introduce approximately 200 μg or 500 μg protein for an analytical gel or preparative gel respectively, to an immobilised pH gradient (IPG) gel using the rehydration loading method *(5)*. Conduct 2-DE using standard protocols or as described elsewhere *(1)*.

4. Notes

1. The pH of 100 mM sodium carbonate is approximately 11. There is no need to adjust the pH by titration. The solution should be stored at 4°C for at least 1 h before use. The extraction can be conducted in an ice bath.
2. ASB14 is available from Calbiochem. Biolytes are available from Bio-Rad, although they can be substituted with carrier ampholytes from other suppliers. The ampholytes pH range should match the pH of the isoelectric focusing gel.
3. Following carbonate extraction, approximately 6% of the cellular protein is recovered for analysis. 20 mg starting material represents a good starting quantity of protein for 2-DE gel analysis. Techniques such as amino acid analysis or a protein assay using a replicate sample set are used to determine protein quantities.
4. Examine the unbroken-cell pellet size after French pressing. If the cell pellet is large, resuspend and repeat the French press procedure.
5. Some membranes can be collected by using a microcentrifuge operated at full speed (approximately 14,000 rpm). This can be a useful alternative to ultracentrifugation for preliminary studies. However, it should be cautioned that not all membranes will pellet at this speed and that for comprehensive analyses, ultracentrifugation should be used to collect cell membranes.
6. It is important to thoroughly wash the membranes to remove sodium carbonate that would otherwise interfere with isoelectric focusing.
7. To aid solubilization, use an ultrasonic probe tip with the sample chilled on ice between pulses. Care should be taken to avoid heating urea solutions above 30°C.

References

1. Molloy, M. P., Herbert, B. R., Slade, M. B., Rabilloud, T., Nouwens, A. S., Williams, K. L., Gooley, A. A. (2000) Proteomic analysis of the Escherichia coli outer membrane. *Eur. J. Biochem.* 267, 2871–2881.
2. Fujiki, Y., Hubbard, A. L., Fowler, S., Lazarow, P. B. (1982) Isolation of intracellular membranes by means of sodium carbonate treatment: application to endoplasmic reticulum. *J. Cell. Biol.* 93, 97–102.

3. Molloy, M. P., Phadke, N. D., Maddock, J. R., Andrews, P. C. (2001) Two-dimensional electrophoresis and peptide mass-fingerprinting of bacterial outer membrane proteins. *Electrophoresis* 22, 1686–1696.
4. Nouwens, A. S., Cordwell, S. J. Larsen, M. R., Molloy, M. P., Gillings, M., Willicox, M. D. P., Walsh, B. J. (2000) Complementing genomics with proteomics: the membrane sub-proteome of Pseudomonas aeruginosa PA01. *Electrophoresis* 21, 3797–3809.
5. Sanchez, J-C., Hochstrasser, D., Rabilloud, T. (1999) In-gel sample rehydration of immobilized pH gradient, in *Methods in molecular biology, vol. 112: 2D Proteome Analysis Protocols* (Link, A.J., ed.), Humana Press, Totowa, NJ. pp. 221–225.

31

Enrichment of Membrane Proteins by Partitioning in Detergent/Polymer Aqueous Two-Phase Systems

Henrik Everberg, Niklas Gustavsson, and Folke Tjerneld

Summary

Methods that combine efficient solubilization with enrichment of proteins and intact protein complexes are of central interest in current membrane proteomics. We have developed methods based on nondenaturing detergent extraction of yeast mitochondrial membrane proteins followed by enrichment of hydrophobic proteins in aqueous two-phase system. Combining the zwitterionic detergent Zwittergent 3–10 and the nonionic detergent Triton X-114 results in a complementary solubilization of proteins, which is similar to that of the anionic detergent sodium dodecyl sulfate (SDS) but with the important advantage of being nondenaturing. Detergent/polymer two-phase system partitioning offers removal of soluble proteins that can be further improved by manipulation of the driving forces governing protein distribution between the phases. Integral and peripheral membrane protein subunits from intact membrane protein complexes partition to the detergent phase while soluble proteins are found in the polymer phase. An optimized solubilization protocol is presented in combination with detergent/polymer two-phase partitioning as a mild and efficient method for initial enrichment of membrane proteins and membrane protein complexes in proteomic studies.

Key Words: Aqueous two-phase systems; detergents; enrichment; membrane proteins; proteomics; solubilization.

1. Introduction

A membrane proteome contains both integral membrane proteins and peripheral proteins associated with the integral proteins or the membrane surface by covalent or noncovalent interactions. Traditionally, protein separation for proteome analysis is performed by two-dimensional gel electrophoresis (2DE),

From: *Methods in Molecular Biology, vol. 424: 2D PAGE: Sample Preparation and Fractionation, Volume 1*
Edited by: A. Posch © Humana Press, Totowa, NJ

combining first dimension iso-electric focusing (IEF) with second dimension sodium dodecyl sulfate poly-acrylamide gel electrophoresis (SDS-PAGE). However, membrane proteins are poorly compatible with this methodology because of their hydrophobic nature and are known to aggregate and precipitate when focused at their iso-electric point in the first dimension *(1,2)*. Thus, transfer to the second dimension SDS-PAGE will be impeded and consequently membrane proteins will not be quantitatively, if at all, represented on the final 2DE gel. Currently used strategies in membrane proteomics are often based on high-resolution separation of peptides by liquid chromatography in combination with tandem mass spectrometry (LC-MS/MS) *(3–5)*. Although this approach has been successful in identifying large numbers of membrane proteins, a major drawback derived from the peptide-based nature of the approach is that protein-specific information such as size, iso-electric point, isoforms, post-translational modifications, and protein-protein interactions such as for multi-subunit membrane protein complexes, is lost in the proteolysis step *(6)*. Other separation methods relying on chromatography or one-dimensional SDS-PAGE suffer from low resolution and require sample prefractionation for successful characterization of membrane proteomes. Alternative separation methods, at the protein level, are therefore needed to facilitate and develop membrane proteomic research.

Here we describe a protocol for initial extraction of membrane proteins (solubilization) which, using a zwitterionic detergent (Zwittergent 3–10) combined with a nonionic detergent (Triton X-114) and after optimizing experimental conditions in terms of detergent/protein ratio, is both quantitatively and qualitatively comparable to the solubilization efficiency of sodium dodecyl sulfate (SDS), but with the important advantage of being nondenaturing. Thus, it will be possible to obtain information about protein-protein interactions and to further prefractionate enriched protein complexes after using the solubilization protocol described here.

The optimized membrane solubilization is compatible with membrane protein enrichment by detergent/polymer aqueous two-phase partitioning (Fig. 1). Many nonionic detergents will form aqueous two-phase systems when mixed with a polymer in sufficient concentrations and proteins are distributed between them depending on inherent properties of the proteins and the phase components *(7,8)*. Detergent/polymer two-phase systems can be designed to increase the partitioning of soluble proteins to the polymer phase by manipulating the thermodynamic driving forces governing protein partitioning between the phases. In this protocol electrostatic driving force is employed to influence the partitioning of soluble proteins whereas hydrophobic interaction governs the partitioning of membrane proteins. This results in an enrichment of membrane

Enrichment of Membrane Proteins by Partitioning

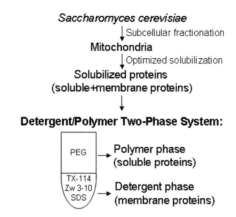

Fig. 1. Enrichment strategy. Mitochondria from *S. cerevisae* were isolated by subcellular fractionation. The isolated mitochondria were solubilized and fractionated in a detergent/polymer aqueous two-phase system formed by addition of polymer and salt to the detergent solubilized mitochondrial membranes. Membrane proteins partitions to the detergent (bottom) phase and the bulk of soluble proteins to the polymer (top) phase at the chosen conditions.

proteins in the detergent phase *(7–9)* and the detergent/polymer two-phase system we have developed for this application will be described below.

The method described in this chapter has been utilized for initial extraction and enrichment of membrane proteins and membrane protein complexes in mitochondria from the yeast *Saccharomyces cerevisiae*. The method can also be combined with further prefractionation by ion-exchange chromatography *(7)*, separation by SDS-PAGE and LC-MS/MS analyses *(9)*.

2. Materials

The protocol described here applies to preisolated mitochondria from commercially processed *S. cerevisiae*. This membrane system represents a relatively small proteome from a fully sequenced organism that contains, besides both soluble and membrane-bound proteins, a number of well-known multi-subunit membrane protein complexes.

2.1. Preparation of Mitochondria from S. cerevisiae

1. Mitochondria isolation buffer (MIB): 20 mM HEPES-KOH, pH 7.4, 0.6 M mannitol, Complete protease inhibitor cocktail, (Boehringer Mannheim GmbH, Germany), 1 tablet/ 50 ml buffer. Prepare MIB fresh at use.
2. Dounce homogeniser with loose fitting plunger

3. Glass beads 450–600 microns (Sigma-Aldrich, St Louis, MO, USA).
4. Beadbeater homogenizer (BioSpec Products Inc. Bartlesville, OK, USA).

2.2. Solubilization

1. Detergent stock solutions (% w/w): 20% Zwittergent 3-10 (Calbiochem, San Diego, CA, USA) and 2% Triton X-114 (Sigma-Aldrich, St Louis, MO, USA), (*see* Note 1). Store at +4 °C.
2. 100 mM Tris-HCl, pH 9.0, (stock solution).
3. Refrigerated centrifuge with rotor for 200 µl and capacity for 100000×g.
4. Sodium dodecyl sulfate (stock solution 10% w/w) (Merck, Darmstadt, Germany). Store at room temperature.
5. Bicinchoninic acid (BCA) protein determination assay reagent A and B (Pierce, Rockford, Il, USA).

2.3. Detergent/polymer two-phase partitioning

1. Poly-(ethylene glycol) (PEG) glycol (M_r: 40000, Serva, Heidelberg, Germany) 25% (w/w) stock solution. Store at +4 °C.
2. Detergent stock solutions: 100% (w/w) Triton X-114, 20% (w/w) (Sigma-Aldrich, St Louis, MO, USA), Zwittergent 3–10 (Calbiochem, San Diego, CA, USA), 100 mM SDS (Merck, Darmstadt, Germany).
3. 200 mM glycine-NaOH buffer, pH 10.0 (stock solution).
4. Maximum recovery 1.5-ml tubes (Axygen, Union City, CA, USA) (*see* Note 2).
5. Refrigerated centrifuge for Eppendorf tubes (1000×g).

2.4. Sample Concentration and Clean-Up

The isolated phases after two-phase partitioning require concentration and removal of polymers and detergents before separation by SDS-PAGE (*see* Note 3).

1. SDS-PAGE clean-up kit (GE Healthcare, Uppsala, Sweden).
2. Dithiothreitol (Sigma-Aldrich, St Louis, MO, USA).

3. Methods

3.1. Preparation of Mitochondria from S. cerevisiae

This protocol is adapted from McAda et al. *(10)* and all experiments are performed at +4 °C.

1. Pressed yeast (400 g) is suspended in 200 ml ice cold MIB, using a Dounce homogenizer.
2. The yeast suspension is homogenized 5 × 20 sec with 1 min rest on ice between each treatment, using a Beadbeater homogenizer filled to 2/3 with glass beads.

3. The homogenized cell suspension is pooled and the suspension is centrifuged at 3,500g for 10 min and the pellet is discarded. This is performed three times in total to remove cell debris.
4. The supernatant is centrifuged for 20 min at 17,000g, to pellet the mitochondria.
5. The pellet is washed once in ice cold MIB and step 4 is repeated.
6. The mitochondrial pellet is resuspended in a minimal volume of MIB (~5 ml) and stored in aliquots at −80 °C until use.

3.2. Solubilization

1. Mitochondria (0.5 mg total protein) are homogenized by freeze-thawing twice (−80 °C) to disrupt membrane structures and release the soluble fraction into the solution (*see* Note 4).
2. Add 20 µl Tris-HCl buffer from stock solution (10 mM final concentration).
3. Add 60 µl Zwittergent 3-10 from stock solution (20% w/w) to obtain the desired detergent/protein ratio (30 mmol detergent/g protein) (*see* **Note 5**).
4. Add 56 µl TX-114 from stock solution (2% w/w) to obtain the desired detergent/protein ratio (5 mmol detergent/g protein).
5. Add 64 µl water to obtain the right final detergent/protein ratio and a solubilization volume of 200 µl.
6. Incubate with gentle agitation at +4 °C for 30 min.
7. Centrifuge at 100,000g for 45 min to pellet any unsolubilized material and recover the supernatant.
8. The pellet is resolubilized by vortexing in 50 µl 2% SDS at room temperature until totally dissolved.
9. The amount of protein in the supernatant (and the pellet) is determined using the BCA protein assay (*see* Note 6).

3.3. Detergent/Polymer Two-Phase Partitioning

The supernatant after optimized detergent solubilization is transferred to a two-phase system premixed in Maximum recovery 1.5-ml tubes (Axygen, Union City, CA, USA) and the total weight of each system is 0.5 g (*see* Note 7).

1. Add 0.075 g TX-114 (100% stock solution) to obtain final concentration of 15% (w/w).
2. Add 0.080 g PEG 40000 (25% stock solution) to obtain final concentration of 4% (w/w).
3. Add 0.025 g glycine-NaOH, pH 10 (200 mM stock solution) to a final concentration of 10 mM (*see* Note 8).
4. Add 0.015 g SDS (100 mM stock solution) to a final concentration of 3 mM (*see* Note 9).
5. Add 0.138 g water.

6. Finally add 0.167 g of the supernatant after solubilization (6% (w/w) Zwittergent 3–10) to reach a final concentration of 2% (w/w) Zwittergent 3–10) (*see* Note 10) and the desired total system weight of 0.5 g.
7. Mix thoroughly by inverting the tube until the system is opaque and phase separation is achieved.
8. Incubate for 30 min at +4 °C with gentle agitation.
9. Speed up phase separation by centrifugation at 1,600g for 15 min.
10. Isolate the phases using a Pasteur pipet (*see* Note 11).
11. The protein amount in each phase is determined by the BCA-assay (*see* Note 6) and the partitioning coefficient is calculated as the total protein concentration in the top phase divided by the total protein concentration in the bottom phase.

3.4. Sample Concentration and Clean Up

1. Perform clean up of sample to remove polymers and detergents using SDS-PAGE clean up kit (Amersham Biosciences) (*see* Note 12). Concentrated and cleaned samples can be used immediately or stored at −20 °C.
2. Sample is then ready for SDS-PAGE, and visualization and analysis as required (*see* Note 13).

4. Notes

1. As solubilization efficiency might not be comparable between different membrane systems, an initial screening of detergents and combinations of detergents should be performed. We have studied the quantitative solubilization efficiency of one zwitterionic (Zw 3–10), one nonionic (TX-114), one anionic detergent (SDS) (Fig. 2a) and the qualitative solubilization efficiency of Zw 3–10 and TX-114 combined at optimized conditions (Fig. 2b). It is essential that tubes used during screening for optimal solubilization conditions are absolutely clean by washing with 70% EtOH and Millipore water to avoid contamination of the sample with other detergents giving false results of solubilization efficiency.
2. To minimize the loss of protein because of adsorption to the test tube walls.
3. This step is performed to be able to transfer as much of the protein content in the isolated samples to the SDS-PAGE and to remove polymers and detergents disturbing protein migration in the gels.
4. As the important parameter is detergent/protein ratio, volumes, or amounts should be scaled accordingly.
5. The relatively high detergent/protein ratio of ∼30 mmol/g needed to reach the maximum level of protein extraction for Zw 3–10 (Fig 2a) is probably because of its high critical micell concentration (CMC) (∼40 mM).
6. To minimize the interference of membrane lipids and detergents, all samples are diluted with SDS stock solution to reach a final concentration of 2% SDS. Standard curves are made from known concentrations of BSA including 2%

Enrichment of Membrane Proteins by Partitioning

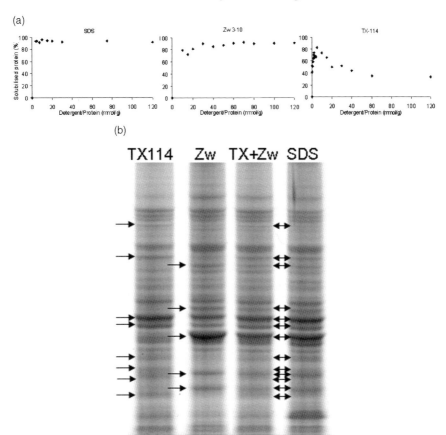

Fig. 2. Optimization of solubilization efficiency. (a) The quantitative solubilization efficiency was investigated for one ionic (SDS), one zwitterionic (Zw 3–10) and three nonionic (here represented by TX-114) detergents. The detergent/protein molar ratio was screened and the amount of protein in the supernatant after incubation and centrifugation was measured. (b) Complementary detergent solubilization. After incubation with detergent and centrifugation the protein content of the supernatants was separated by SDS-PAGE. The arrows indicate a number of membrane proteins more efficiently solubilized by either Triton X-114 (TX114) or Zwittergent 3–10 (Zw). By mixing the detergents (TX+Zw) a complementary protein pattern was obtained. This pattern was also found comparable to the protein pattern obtained after solubilization with SDS.

SDS and 1% of the detergent used for solubilization (provided that a dilution series has been made to ensure that increased detergent concentration do not result in increased absorption). The solubilization efficiency is presented in Figure 2a. It is defined (Eq. 1) as the fraction of solubilized membrane protein found in the supernatant of the total amount of membrane protein in pellet plus supernatant.

$$\text{Detergent solubilization efficiency (\%)} \frac{(A-B)}{C+(A-B)} \times 100 \quad (1)$$

Here, A stands for total protein amount in the supernatant; B for soluble protein amount in the supernatant; and C for protein amount in the pellet. The soluble protein amount was determined as the amount of protein found in the supernatant when no detergent was added.

7. It is important to get the right proportions of the phase components for successful phase separation and phase volume ratio. Thus, because of the viscosity of the polymers and detergents they are weighed in to obtain the right system composition.
8. The ionic strength of the system should be kept to a minimum for optimal electrostatic repulsion effect discussed in Note 9.
9. Increased partitioning of proteins to the polymer phase is obtained by addition of a negatively charged detergent, SDS, at a concentration below its CMC (10 mM). Monomers of SDS get incorporated in the mixed micelles of nonionic and zwitterionic detergents, thus introducing negative charges *(7–9,11)*. This results in repulsion of negatively charged soluble proteins from the negatively charged mixed micelles in the detergent phase. The denaturing properties of SDS will be insignificant because of the low concentration of SDS monomers in the solution. By raising the pH, increased negative protein net charge leads to stronger repulsion between proteins and negatively charged mixed micelles,

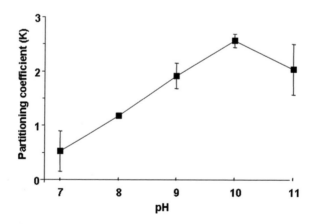

Fig. 3. Effect of pH on protein partitioning in the TX-114/PEG aqueous two-phase system. By increasing the pH of the system including SDS, the partitioning coefficient (K = [protein concentration in the top phase)/(protein concentration in the bottom phase]) is shifted from a partitioning predominantly to the detergent phase (K = 0.5) at pH 7, to a more pronounced partitioning to the polymer phase (K = 2.5) at pH 10.

which enhances the partitioning of soluble proteins to the polymer-enriched top phase. Membrane proteins are not affected because the hydrophobic interaction between membrane protein and the hydrophobic core of the mixed micelles govern the partitioning for these proteins. The results shown in Figure 3 demonstrates an increased partitioning of proteins to the polymer phase as an effect of including SDS and increasing the pH of the system from 7 to 10.

10. The concentration of Zwittergent 3–10 should be kept to a minimum but above the CMC to avoid loosing membrane protein solubility.
11. Be sure to leave the interphase undisturbed to minimize contamination from the bottom phase as the top phase is withdrawn. Discard the interphase and collect the bottom phase with a clean Pasteur pipet
12. The concentration and clean up is performed according to the manufacturers instructions but instead of the final heating step (95 °C, 5 min.), the samples are incubated at room temperature for 30 min. to avoid aggregation of hydrophobic proteins. The pellet was totally dissolved in 25 μl loading buffer by vortexing and incubation in a sonicator bath.
13. We used NuPAGE Bis-Tris precast gradient gels (4–12%) from Novex (San Diego, CA, USA) for SDS-PAGE according to the manufacturer's instructions, stained in colloidal Coomassie brilliant blue and destained in water, followed by tryptic in-gel digestion and LC-MALDI TOF/TOF analysis.

Acknowledgments

This work is funded by the Swedish Research Council (FT) and the Swedish Foundation for Strategic Research (NG).

References

1. Santoni, V., Molloy, M. and Rabilloud, T. (2000) Membrane proteins and proteomics: un amour impossible? *Electrophoresis* 21, 1054–70.
2. Schluesener, D., Fischer, F., Kruip, J., Rogner, M. and Poetsch, A. (2005) Mapping the membrane proteome of Corynebacterium glutamicum. *Proteomics* 5, 1317–30.
3. Washburn, M. P., Wolters, D. and Yates, J. R., 3rd. (2001) Large-scale analysis of the yeast proteome by multidimensional protein identification technology. *Nat Biotechnol* 19, 242–7.
4. Wu, C. C., MacCoss, M. J., Howell, K. E. and Yates, J. R., 3rd. (2003) A method for the comprehensive proteomic analysis of membrane proteins. *Nat. Biotechnol.* 21, 532–8.
5. Zhao, Y., Zhang, W. and Kho, Y. (2004) Proteomic analysis of integral plasma membrane proteins. *Anal Chem* 76, 1817–23.
6. Nesvizhskii, A. I. and Aebersold, R. (2005) Interpretation of shotgun proteomic data: the protein inference problem. *Mol Cell Proteomics* 4, 1419–40.
7. Everberg, H., Sivars, U., Emanuelsson, C., Persson, C., Englund, A. K., Haneskog, L., Lipniunas, P., Jornten-Karlsson, M. and Tjerneld, F. (2004) Protein

pre-fractionation in detergent-polymer aqueous two-phase systems for facilitated proteomic studies of membrane proteins. *J Chromatogr A* 1029, 113–24.
8. Sivars, U. and Tjerneld, F. (2000) Mechanisms of phase behaviour and protein partitioning in detergent/polymer aqueous two-phase systems for purification of integral membrane proteins. *Biochim Biophys Acta* 1474, 133–46.
9. Everberg, H., Leiding, T., Schioth, A., Tjerneld, F. and Gustavsson, N. (2006) Efficient and non-denaturing membrane solubilization combined with enrichment of membrane protein complexes by detergent/polymer aqueous two-phase partitioning for proteome analysis. *J Chromatogr A* 1122, 35–46.
10. McAda, P. C. and Douglas, M. G. (1983) Biomembranes. in *Methods in Enzymology* (Fleischer, S. & Fleischer, B., eds.), Vol. 97. Academic press, New York, pp. 337.
11. Sivars, U., Bergfeldt, K., Piculell, L. and Tjerneld, F. (1996) Protein partitioning in weakly charged polymer-surfactant aqueous two-phase systems. *J Chromatogr B* 680, 43–53.

32

The Isolation of Detergent-Resistant Lipid Rafts for Two-Dimensional Electrophoresis

Ki-Bum Kim, Jae-Seon Lee, and Young-Gyu Ko

Summary

Because lipid rafts are plasma membrane platforms mediating various cellular events such as in signal transduction, immunological response, pathogen invasion, and neurodegenerative diseases, protein identification in the rafts could provide important information to study their function. Here, we present an optimized method to isolate detergent-resistant lipid rafts that are subsequently analyzed by two-dimensional electrophoresis (2-DE). Lipid rafts were isolated based on their two distinct biochemical properties such as Triton X-100 insolubility and low density. To solubilize completely the proteins embedded in lipid rafts, sample lysis buffer (9 M urea, 2 M thiourea, 100 mM DTT, 2% CHAPS (w/v), 60 mM n-octylβ-D-glucopyranoside, 2% IPG buffer) was applied to the isolated rafts. This method was found to be the most suitable choice for obtaining 2-DE profile of lipid raft proteome from various cells and tissues. We expect that this method could provide the way to dissect the function of raft-associated proteins and to gain a comprehensive insight upon various cellular events mediated through lipid rafts, the specialized domains in cell surface.

Key Words: detergent-resistant lipid rafts; Two-dimensional electrophoresis.

1. Introduction

Specific membrane compartments in the plasma membrane concentrate and organize signal molecules for efficient and rapid cellular signal transductions. Because these specific membrane compartments are rich in cholesterol and glycosphingolipids with saturated long fatty acid chains, they exist in the liquid-ordered phase (l_o). Conversely, other plasma membrane regions mainly contain phospholipids with unsaturated short fatty acid chains, so that they show a liquid-disordered phase (l_d) *(1)*. The l_o and l_d membrane domains are present separately in the model membranes and the cellular plasma membrane *(2–5)*. Because l_o domains are among the "sea" of phospholipids (l_d domains) present in the plasma membrane, they are referred to as lipid rafts. Because of their tight molecular packing, lipid rafts are resistant to extraction by the nonionic detergents, which include Triton X-100, Lubrol XW, and Brij *(6–8)*. Thus, the detergent-resistant lipid rafts can be isolated based on their detergent insolubility and low density using sucrose gradient ultracentrifugation.

Detergent-resistant lipid rafts have been known to concentrate transmembrane proteins and lipid-modified proteins, but the means by which these proteins are incorporated into the lipid rafts has not yet been elucidated *(9)*. The detergent-resistant lipid rafts contain different kinds of receptors and downstream signaling molecules that mediate various cellular signal transductions. For example, epidermal growth factor (EGF) receptors are predominantly enriched into the plasma membrane lipid rafts before the stimulation of EGF *(10)*. After EGF stimulation, Ras, Raf, and MAPK are rapidly recruited into the lipid rafts to operate EGF signaling *(10)*. In addition to the mediation of various cellular signal transductions, the lipid rafts are required for pathogen entry, immunological synapse formation, cellular migration, endocytosis, and cholesterol homeostasis *(11–15)*.

To identify lipid raft proteins involved in specific cellular events, detergent-resistant lipid raft proteins have been identified via proteomic analysis in a variety of mammalian organs or cell types *(8,15–18)*. However, relatively few differential proteomic analyses have been performed for lipid rafts. Although differential quantitative proteomic techniques, including stable isotope-labeling with amino acids in cell cultures (SILAC) *(19)* and isotope-coded affinity tag (ICAT) techniques *(20)*, have recently been developed, two-dimensional electrophoresis (2-DE) remains the most powerful technique in differential proteome analysis. In addition, protein spots in two-dimensional gels can be visualized relatively easily by silver staining or fluorescent dyes, and these spots can be analyzed via matrix-assisted laser desorption ionization and time-of-flight spectrometry (MALDI-TOF)*(21)*, or by electrospray ionization tandem mass spectrometry (ESI-MS/MS) *(22)*. Here, we describe an optimized method to obtain detergent-resistant lipid rafts suitable for analysis by 2-DE.

2. Materials
2.1. Preparation of Lipid Rafts from Cultured Cells
1. 100 mM phenylmethylsulfonyl fluoride (PMSF) stock solution: Dissolve in methanol and store at room temperature. Add PMSF stock solution at 1 mM of working concentration just before use.

2. Protease inhibitor cocktail: Commercially purchased (Roche Diagnostics, Mannheim, Germany; Complete Mini or equivalent). One tablet of protease inhibitor cocktail per 10 ml buffer (*see* Note 1).
3. Phosphate buffered saline (PBS): 8 mM Na_2HPO_4, 15 mM KH_2PO_4, 27 mM KCl, 137 mM NaCl, pH 7.4. Prepare 10× and store 4 °C.
4. Modified HEPES buffer: 25 mM HEPES-HCl, pH 6.5, 150 mM NaCl, 1 mM EDTA, 1 mM PMSF, protease inhibitor cocktail. Prepare 2× and store at 4 °C (*see* Note 2).
5. 1% Triton X-100 lysis buffer: 1% (v/v) Triton X-100 in 10 ml modified HEPES buffer. Store at –20 °C (*see* Note 3).
6. Sucrose cushion: 80%, 30%, 5% (w/v) sucrose in modified HEPES buffer (without PMSF and protease inhibitor cocktail) for discontinuous gradient ultra centrifugation. Store at 4 °C.
7. Miscellaneous: Teflon-coated Dounce homogenizer (Wheaton Science Products, Millville, USA), ultracentrifuge tubes (Beckman Instruments, Palo Alto, USA).

2.2. Preparation of Lipid Rafts from Various Mouse Tissues

1. Male mouse (160–180 g).
2. Nylon screen (pore size: 300 mesh).
3. Motor-driven tissue homogenizer: Brinkmann homogenizer (Brinkmann Instruments, Westbury, USA; or equivalent).
4. Dissecting knives and scissors for organ extraction and mincing procedure.
5. All materials described in Section 2.1 are also necessary for the preparation of lipid rafts from mouse tissue.

2.3. Fractionation of Lipid Rafts for Immunoblotting

1. RIPA buffer: 25 mM Tris-HCl, pH 7.2, 150 mM NaCl, 0.1% (w/v) SDS, 1.0% Triton X-100, 1% deoxycholate, 5 mM EDTA. Store at room temperature.
2. 5× SDS sample buffer: 60 mM Tris-HCl, pH 6.8, 25% (v/v) glycerol, 2% (w/v) SDS, 14.4 mM 2-mercaptoethanol, 0.1% (w/v) bromophenol blue. Store at –20 °C.
3. Peristaltic pump (Biorad, Hercules, USA; Econo Pump or equivalent) and its adjunct silicon tubes (Tube diameter: 1.6 mm), mirocapillary tubes (Sigma # P1049).
4. Heat block (Bioneer, Seoul, Korea; My genie 32 thermal block or equivalent)
5. Sonicator (Branson Ultrasonic Corporation, Danbury, USA; Sonifier 250 or equivalent)

2.4. Sample Preparation of Isolated Lipid Rafts for Proteomic Analysis

1. 2-DE sample lysis buffer: 9 M urea, 2 M thiourea, 100 mM DTT, 2% CHAPS (w/v), 60 mM *n*-octyl β-D-glucopyranoside, 2% IPG buffer (Amersham Biosciences, Piscataway, USA; IPG buffer, pH 3-10), protease inhibitor cocktail. Store at –20 °C.
2. Kit for protein quantification: Bio-Rad protein assay kit (Biorad, Hercules, USA) or equivalent.

3. Methods
3.1. Isolation of lipid rafts from cultured cells

1. Prepare cells grown to 90% confluence in 150 mm culture plates (4 to 5 plates or equivalent to 300 μl packed cells) (*see* Note 4).
2. Wash cells with ice-cold PBS buffer twice and place the culture plates on ice.
3. Detach cells with cell-scrapper in several ml of PBS buffer, and suspend well by gentle pipeting (*see* Note 5).
4. Centrifuge at 288g for 5 min and remove supernatant.
5. Apply 700 μl 1% Triton X-100 lysis buffer, and then gently suspend the cells by using a pipet.
6. Homogenize sample-detergent mixture (lysates) using a Teflon-coated Dounce homogenizer (20–30 strokes), and incubate at 4 °C (leave the tube containing the homogenate on ice) for 30 min (*see* Note 6).
7. After incubation, mix lysates with same volume (1 ml) of 80% sucrose cushion solution to yield a mixture at a final 40% sucrose gradient, and then transferred into a 12 ml polyallomer ultracentrifuge tube (for an SW 41 rotor, Beckman Instruments).
8. At the top of the sample-sucrose mixture, overlay 6.5 ml of 30% and 3.5 ml of 5% sucrose cushion, respectively.
9. Ultracentrifuge at 187, 813g, 20 h, 4 °C using an SW41 rotor (*see* Note 7).
10. After centrifugation, the floating opaque band corresponding to the detergent-resistant membrane fraction can be found at the interface between the 30% and 5% sucrose gradients (*see* Fig. 1).

The procedure of lipid raft isolation is basically completed at step 10. A different method of sample preparation can be applied depending on the type of analysis.

3.2. Isolation of Lipid Rafts from Various Mouse Tissues

1. Dissect the tissue from a mouse and rinse with ice-cold PBS buffer.
2. Mince the tissue with scissors and a knife, and briefly grind with a motor-driven mechanical grinder (*see* Note 8).
3. Apply 700 μl of 1% Triton X-100 lysis buffer to 300 μl of tissue sample, and suspend well by gentle pipeting.
4. Homogenize lysate with a Dounce homogenizer (up to 30 strokes), and incubate at 4 °C for 30 min.
5. The subsequent sucrose gradient ultracentrifugation procedures are identical to the steps performed in the preparation of lipid rafts from cultured cells (*see* Note 9).

3.3. Preparation of Lipid Rafts for Immunoblotting (see Note 10).

1. Once the ultracentrifugation step is completed, fractionate the sucrose gradient into twelve 1 ml fractions from the bottom to the top using a capillary tube that is connected with a peristaltic pump (*see* Fig. 2). Fractions corresponding to the

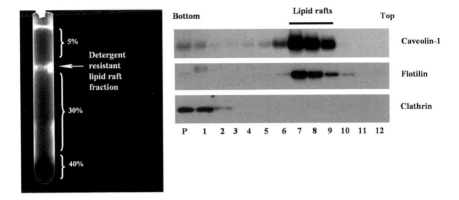

Fig. 1. Isolation of detergent-resistant lipid raft fractions from mouse liver followed by quality assessment of isolated lipid raft fractions by immunoblotting. (**A**) Lipid rafts from mouse liver were prepared based on their detergent insolubility and low density. (**B**) After density gradient ultracentrifugation, a peristaltic pump was used to fractionate the sucrose gradients from the bottom to the top with 13 fractions (*see* **Fig. 2** for detailed description), and the fractions were then analyzed via immunoblotting with anti-caveolin-1 and anti-flotillin-1 antibodies. As shown in the figure, markers for lipid rafts (caveolin-1, flotillin) appeared predominantly in the fractions containing lipid rafts (#7-9), whereas the nonraft marker, clathrin, was mostly found in the bottom of the fraction consisting of 40% sucrose. P indicates pellet fraction.

5–30% sucrose interface (fractions #7 and 8) are usually referred as lipid raft fractions.
2. To create the pellet fraction, suspend the pellet remaining at the bottom of the tube with 1 ml of RIPA buffer. Sonicate sample to enhance solubilization.
3. Mix each fraction with 5× SDS-PAGE sample buffer and boil at 95 °C for 5 min in a heat block (*see* Note 11).

3.4. Preparation of Isolated Lipid Rafts for 2-D Electrophoresis (see Note 12).

1. After ultracentrifugation, collect the buoyant band at the interface between the 5% and 30% sucrose gradients using a pipet.
2. Dilute raft fractions to 10–12 ml with modified HEPES buffer, and ultracentrifuge at 49,392g, 30 min, 4 °C to obtain the lipid raft pellet.
3. Gently discard supernatant and dissolve the pellet with 100–150 µl of 2-DE sample lysis buffer with brief vortexing (*see* Note 13).
4. Centrifuge at 20,817g, 10 min, 4 °C, then secure supernatant for subsequent 2-D gel analysis (*see* Note 14 and 15). Fig. 3 shows the lipid raft protein profiles on 2-dimensional electrophoresis gels.

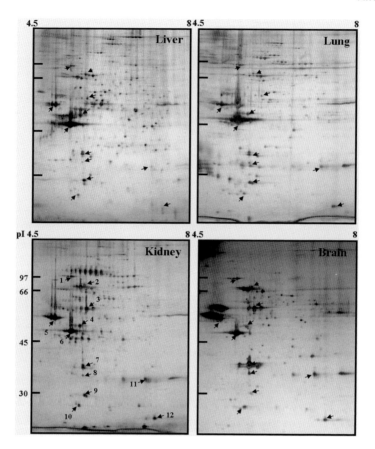

Fig. 2. Lipid raft fractionation followed by density gradient ultracentrifugation. During fractionation, all samples are placed on ice. The remaining pellet at the bottom of the ultracentrifuge tube after fractionation is separately prepared with 1 ml RIPA buffer for the "Pellet" fraction (*see* Section **3.3.2**).
Source: (From Kim, K.-B. *et al.*, *Proteomics*, 6, 2447, 2006, with permission).

4. Notes

1. Instead of a protease inhibitor cocktail, other reagents that have the same effect of protease inhibition, such as leupeptine (serine and cystein protease inhibitor), pepstatin (aspartyl protease inhibitor), and bestatin (aminopeptidase inhibitor), can be used at concentrations of 2–20 μg/ml.
2. This buffer is used in many steps of lipid raft isolation, as described in the *Materials* section.
3. Although Triton X-100 is the detergent that is most commonly used for the isolation of lipid rafts, several other nonionic detergents, including Lubrol WX,

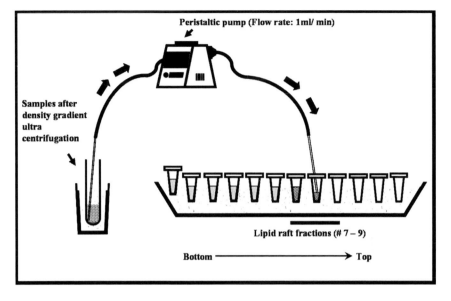

Fig. 3. Lipid raft protein profiles on 2-DE gels. The lipid raft samples prepared from mouse liver, lung, kidney, and large brain were solubilized with 300 μl of 2-DE sample lysis buffer. An equal concentration (150 μg) of raft proteins was subjected to 2-D gel electrophoresis, and spots were visualized via silver staining. The black arrows and numbers indicate protein spots which were found ubiquitously in all lipid rafts prepared from the liver, lung, kidney, and cerebrum.

Brij 35, Brij98, and CHAPS, are suited for lipid raft preparations *(7,8,23,24)*. However, it is recommended that applications of new types of detergents for the preparation of lipid rafts should be performed with caution.
4. The adequate amount of cells for lipid raft isolation can vary depending on the cell type. In most cases, lysis of 300 μl packed cells with 700 μl 1% Triton X-100 lysis buffer is sufficient for partitioning the detergent-resistant raft fractions from the rest of the cell compartments.
5. This protocol is optimized for lipid raft isolation with adherent cells. In the case of suspension cells, centrifugation at 288g, 4 °C, 5 min, followed by washing with PBS buffer can replace steps 1–3.
6. Incubation time is important for this step. An incubation time shorter than 30 min may result in incomplete lipid raft partitioning, which generally results in contamination from other subcellular organelles, i.e., endoplasmic reticulum and mitochondria.
7. For the same reason described in Note 5, enough time should be allocated to ultracentrifugation. Twenty hours of ultracentrifugation was proven to be sufficient for separation of lipid rafts in our experiments. This period can be prolonged to 24 h without causing significant damage to the sample.

8. Homogenization should be carried out on ice to avoid the tissue damage that may result from heating.
9. In our experiment, this method was applied successfully to many other rat organs, including the cerebrum, liver, lung, and kidney, as well as those of the mouse.
10. Immunoblotting analysis following lipid raft fractionation is a frequently applied experimental procedure used for studying lipid rafts and their associated proteins. As part of a preliminary experiment, it is possible to validate whether raft isolation is successfully performed by using antibodies for both positive and negative raft marker proteins (see Fig. 1B). It is also possible to determine whether the protein of interest is associated with these membrane micro domains by using specific antibodies. This procedure can also be adopted to detect any changes in the interaction of certain proteins with lipid rafts after various stimuli or treatments are applied.
11. In general, apply 100 μl of 5 × SDS sample buffer to 400 μl of each fraction (from bottom to top) for the preparation of the immunoblotting sample.
12. This sample preparation method is designed to analyze raft-associated proteomes by 2-dimensional electrophoresis (2-DE) with a large gel size (25.5 cm × 19.6 cm × 1 mm). Thus, the method may not be appropriate for other proteome analysis procedures such as stable isotope-labeling with amino acids in cell cultures (SILAC) (19), which does not routinely require the use of polyacrylamide gels.
13. The amount of 2-DE lysis buffer for solubilization of lipid rafts can be adjusted depending on the yield of lipid raft proteome. It is important to note that more than 100 μg of lipid raft proteins are required to perform large format 2-DE. Thus, in most cases, less than 100 μL is sufficient for the primary solubilization of the sample. One can then increase the amount of lysis buffer depending on the needs or the result of protein quantitation.
14. In most cases, brief vortexing in 2-DE sample lysis buffer is sufficient for the complete solubilization of lipid raft fractions. However, if there is some difficulty in solubilization of the raft pellet, sonication (25 cycles, 5–6 bursts each) can enhance the solubilization of the sample without causing significant damage to the proteins.
15. The prepared samples can be stored for several months at –20 °C.

Acknowledgments

This work was supported by grants awarded to YK from KOSEF (R01-2004-000-10765-0), Korea Research Foundation (KRF-2006-311-C00407), and Top Brand Project of Korea Basic Science Institute.

References

1. Munro, S. (2003) Lipid rafts: elusive or illusive? Cell 115, 377–388.
2. Edidin, M. (2003) The state of lipid rafts: from model membranes to cells. Annu. Rev. Biophys. Biomol. Struct. 32, 257–283.

3. Lagerholm, B. C., Weinreb, G. E., Jacobson, K., and Thompson, N. L. (2005) Detecting microdomains in intact cell membranes. Annu. Rev. Phys. Chem. 56, 309–336.
4. London, E. (2005) How principles of domain formation in model membranes may explain ambiguities concerning lipid raft formation in cells. Biochim. Biophys. Acta. 1746, 203–220.
5. Mukherjee, S., Maxfield, F.R. (2004) Membrane domains. Annu. Rev. Cell. Dev. Biol. 20, 839–866.
6. Brown, D. A., Rose, J. K. (1992) Sorting of GPI-anchored proteins to glycolipid-enriched membrane subdomains during transport to the apical cell surface. Cell 68, 533–544.
7. Roper, K., Corbeil, D., Huttner, W. B. (2000) Retention of prominin in microvilli reveals distinct cholesterol-based lipid micro-domains in the apical plasma membrane. Nat. Cell Biol. 2, 582–592.
8. Kim, K. B., Kim, S. I., Choo, H. J., and Ko, Y. G. (2004) Two dimensional electrophoretic analysis reveals that detergent resistant lipid rafts in physiological condition. Proteomics 4, 3527–3535.
9. Anderson, R. G. (1998) The caveolae membrane system. Annu. Rev. Biochem. 67, 199–225.
10. Mineo, C., Gill, G. N. and Anderson, R. G. (1999) Regulated migration of epidermal growth factor receptor from caveolae. J. Biol. Chem. 274, 30636–30643.
11. Simons, K., Toomre, D. (2000) Lipid rafts and signal transduction. Nat. Rev. Mol. Cell Biol. 1, 31–39.
12. Manes, S., del Real, G., and Martinez-A, C. (2003) Pathogens: raft hijackers. Nat. Rev. Immunol. 3, 557–568.
13. Gaus, K., Rodriguez, M., Ruberu, K. R., Gelissen, I., Sloane, T. M., Kritharides, L., Jessup, W. (2005) Domain-specific lipid distribution in macrophage plasma membranes. J. Lipid Res. 46, 1526–1538.
14. Nabi, I. R., Le, P. U. (2003) Caveolae/raft-dependent endocytosis. J. Cell Biol. 161, 673–677.
15. Bae, T. J., Kim, M. S., Kim, J. W., Kim, B. W., Choo, H. J., Lee, J. W., Kim, K. B., Lee, C. S., Kim, J. H., Chang, S. Y., Kang, C. Y., Lee, S. W., Ko, Y. G.. (2004) Lipid raft proteome reveals ATP synthase complex in the cell surface. Proteomics 4, 3536–3548.
16. Foster, L. J., De Hoog, C. L., Mann, M. (2003) Unbiased quantitative proteomics of lipid rafts reveals high specificity for signaling factors. Proc. Natl. Acad. Sci. USA 100, 5813–5818.
17. Calvo, M., Enrich, C. (2000) Biochemical analysis of a caveolae-enriched plasma membrane fraction from rat liver. Electrophoresis 21, 3386–3395.
18. Kim, K. B., Lee, J. W., Lee, C. S., Kim, B. W., Choo, H. J., Jung, S. Y., Chi, S. G., Youn, Y. S., Yoon, G.. S., and Ko, Y. G. (2006) Oxidation-reduction respiratory chains and ATP synthase complex are localized in the detergent-resistant lipid rafts. Proteomics 6, 2444–2453.

19. Blagoev, B., Kratchmarova, I., Ong, S. E., Nielsen, M., Foster, L. J., and Mann, M. (2003) A proteomics strategy to elucidate functional protein-protein interactions applied to EGF signaling. Nat. Biotechnol. 21, 31–318.
20. Gygi, S. P., Rist, B., Gerber, S. A., Turecek, F., Gelb, M. H., and Aebersold, R. (1999) Quantitative analysis of complex protein mixtures using isotope-coded affinity tags. Nat. Biotechnol. 17, 994–999.
21. Griffin, T. J., Tang, W., Smith, L. M. (1997) Genetic analysis by peptide nucleic acid affinity MALDI-TOF mass spectrometry. Nat. Biotechnol. 15, 1368–1372.
22. Miranker, A., Robinson, C. V., Radford, S. E., Aplin, R. T., and Dobson, C. M. (1993) Detection of transient protein folding populations by mass spectrometry. Science 262, 896–900.
23. Schuck, S., Honsho, M., Ekroos, K., Shevchenko, A., and Simons, K. (2003) Resistance of cell membranes to different detergents. Proc. Natl. Acad. Sci. USA 100, 5795–5800.
24. Chamberlain, L. H. (2004) Detergents as tools for the purification and classification of lipid rafts. FEBS lett. 559, 1–5.

33

Isolation of Membrane Protein Complexes by Blue Native Electrophoresis

Veronika Reisinger and Lutz A. Eichacker

Summary

Blue native PAGE is a discontinuous electrophoretic system that allows the separation of membrane protein complexes in a native, enzymatically active state with high resolution. Membrane protein complexes are solubilized by neutral, nondenaturing detergents like n-dodecyl-β-D-maltoside. After addition of Coomassie G250 that binds to the surface of the proteins, separation of the negatively charged complexes according to molecular mass is possible. After electrophoresis the structure and function of the isolated protein complexes can be investigated.

Key Words: BN PAGE; Coomassie G250; electrophoresis; functional complex; membrane protein; n-dodecyl-β-D-maltoside.

1. Introduction

In proteomic research high resolution separation of proteins is achieved by two-dimensional polyacrylamide gel electrophoresis. This denaturing technique separates hundreds to thousands of proteins in complex samples (1). However, hydrophobic membrane proteins are hardly detectable after IEF/SDS-PAGE (2). Therefore global analysis of membrane proteomes has not been started, despite the importance of membranes for the living cell.

An alternative strategy to separate membrane proteins with high resolution and maintenance of their enzymatic function is presented by blue native PAGE (3, 4). It is mainly used in the investigation of organelle membrane complexes like mitochondria and chloroplasts but other fields of application come up

(5–7). The technique relies on the solubilization of protein complexes from the membrane with mild neutral detergents and the transfer of negative charges to the complex by binding Coomassie blue G250 to its surface. The high charge to mass ratio allows the protein complexes to migrate to the anode like in SDS-PAGE. High resolution mass separation is achieved by electrophoresis of the protein complexes into an acrylamide gradient with decreasing pore sizes. Hereby the protein complexes are focused at the corresponding pore size limit.

2. Materials

2.1. Sample Preparation

1. Sample buffer: 750 mM ε- amino caproic acid, 50 mM Bis-Tris-HCl pH 7.0, 0.5 mM EDTA-Na_2. Store at –20 °C.
2. Detergent solution: 10 % (w/v) n-dodecyl-β-D-maltoside. Store at –20 °C.
3. Loading buffer: 750 mM ε- amino caproic acid, 5 % (w/v) Coomassie G 250 (*see* Note 1). Store at –20 °C.

2.2. Casting of Gradient Gels

1. Gel buffer (6×): 3 M ε- amino caproic acid, 0.3 M Bis-Tris-HCl pH 7.0. Store at 4 °C.
2. Acrylamide solution: 30% (w/v) acrylamide/bis acrylamide solution (37.5:1, 2.6%C), acts in unpolymerized state as a neurotoxin. Store at 4 °C.
3. Glycerol (100%). Store at room temperature.
4. TEMED: *N,N,N,N'*-tetramethyl-ethylenediamine. Store at room temperature.
5. APS: ammonium persulfate: 10% (w/v) solution. Stable at 4 °C for up to 2 weeks.
6. Water-saturated isobutanol: 50% (v/v) isobutanol. Store at room temperature.

2.3. Electrophoresis

1. Running buffer cathode (10×):
   a. Blue cathode buffer: 500 mM Tricine, 150 mM Bis-Tris-HCl pH 7.0, 0.2 % Coomassie 250 G (*see* Note 1). Store at 4 °C.
   b. Colorless cathode buffer: 500 mM Tricine, 150 mM Bis-Tris-HCl pH 7.0. Store at 4 °C.
2. Running buffer anode: 500 mM Bis-Tris-HCl pH 7.0. Store at 4 °C.

3. Methods

Blue native PAGE separates membrane protein complexes in a native state. It is therefore advisable to carry out all steps of sample preparation on ice. This procedure avoids protein degradation as well as a loss of protein subunits with only weak interactions to the complex (*see* Note 2). With regard to an optimal result, sample preparation is the most critical step. As the extent of protein complex solubilization and the stability of the solubilized proteins depends as well on the nature of the single sample as on the detergent type and concentration these variables have to be tested for the sample of interest. Indications for sample preparation given below are optimized for 400 μg thylakoid membrane proteins but the stated detergent concentration provides an appropriate initial value for a lot of biological tissues in general.

As blue native PAGE separates protein complexes depending on the volume/size of the single protein complexes the acrylamide concentration of the gel is responsible for the separation range. In practice the separation of membrane protein complexes in the mass range from 10 to 10,000 kDa is possible (8), (see Note 3). Instructions stated below correspond to a 6–12% linear gradient separating gel and a 4% stacking gel. This set up allows the separation of protein complexes in the molecular mass range from ~ 50 kDa to 700 kDa. The entire experiment was carried out in a Hoefer SE 660 (GE Healthcare) that has a gel size of $18 \times 24 \times 0.075$ cm.

3.1. Sample preparation

1. Pellet the membrane fraction of your sample by centrifugation. Remove the supernatant containing all soluble and peripheral proteins. To be sure that all peripheral proteins are removed, repeat this step at least one time (see Note 4).
2. Resuspend the pellet in 70 μl of sample buffer.
3. Add 10 μl of detergent solution to each sample and mix the samples gently. For the subsequent loading of the samples on the gel, it is important to avoid foam formation. In case of foam it can be difficult to underlay the complete sample into the wells.
4. Incubate the sample on ice for at least 10 min to solubilize the membrane protein complexes. Depending on the sample it can be necessary to extend this step to up to one hour to achieve complete solubilization of the membrane protein complexes.
5. Centrifuge for 10 min at 16,000g at 4 °C in a microfuge to pellet the unsolubilized material (see Note 5). Unsolubilized material can affect the subsequent electrophoretic separation of the protein complexes in a negative way.
6. Add 5 μl of loading buffer to the bottom of a new cap and transfer the supernatant of step 5 to the cap. Mix the sample gently to avoid foaming.

3.2. Casting of Gradient Gels

3.2.1. Separating Gel

1. Label two filtering flasks and prepare the gel solutions (see Note 6) stated in Table 1 in the labelled flasks (see Note 7 and 8).
2. Degas both solutions for 2 min to get rid of dissolved oxygen.

Table 1
Solutions for the preparation of a BN gradient gel

	12%	6%
acrylamide (30/0.8)	4.60ml	2.30ml
6× gel buffer	1.92ml	1.92ml
glycerol	2.30g	–
ddH$_2$O	3.29ml	7.28ml
Σ	11.5ml	11.5ml

3. Clean glass plates and spacers with denatured ethanol (100%), assemble the glass plate sandwich, fix it on the casting stand and adjust the assembled casting stand by a spirit level.
4. Place the gradient mixer on the magnetic stirrer and connect the pipet tip with the casting stand. The gradient mixer has to show a moderate incline towards the casting stand to achieve a directed flow of the acrylamide solutions (*see* Note 9). Ensure that all ports are closed: The tube that connects the gradient mixer with the glass plates is closed by a tube clip and fixed in the middle of the assembled glass plates. The tube ends in a cut pipet tip. The valve between the two chambers is closed (**Fig. 1**).
5. Place a magnetic stirring rod in chamber 1 of the gradient mixer and fill in the 12% gel solution carefully to avoid air bubbles.
6. Open the valve between the two chambers shortly to fill the valve with gel solution.
7. Pour the 6% gel solution in chamber 2 of the gradient mixer and add a nonmagnetic rod to balance the solution levels in the both chambers.
8. Add 5.5 µl of TEMED and 20 µl of APS to both chambers and mix the solutions gently.
9. Open the valve between the two chambers and remove the tube clip. Let the solutions rinse between the glass plates.
10. Overlay the cast gel with water-saturated isobutanol. The gel should polymerize within 90 min.
11. Rinse the gradient mixer immediately with distilled water to prevent the polymerization of the residual acrylamide solution in the tubes.

3.2.2. Stacking Gel

1. After polymerization of the separating gel remove the isobutanol by distilled water (*see* Note 10).
2. Dry the area above the separating gel completely with Whatman paper.
3. Clean a 10-well comb with denatured ethanol (100%) and insert the comb in the glass plate sandwich (*see* Note 11).

Isolation of Membrane Protein Complexes

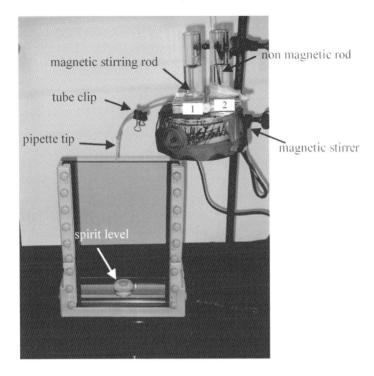

Fig. 1. Assembled casting stand for gradient gels.

4. Prepare the gel solution as stated in Table 2 in a filtering flask.
5. Mix the solution and degas it for 2 min to get rid of the dissolved oxygen.
6. Add 10 µl of TEMED and 100 µl of APS to the solution and pipet the gel solution quickly between the glass plates up to the top of the glass plate. The gel should polymerize in about 20 min (*see* Note 12).

Table 2
Solutions for the preparation of a BN stacking gel

	4%
acrylamide (30/0.8)	1.35 ml
6× gel buffer	1.67 ml
ddH$_2$O	6.98 ml
Σ	10.00 ml

3.3. Electrophoresis

1. Prepare the cathode running buffer by diluting 50 ml of the 10× blue cathode buffer with 450 ml of double distilled water. Pour the blue 1 × cathode buffer into the upper chamber.
2. As soon as the stacking gel has set, remove carefully the comb from the blue native gel. Rinse the wells with buffer using a 100 μl microsyringe to remove the residues of acrylamide.
3. Underlay the samples into the wells using the microsyringe. Rinse the microsyringe with buffer before applying a new sample (*see* Note 13 and 14).
4. Dilute 600 ml of 10× anode buffer with 5,400 ml of double distilled water and pour the 1 × anode buffer in the lower buffer chamber.
5. Fit the upper buffer chamber assembly into the lower buffer chamber and connect the tubes of the thermostatic circulator to the upper buffer chamber. Start the thermostatic circulator. Ensure that the electrophoretic run is carried out at 4 °C (*see* Note 15).
6. Assemble the electrophoresis unit completely and connect to a power supply.
7. Set the power supply to 12 mA, 1,000 V and 24 W and start the electrophoresis.
8. Prepare the colorless cathode buffer by diluting 50 ml of the 10× colorless cathode buffer with 450 ml of double distilled water.
9. As soon as the blue Coomassie dye front has reached half of the separating gel, pause the electrophoretic run and remove the blue cathode buffer from the upper buffer chamber via filtering flask or syringe. Dispose the buffer in the sink.
10. Pour colorless blue native cathode buffer in the upper buffer chamber and continue the electrophoretic run.
11. Stop electrophoresis when the blue cathode buffer front has reached the bottom of the separating gel (about 3.5 hours).
12. Stop the thermostatic cooler and disassemble the buffer chamber assembly. Disassemble the glass plate sandwich and remove the stacking gel.

The separating gel can be used for different applications. The single complexes can be visualized both by Coomassie or silver staining and by antibody detection after immunoblot (Fig. 2), (9). Furthermore the gel can be used for "in gel" activity assays e.g., for measuring the activity of mitochondrial enzyme complexes (10). The subunits of the protein complexes can be resolved in a subsequent electrophoretic step. Therefore the protein complexes are denatured with SDS/β-mercaptoethanol and transferred to SDS-PAGE for separating the complex subunits according to their molecular mass. After SDS-PAGE protein subunits can be visualized by Coomassie staining and identified by mass spectrometry (9).

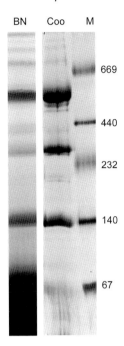

Fig. 2. Blue native PAGE of thylakoid membranes. Four hundred micrograms protein of thylakoid membranes isolated from barley chloroplasts were separated by a linear gradient gel (6–12%). After blue native PAGE the gel was stained with Coomassie blue. "BN" represents the gel directly after the electrophoretic run, "Coo" the Coomassie stained gel, and "M" the molecular mass standard with the associated molecular masses in kDa.

4. Notes

1. It is crucial to use Coomassie 250 G for all buffers in blue native PAGE to yield the best resolution of the protein complexes in the aqueous environment of blue native PAGE.
2. It is advisable to test the integrity of the entire complex at least one time by using freshly prepared source material because in case of frozen material the loss of weakly associated complex subunits could occur.
3. If acrylamide gels of lower percentages are used it is recommended to stabilize the polyacrylamide gel matrix by the use of gel strengtheners like Rhinohide Gel Strengthener (Molecular Probes) or the use of supporting films suitable for IEF electrophoresis.
4. In case of nuclear proteins or tissues that contain a lot of nucleic acid, it is advisable to remove the nucleic acid by DNAse or/and RNAse. Nucleic acids interfere with the electrophoretic run.

5. The completeness of solubilization can be tested by resuspension of the unsolubilized material in SDS sample buffer. After SDS electrophoresis proteins are Coomassie stained. If any bands are detected after Coomassie staining solubilization of the protein complexes was incomplete and has to be optimized.
6. Acrylamide acts as a neurotoxin in the unpolymerized state. Therefore it is strongly recommended to wear gloves during all steps of gel casting and electrophoresis.
7. The simplest way to transfer the glycerol in the filtering flask is to pipet it with a cut 1 ml pipet tip.
8. The separating gel will fill approximately 80% of volume between the glass plates.
9. It is also possible to cast the gradient gel from the bottom of the gel. For the first attempt it is recommended to cast the gel from the top because the other approach is more complicated: a peristaltic pump has to be installed, the time for pouring the gel is prolonged, the amount of radical starter has to be adapted and the risk of air bubbles in the gel during the casting process is higher.
10. Polymerisation of the separating gel is completed when a layer of water is formed between the upper edge of the gel and the *n*-butanol layer.
11. The maximum capacity of sample loading depends on the dimension of the gel concerning the size and the thickness of the gel and the dimension of the single wells. Gels as described above are able to resolve about 500 µg of total protein that can be loaded in up to 100-µl fractions per well. If the amount of protein loaded should be increased it is advisable to use spacers with 1 mm or 1.5 mm thickness.
12. Gels can be stored up to one week at 4 °C wrapped in wet tissues in a plastic bag.
13. Beside the use of microsyringes it is also possible to use gel loader tips to underlay the samples into the wells. As gel loader tips are much more flexible than a microsyringe and end in a capillary tip it is easier to apply the samples between the glass plates. On the other hand gel loader tips are disposables and can therefore only be used one time.
14. There are marker proteins purchasable for native PAGE but they only cover a part of the mass range that occurs in protein complex separation.
15. The electrophoretic run can also be performed in a cold room at 4 °C if no thermostatic cooler is existent. It is advisable to cool the gel during electrophoresis to keep the complexes intact, maintain their structure and protect them against proteolysis.

References

1. Righetti, P. G. (2005) Electrophoresis: the march of pennies, the march of dimes. *J. Chromatogr. A.* **1079**, 24–40.
2. Zahedi, R. P., Meisinger, C. and Sickmann, A. (2005) Two-dimensional benzyldimethyl-n-hexadecylammonium chloride/SDS-PAGE for membrane proteomics. *Proteomics*, **5**, 3581–3588.

3. Schägger, H. and von Jagow, G. (1991) Blue native electrophoresis for isolation of membrane protein complexes in enzymatically active form. *Anal. Biochem.* **166**, 223–231.
4. Schägger, H., Cramer, W. A. and von Jagow, G. (1994) Analysis of molecular masses and oligomeric states of protein complexes by blue native electrophoresis and isolation of membrane complexes by two dimensional native electrophoresis. *Anal. Biochem.* **217**, 220–230.
5. Brouillard, F., Bensalem, N., Hinzpeter, A., Tondelier, D., Trudel, S., Gruber, A. D., Ollero, M. and Edelman, A. (2005) Blue native-SDS PAGE analysis reveals reduced expression of the mClCA3 protein in cystic fibrosis knock-out mice. *Mol Cell Proteomics.* **4**, 1762–1775.
6. Eubel, H., Heinemeyer, J., Sunderhaus, S. and Braun, H. P. (2004) Respiratory chain supercomplexes in plant mitochondria. *Plant Physiol Biochem.* **42**, 937–942.
7. Komenda, J., Reisinger, V., Müller, B. C., Dobakova, M., Granvogl, B. and Eichacker, L. A. (2004) Accumulation of the D2 protein is a key regulatory step for assembly of the photosystem II reaction center complex in Synechocystis PCC 6803. *J. Biol. Chem.* **279**, 48620–48629.
8. Schägger, H. (2003) Blue Native Electrophoresis, in *Membrane Protein Purification and Crystallization* (Hunte, C., von Jagow, G. and Schägger, H., eds.), Elsevier Science, USA, pp. 105–130.
9. Darie, C. C., Biniossek, M. L., Winter, V., Mutschler, B. and Haehnel, W. (2005) Isolation and structural characterization of the Ndh complex from mesophyll and bundle sheath chloroplasts of Zea mays. *FEBS J.* **272**, 2705–2716.
10. Jung, C., Higgins C.M. and Xu, Z. (2000) Measuring the quantity and activity of mitochondrial electron transport chain complexes in tissues of central nervous system using blue native polyacrylamide gel electrophoresis. *Anal Biochem.* **286**, 214–223.

34

Tissue Microdissection

Heidi S. Erickson, John W. Gillespie, and Michael R. Emmert-Buck

Summary

Procurement of pure populations of cells from heterogeneous histological sections can be accomplished utilizing tissue microdissection. At present, a variety of different manual and laser-based dissection tools are available and each method has particular strengths and weaknesses. The types of biomolecular analyses that can be performed on microdissected cells depend not only on the method of cell procurement, but also on the effects of upstream tissue handling and processing. Tissue preparation protocols include two major approaches; snap-freezing, or, fixation and embedding. Snap-freezing generally provides the best quality tissue for subsequent study, including proteomic analyses such as two-dimensional polyacrylamide gel electrophoresis (2D-PAGE). Tissue fixatives include either precipitating reagents or biomolecular cross-linkers. The fixed samples are then further processed and embedded in a wax medium. In general, the biomolecules recovered from fixed and embedded tissue specimens are lower in both quantity and quality than those from snap-frozen specimens, although they are useful for certain types of analyses. The protocols provided here for tissue handling and processing, preparation of tissue sections, and microdissection are derived from our experience at the Pathogenetics Unit of the National Cancer Institute.

Key Words: Laser capture microdissection; LCM; microdissection; slide preparation; tissue.

1. Introduction

The tissue microdissection field has evolved over the past several decades from rudimentary techniques used to study relatively large and impure samples to sophisticated laser-based technologies that permit precise procurement of specific target cells. The advent of new PCR and RT-PCR techniques has been a particularly important driving force in this process as these methods enable sophisticated genomic analyses of small samples. A similar change is occurring in the field of proteomics as new and improved technologies are being developed for large-scale protein analysis, although the inability to amplify proteins is clearly a limiting factor. Nonetheless, more sensitive proteomic technologies and detection methods are pushing the limits for the study of protein content of dissected cells. These protein-based efforts are an important complement to the established genomic approaches, and permit dissected samples to be investigated at the levels of the genome, transcriptome, and proteome.

Tissue microdissection began with relatively crude gross dissection tissue block trimming methods and then progressed to targeted scraping of tissue from histological sections. Both of these approaches were effective for analyzing large anatomic regions of a specimen, but were not sufficient for procuring specific cell types, particularly those in a complex histological pattern. The field then evolved to the use of more refined manual dissection tools such as micromanipulators and other such devices. These methods enabled precise dissections for the first time, but were generally tedious and labor-intensive to employ. For example, one approach to manual microdissection (MD) is to use positive selection of the desired area using microscopic visualization and a small gauge needle (*1–7*). Although effective, this method can be slow, tedious, operator dexterity dependent, and studies are typically limited to cases where infiltrative patterns are easier to dissect (*8*). However, manual dissection is a good option for studies where minute precision is not needed, or in pilot studies where a regional dissection is sufficient.

Laser-based microdissection techniques, which are used for most tissue microdissection studies today, provide pure cell populations from heterogeneous tissue specimens. As an example, laser capture microdissection (LCM) (*9,10*) was invented at the National Institutes of Health (NIH) and developed through a cooperative research and development agreement (CRADA) with Arcturus Engineering, Inc. (www.arcturus.com), which has now been acquired by Molecular Devices, Inc. (www.moldev.com). LCM is an operator-dependent method in which the tissue area of interest is first identified morphologically and then targeted with the laser (Figs. 1 and 2). Currently, there are three laser dissection systems available and each operates on a slightly different principle: 1) the Arcturus developed laser capture microdissection system transfers cells onto a film, 2) the PALM micro-laser system (www.palm-microlaser.com) catapults the cells into a collection tube, and 3) the Leica AS LMD system (www.light-microscopy.com) cuts and gravitationally deposits the cells into a test tube. Unlike manual microdissection, laser dissection systems are simple, reliable, quick, and the technique is easy to master.

Given that each cell type has its own unique expression profile, microdissection allows investigators to isolate pure populations of cells that are representative of the disease or biological process that a researcher is

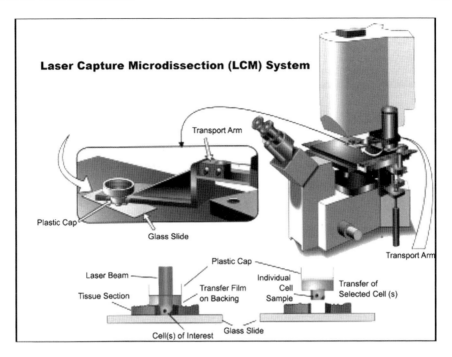

Fig. 1. NIH laser capture microdissection. Microdissection is done by placing a cap with a membrane on the tissue and melting the membrane overlying the tissue of interest with the laser. Next, the cap is lifted that separates the selected cells from the remaining tissue. The selected cells can then be used immediately for protein analysis.

interested in examining. For example, early studies using LCM focused on gene and gene expression studies and enabled investigators to gain insights into the molecular makeup of disease states (*11–18*) by isolating specific target and successfully analyzing mRNA transcripts using polymerase chain reaction (*1,3,4, 9,10,19–21*) and gene expression analysis (*1,2,22–35*). More recently, microdissection was applied to analyses of cellular proteins (*1,9,22–29, 31,36,37*), including efforts to identify new markers for potential drug targets using clinical tissue samples as the discovery platform (*38–46*). Of note, dissection of specific cells of interest is critical in accurately determining target cell proteomes uncompromised by other cell types in the specimen. A comparison of prostate epithelium and stroma showed that less than 45% of the proteins were shared between the two cell types (*47*). In addition, a comparison of in vivo and in vitro prostate cancer cells demonstrated less than 20% shared proteins between the cell types, with cell culture conditions shown to introduce protein expression artifacts (*47*). These data underscore the value of including microdissected cells from clinical samples in the overall effort to define cellular proteomes.

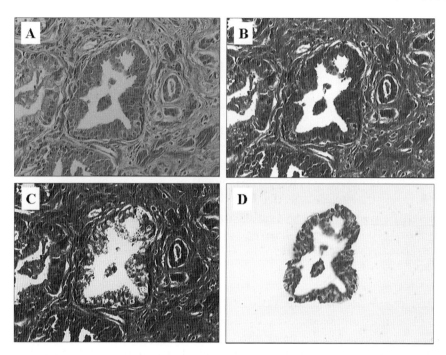

Fig. 2. LCM procurement sequence of epithelial components of a prostatic intraepithelial neoplasia (PIN) focus. **A** H&E roadmap image. **B** Hematoxylin stained frozen tissue section before LCM. **C** Tissue left on section after lifting cap. **D** Microdissected epithelial cells lifted on cap.

Tissue handling and processing is an important factor in biomolecular analyses after microdissection. It is imperative that one does not overlook the steps that the tissue takes from the time it leaves the patient *(48)* or animal model to the isolation and extraction of proteins at the bench. The concept of "garbage in equals garbage out" applies to all types of "–omic" profiling. Of particular importance are the fixative and the processing method. There are essentially two different types of preparations for tissue preservation; snap frozen, or, fixed and embedded. Frozen tissue allows for optimum recovery of biomolecules for downstream analysis, but can often compromise histological detail due to freezing artifact. Moreover, the preparation, use, and storage of frozen tissue blocks and slides are resource-intensive as compared to standard fixed and embedded specimens. Nonetheless, frozen tissues are the optimal tissue type for proteomic study, and have successfully been used to analyze proteins from dissected cells via numerous techniques *(38,39,46,47,49–54)*.

Tissue fixation and processing methods were initially developed to preserve the histological characteristics of the cells for microscopic and morphological analysis, and were not designed with today's sophisticated molecular analysis technologies in mind. There are a variety of fixatives available, including cross-linking and non cross-linking agents. Formalin is the most widely used fixative for tissue in pathology departments, and the traditional embedding agent is paraffin wax. Low-melt polyester wax may also be used as an embedding medium to better preserve the molecular content of samples; however, these embedding compounds are more difficult to handle than standard paraffin and thus are not routinely used. Archival formalin-fixed and paraffin-embedded tissues are the most widely available specimens worldwide. For the most part, it has not been possible to efficiently recover proteins from microdissected cells from these specimens, although this is an active area of investigation for many groups. As an alternative to formalin, we investigated the feasibility of using alcohol-based fixation of specimens followed by standard paraffin embedding, and found it to be a useful method for molecular profiling studies *(55,56)*. Sample protein content was analyzed by one-dimensional gel electrophoresis, immunoblot, two-dimensional gel electrophoresis, and layered expression scanning. Ethanol-fixed tissues produced results similar to that obtained from snap-frozen specimens, although the protein quantity and quality was somewhat decreased. However, overall protein content and profiles obtained resolved by two-dimensional gel electrophoresis (2-D PAGE) and stained by standard methods, were successfully analyzed and were found to be similar to those observed from frozen tissues *(54,57)*, in contrast to the poor protein recovery from formalin-fixed material *(57)*.

The amount of material needed for most proteomic technologies, including electrophoresis, mass spectroscopy, and isotope-coded affinity tagging has limited their application to microdissected samples. Dissection of the number of cells needed for this task is extremely labor intensive and often impractical, thus development of better tissue microanalysis technologies will be important to the future of the field *(57–62)*. One attempt to solve this issue was the creation of an operator-independent dissection system, termed expression microdissection (×MD) *(63)*. ×MD is performed by conducting immunohistochemistry (IHC) on a tissue section to target specific cells, applying a membrane over the tissue, irradiating the tissue with a laser (which specifically melts the membrane only where the IHC staining substrate is located), and then lifting the membrane to selectively recover the cells of interest. Although still in prototype form, ×MD has been successfully used to dissect fine targets (e.g., individual endothelial cells) and to procure larger quantities of isolated cells *(64,65)* in a shorter period of time than is possible with standard laser microdissection methods. Another solution to potentially provide for high throughput assessments using small amounts of patient material is to employ mass spectrometry (MS) directly to tissue samples, thus obviating the need for microdissection. This approach is applicable for development of clinical diagnostics and biomarker discovery. MS has become a major focus in the tissue analysis field *(66–74)*.

2. Materials

2.1. Manual Microdissection Slide Preparation

1. 70%, 95%, and 100% ethanol.
2. Xylenes, mixed, ACS reagent (Sigma).
3. Glycerol, ultra pure (Gibco).
4. Deionized water.
5. Hematoxylin solution, Mayer's (Sigma).
6. Eosin Y solution (Sigma).
7. Complete, mini protease inhibitor cocktail tablets (Roche Corp.) (*see* Note 1).

2.2. Laser Capture Microdissection (LCM) Slide Preparation

1. 70%, 95%, and 100% ethanol.
2. Xylenes, mixed, ACS reagent (Sigma).
3. Deionized water.
4. Hematoxylin solution, Mayer's (Sigma).
5. Eosin Y solution (Sigma).
6. Complete, mini protease inhibitor cocktail tablets (Roche Corp.) (*see* Note 1).

2.3. Manual Microdissection

1. Standard inverted microscope.
2. Microdissecting tool (30 gauge needle on a syringe).

2.4. Laser Capture Microdissection

1. LCM machine (e.g., PixCell II, Arcturus, Inc.).
2. LCM caps.
3. Adhesive pad.
4. Microcentrifuge tubes.

3. Methods

3.1. Manual Microdissection Slide Preparation

3.1.1. Storage of Tissue Sections Before Use

1. Store recut paraffin sections at or below room temperature. Deparaffinize only immediately before microdissection.
2. Store low-melt polyester sections at 4 °C.
3. Store frozen sections at −80 °C or below.

3.1.2. Paraffin-Embedded and Frozen Sections

For paraffin-embedded sections, start at step 1 and for frozen-embedded sections, start at Step 4. Place the sections in the following solutions (*see* Note 2):

1. Fresh xylenes (to depariffinize the sections) -5 min.
2. Fresh xylenes - 5 min.
3. 100% ethanol - 30 sec.
4. 95% ethanol - 30 sec.
5. 70% ethanol - 30 sec.
6. Deionized water - 30 sec.
7. Mayer's Hematoxylin - 30 sec.
8. Deionized water - rinse 15 sec (×2).
9. 70% ethanol - 30 sec.
10. Eosin Y - 15 sec.
11. Deionized water - 30 sec (x 2).
12. 3% glycerol in deionized water - 30 sec (*see* Note 3).
13. Shake the slide in the air to remove the layer of glycerol/water. (*see* Note 4).

3.1.3. Low-Melt Polyester-embedded Sections

Place the sections in the following solutions (*See* Note 5):

1. 100% ethanol - 5 min.
2. 100% ethanol - 5 min.
3. 95% ethanol - 30 sec.
4. 70% ethanol - 30 sec.
5. Deionized water - 30 sec.
6. Mayer's hematoxylin - 30 sec.
7. Deionized water - 30 sec.
8. 70% ethanol - 30 sec.
9. Eosin Y - 15 sec.
10. Deionized water - 30 sec (×2).
11. 3% glycerol in deionized water - 30 sec (*see* Note 6).
12. The tissue is now ready for microdissection.

3.2. Laser Capture Microdissection (LCM) Slide Preparation

3.2.1. Storage of Tissue Sections Before Use

1. Store recut paraffin sections at or below room temperature. Deparaffinize immediately before microdissection.
2. Store low-melt polyester sections at 4 °C.
3. Store frozen sections at –80 °C or below.

3.2.2. Paraffin-embedded and Frozen Sections

For paraffin-embedded sections, start at step 1. For frozen-embedded sections, melt the section gently (e.g., on the back of the hand) for approximately 30 sec after removal from the freezer. This creates a "rougher" tissue surface and allows for better adhesion to the LCM cap. Start at Step 4.

Place the sections in the following solutions (S*ee* Note 2):

1. Fresh xylenes (to depariffinize the sections) - 5 min.
2. Fresh xylenes - 5 min.
3. 100% ethanol - 15 sec.
4. 95% ethanol - 15 sec.
5. 70% ethanol - 15 sec.
6. Deionized water - 15 sec.
7. Mayer's Hematoxylin - 30 sec.
8. Deionized water - rinse (x 2) -15 sec.
9. 70% ethanol - 15 sec.
10. Eosin Y - 5 sec.
11. 95% ethanol - 15 sec.
12. 95% ethanol - 15 sec.
13. 100% ethanol - 15 sec.
14. 100% ethanol - 15 sec.
15. Xylenes (to ensure dehydration of the section)- 60 sec.
16. To completely remove xylenes, air-dry for approximately 2 min or gently use air gun.
17. The tissue is now LCM ready.

3.2.3. Low-melt Polyester-embedded Sections

Place the sections in the following solutions (*see* Note 5):

1. 100% ethanol (to remove polyester wax) - 5 min.
2. 100% ethanol - 5 min.
3. 95% ethanol - 15 sec.
4. 70% ethanol - 15 sec.
5. Deionized water - 15 sec.
6. Mayer's hematoxylin - 30 sec.
7. Deionized water - 15 sec.
8. 70% ethanol - 30 sec.
9. Eosin Y - 5 sec.
10. 95% ethanol - 15 sec.
11. 95% ethanol - 15 sec.
12. 100% ethanol - 15 sec.
13. 100% ethanol - 15 sec.
14. 50:50, xylenes:100% ethanol - 10 sec (*see* Note 7).
15. The tissue is now LCM ready (*see* Note 8).

3.3. Manual Microdissection

1. Dissection is performed on a standard inverted microscope using a 30 gauge needle on a syringe as the microdissecting tool (*see* Note 9).
2. The cell population of interest should be gently scraped with the needle while viewing the tissue through the microscope. Dissected cells become detached from the slide and form small dark clumps of tissue that can be collected on the needle by electrostatic attraction. Several small tissue fragments can be procured simultaneously. Collect an initial fragment on the tip of the needle to assist in procuring subsequent tissue. The tip of the needle with the procured tissue fragments should be carefully placed into a small PCR tube containing the appropriate buffer. Gentle shake the tube to ensure the tissue detaches from the tip of the needle (*see* Note 10).
3. To help maintain enzyme stability, frozen tissue sections can be placed directly on agarose coated slides. In addition, before and during the microdissection the frozen tissue sections on the agarose gels can be prepared or soaked in custom buffers (e.g., pH, salt concentration, proteinase inhibitors, etc.) that can be varied specifically for a given enzyme. To prepare agarose coated slides for microdissection, place 200 µl of warm agarose on standard uncoated glass slides, cover with a glass coverslip, and allow the gel to polymerize. Remove the coverslip from the slide and the frozen tissue section is immediately placed onto the agarose gel by directly transferring the freshly cut section from the cryostat to the agarose coated slide.
4. Manual microdissection can be performed similar to the step 2 method described above. However, because the tissue remains bathed in the fluid from the gel and can be gently pulled apart the dissector may find it easier to "tease" the tissue apart as tissue will separate along tissue planes, e.g., stroma and epithelium will easily separate from each other. The dissected tissue can either be gently picked up from the slide or both the agarose and the tissue fragment together can be procured by using the needle to physically cut the agarose.

3.4. Laser Capture Microdissection

1. The tissue is ready for microdissection once it has been properly processed, sectioned and stained.
2. Place the tissue section on the stage of the laser microscope.
3. Place a thermoplastic film coated cap onto the tissue.
4. Visualized the tissue section under the microscope and take an initial road map image.
5. Repeatedly fire the laser to encompass the cells to be dissected.
6. Take a predissection image.
7. Lift the cap from the tissue, removing the targeted cells from the tissue section.
8. Take a post-dissection image.
9. Image the cap with the dissected cells.
10. Place the cap onto a microcentrifuge tube containing the appropriate buffer depending on the analysis being performed (*see* Notes 11–14).

4. Notes

1. Dissolve 1 protease inhibitor cocktail tablet per 10 ml of each reagent, except xylene, for all protein analysis.
2. To significantly improve macromolecule recovery, use the minimal amount of staining to visualize the tissue for microdissection. For example, use hematoxylin and eosin at 10% of their standard concentrations. Additionally, "dark" images are typically produced because the slides are microdissected without a coverslip, i.e., the tissue is not index-matched and substantial light scattering occurs. Thus, both image quality and molecular recovery can be improved by decreasing stain concentrations.
3. Preparing the tissue for microdissection by soaking it in 3% glycerol is particularly helpful because it renders the tissue less brittle and dissected tissue fragments are easier to procure.
4. The next 5–10 min are the optimal time for microdissection because tissue is dry, but retains a soft consistency. The tissue will become increasingly brittle and the dissected fragments may be repelled as the needle is brought in proximity to the tissue if the dissection takes more than a few minutes. But, if the tissue does become overly dry, re-soak in the 3% glycerol/water solution for 1–2 min.
5. For tissues embedded in polyester wax, proceed gently when staining sections. The tissue has a tendency to detach from the slide, even though the sections are placed on charged slides. Therefore, the sections should be monitored carefully throughout the staining procedure.
6. After removal from the glycerol/water, it is important to ensure that the thin coat of fluid covering the slide is removed. If the fluid layer is not removed before dissection, potential "contamination" of samples may result due to diffusion of tissue fragments. And, large strips of tissue are produced that are not easily homogenized in the extraction buffers.
7. The end xylenes-ethanol step is critical for subsequent LCM. Depending on the tissue type and goals of the study, the length of time for the xylenes-ethanol step may need to be adjusted. If the xylenes-ethanol step is too short, the tissue may be too strongly bound to the slide and will not dissect. On the other hand, if the xylenes-ethanol step is longer than 10–15 sec, the tissue may detach from the slide during dissection.
8. Ensure that the tissue section is completely dry before LCM. To facilitate drying for efficient microdissection, an Accuduster or similar device may be used.
9. The dissector should prop their elbow on a solid surface adjacent to and at the same height as the stage of the microscope to stabilize the dissecting hand. By resting the ulnar aspect of the dissecting hand on the stage of the microscope and moving the needle into the microscopic field, a few millimeters above the tissue, the dissecting arm and hand can be rested on solid support surfaces.
10. To help detach the tissue from the needle and prevent any fragments from remaining lodged in the barrel of the needle, press down on the shaft of the syringe to inject an air bubble into the extraction solution.

11. It is imperative that there are no irregularities in the tissue surface near or in the area to be microdissected, as wrinkles will decrease the membrane contact during laser activation by elevating the LCM cap away from the tissue surface.
12. A decrease in laser activation spot size signals that there are subtle irregularities on the tissue surface (under the LCM cap) that cannot be visually appreciated. This issue can be partially or completely alleviated by temporarily increasing the laser power or by adding an extra weight to the cap support arm.
13. To remove cells that may have attached nonspecifically to the LCM cap, place the cap on an adhesive pad three separate times and then view it microscopically to ensure all of the nonspecific material has been removed.
14. To ensure that nonspecific transfer is not occurring during microdissection, a cap-alone control is recommended for each experiment. Place an LCM cap on the tissue section being dissected and aim and fire the laser at regions where there are no cells or structures present, e.g., lumens of large vessels, cystic structures, etc. Or, place a portion of the LCM cap "off" the tissue and target this region. Process the cap through the buffer and analysis methodology being utilized in the study and to serve as a negative control.

Acknowledgements

This research was supported by the Intramural Research Program of the Center for Cancer Research, National Cancer Institute, NIH.

References

1. Radford, D. M., Fair, K., Thompson, A. M., Ritter, J. H. et al. (1993) Allelic loss on a chromosome 17 in ductal carcinoma in situ of the breast. *Cancer Res* 53, 2947–9.
2. Emmert-Buck, M. R., Roth, M. J., Zhuang, Z., Campo, E. et al. (1994) Increased gelatinase A (MMP-2) and cathepsin B activity in invasive tumor regions of human colon cancer samples. *Am J Pathol* 145, 1285–90.
3. Koreth, J., Bethwaite, P. B., McGee, J. O. (1995) Mutation at chromosome 11q23 in human non-familial breast cancer: a microdissection microsatellite analysis. *J Pathol* 176, 11–8.
4. Speiser, P., Gharehbaghi-Schnell, E., Schneeberger, C., Eder, S. et al. (1996) Microdissection as a means to verify allelic imbalance in tumour biology samples. *Anticancer Res* 16, 461–4.
5. Zhuang, Z., Merino, M. J., Chuaqui, R., Liotta, L. A. et al. (1995b) Identical allelic loss on chromosome 11q13 in microdissected in situ and invasive human breast cancer. *Cancer Res* 55, 467–71.
6. Chuaqui, R. F., Zhuang, Z., Emmert-Buck, M. R., Liotta, L. A. et al. (1997b) Analysis of loss of heterozygosity on chromosome 11q13 in atypical ductal hyperplasia and in situ carcinoma of the breast. *Am J Pathol* 150, 297–303.

7. Quezado, M. M., Moskaluk, C. A., Bryant, B., Mills, S. E. et al. (1999) Incidence of loss of heterozygosity at p53 and BRCA1 loci in serous surface carcinoma. *Hum Pathol* 30, 203–7.
8. Rubin, M. A. (2002) Tech.Sight. Understanding disease cell by cell. *Science* 296, 1329–30.
9. Emmert-Buck, M. R., Bonner, R. F., Smith, P. D., Chuaqui, R. F. et al. (1996) Laser capture microdissection. *Science* 274, 998–1001.
10. Bonner, R. F., Emmert-Buck, M., Cole, K., Pohida, T. et al. (1997) Laser capture microdissection: molecular analysis of tissue. *Science* 278, 1481, 1483.
11. Zhang, L., Zhou, W., Velculescu, V. E., Kern, S. E. et al. (1997) Gene expression profiles in normal and cancer cells. *Science* 276, 1268–72.
12. Golub, T. R., Slonim, D. K., Tamayo, P., Huard, C. et al. (1999) Molecular classification of cancer: class discovery and class prediction by gene expression monitoring. *Science* 286, 531–7.
13. Klose, J. (1999) Genotypes and phenotypes. *Electrophoresis* 20, 643–52.
14. Alizadeh, A. A., Eisen, M. B., Davis, R. E., Ma, C. et al. (2000) Distinct types of diffuse large B-cell lymphoma identified by gene expression profiling. *Nature* 403, 503–11.
15. Perou, C. M., Sorlie, T., Eisen, M. B., van de Rijn, M. et al. (2000) Molecular portraits of human breast tumours. *Nature* 406, 747–52.
16. Strausberg, R. L., Buetow, K. H., Emmert-Buck, M. R., Klausner, R. D. (2000) The cancer genome anatomy project: building an annotated gene index. *Trends Genet* 16, 103–6.
17. Hedenfalk, I., Duggan, D., Chen, Y., Radmacher, M. et al. (2001) Gene-expression profiles in hereditary breast cancer. *N Engl J Med* 344, 539–48.
18. Swalwell, J. I., Vocke, C. D., Yang, Y., Walker, J. R. et al. (2002) Determination of a minimal deletion interval on chromosome band 8p21 in sporadic prostate cancer. *Genes Chromosomes Cancer* 33, 201–5.
19. Emmert-Buck, M. R., Vocke, C. D., Pozzatti, R. O., Duray, P. H. et al. (1995) Allelic loss on chromosome 8p12–21 in microdissected prostatic intraepithelial neoplasia. *Cancer Res* 55, 2959–62.
20. Thiberville, L., Payne, P., Vielkinds, J., LeRiche, J. et al. (1995) Evidence of cumulative gene losses with progression of premalignant epithelial lesions to carcinoma of the bronchus. *Cancer Res* 55, 5133–9.
21. Chuaqui, R. F., Englert, C. R., Strup, S. E., Vocke, C. D. et al. (1997) Identification of a novel transcript up-regulated in a clinically aggressive prostate carcinoma. *Urology* 50, 302–7.
22. Fearon, E. R., Hamilton, S. R., Vogelstein, B. (1987) Clonal analysis of human colorectal tumors. *Science* 238, 193–7.
23. Shibata, D., Hawes, D., Li, Z. H., Hernandez, A. M. et al. (1992) Specific genetic analysis of microscopic tissue after selective ultraviolet radiation fractionation and the polymerase chain reaction. *Am J Pathol* 141, 539–43.
24. Schena, M., Shalon, D., Davis, R. W., Brown, P. O. (1995) Quantitative monitoring of gene expression patterns with a complementary DNA microarray. *Science* 270, 467–70.

25. Chee, M., Yang, R., Hubbell, E., Berno, A. et al. (1996) Accessing genetic information with high-density DNA arrays. *Science* 274, 610–4.
26. DeRisi, J., Penland, L., Brown, P. O., Bittner, M. L. et al. (1996) Use of a cDNA microarray to analyse gene expression patterns in human cancer. *Nat Genet* 14, 457–60.
27. Going, J. J. and Lamb, R. F. (1996) Practical histological microdissection for PCR analysis. *J Pathol* 179, 121–4.
28. Moskaluk, C. A. and Kern, S. E. (1997) Microdissection and polymerase chain reaction amplification of genomic DNA from histological tissue sections. *Am J Pathol* 150, 1547–52.
29. Schena, M., Heller, R. A., Theriault, T. P., Konrad, K. et al. (1998) Microarrays: biotechnology's discovery platform for functional genomics. *Trends Biotechnol* 16, 301–6.
30. Simone, N. L., Bonner, R. F., Gillespie, J. W., Emmert-Buck, M. R. et al. (1998) Laser-capture microdissection: opening the microscopic frontier to molecular analysis. *Trends Genet* 14, 272–6.
31. Celis, J. E., Kruhoffer, M., Gromova, I., Frederiksen, C. et al. (2000) Gene expression profiling: monitoring transcription and translation products using DNA microarrays and proteomics. *FEBS Lett* 480, 2–16.
32. Leethanakul, C., Patel, V., Gillespie, J., Shillitoe, E. et al. (2000) Gene expression profiles in squamous cell carcinomas of the oral cavity: use of laser capture microdissection for the construction and analysis of stage-specific cDNA libraries. *Oral Oncol* 36, 474–83.
33. Ornstein, D. K., Cinquanta, M., Weiler, S., Duray, P. H. et al. (2001) Expression studies and mutational analysis of the androgen regulated homeobox gene NKX3.1 in benign and malignant prostate epithelium. *J Urol* 165, 1329–34.
34. Best, C. J., Leiva, I. M., Chuaqui, R. F., Gillespie, J. W. et al. (2003) Molecular differentiation of high- and moderate-grade human prostate cancer by cDNA microarray analysis. *Diagn Mol Pathol* 12, 63–70.
35. Best, C. J., Gillespie, J. W., Yi, Y., Chandramouli, G. V. et al. (2005) Molecular alterations in primary prostate cancer after androgen ablation therapy. *Clin Cancer Res* 11, 6823–34.
36. Lockhart, D. J., Dong, H., Byrne, M. C., Follettie, M. T. et al. (1996) Expression monitoring by hybridization to high-density oligonucleotide arrays. *Nat Biotechnol* 14, 1675–80.
37. Zhuang, Z., Bertheau, P., Emmert-Buck, M. R., Liotta, L. A. et al. (1995) A microdissection technique for archival DNA analysis of specific cell populations in lesions < 1 mm in size. *Am J Pathol* 146, 620–5.
38. Emmert-Buck, M. R., Gillespie, J. W., Paweletz, C. P., Ornstein, D. K. et al. (2000) An approach to proteomic analysis of human tumors. *Mol Carcinog* 27, 158–65.
39. Ornstein, D. K., Gillespie, J. W., Paweletz, C. P., Duray, P. H. et al. (2000) Proteomic analysis of laser capture microdissected human prostate cancer and in vitro prostate cell lines. *Electrophoresis* 21, 2235–42.

40. Knezevic, V., Leethanakul, C., Bichsel, V. E., Worth, J. M. et al. (2001) Proteomic profiling of the cancer microenvironment by antibody arrays. *Proteomics* 1, 1271–8.
41. Wulfkuhle, J. D., McLean, K. C., Paweletz, C. P., Sgroi, D. C. et al. (2001) New approaches to proteomic analysis of breast cancer. *Proteomics* 1, 1205–15.
42. Wulfkuhle, J. D., Sgroi, D. C., Krutzsch, H. et al. (2002) Proteomics of Human Breast Ductal Carcinoma in Situ. *Cancer Res* 62, 6740–49.
43. Wulfkuhle, J. D., Liotta, L. A., Petricoin, E. F. (2003) Proteomic applications for the early detection of cancer. *Nat Rev Cancer* 3, 267–75.
44. Ahram, M., Best, C. J., Flaig, M. J., Gillespie, J. W. et al. (2002) Proteomic analysis of human prostate cancer. *Mol Carcinog* 33, 9–15.
45. Craven, R. A., Totty, N., Harnden, P., Selby, P. J. et al. (2002) Laser capture microdissection and two-dimensional polyacrylamide gel electrophoresis: evaluation of tissue preparation and sample limitations. *Am J Pathol* 160, 815–22.
46. Jones, M. B., Krutzsch, H., Shu, H., Zhao, Y. et al. (2002) Proteomic analysis and identification of new biomarkers and therapeutic targets for invasive ovarian cancer. *Proteomics* 2, 76–84.
47. Ornstein, D. K., Englert, C., Gillespie, J. W., Paweletz, C. P. et al. (2000) Characterization of intracellular prostate-specific antigen from laser capture microdissected benign and malignant prostatic epithelium. *Clin Cancer Res* 6, 353–6.
48. Leiva, I. M., Emmert-Buck, M. R., Gillespie, J. W. (2003) Handling of clinical tissue specimens for molecular profiling studies. *Curr Issues Mol Biol* 5, 27–35.
49. Posadas, E. M., Simpkins, F., Liotta, L. A., MacDonald, C. et al. (2005) Proteomic analysis for the early detection and rational treatment of cancer—realistic hope? *Ann Oncol* 16, 16–22.
50. Paweletz, C. P., Charboneau, L., Bichsel, V. E., Simone, N. L. et al. (2001) Reverse phase protein microarrays which capture disease progression show activation of pro-survival pathways at the cancer invasion front. *Oncogene* 20, 1981–9.
51. Simone, N. L., Paweletz, C. P., Charboneau, L., Petricoin, E. F. 3rd et al. (2000) Laser capture microdissection: beyond functional genomics to proteomics. *Mol Diagn* 5, 301–7.
52. Simone, N. L., Remaley, A. T., Charboneau, L., Petricoin, E. F. 3rd et al. (2000) Sensitive immunoassay of tissue cell proteins procured by laser capture microdissection. *Am J Pathol* 156, 445–52.
53. Herrmann, P. C., Gillespie, J. W., Charboneau, L., Bichsel, V. E. et al. (2003) Mitochondrial proteome: altered cytochrome c oxidase subunit levels in prostate cancer. *Proteomics* 3, 1801–10.
54. Zhou, G., Li, H., DeCamp, D., Chen, S. et al. (2002) 2D differential in-gel electrophoresis for the identification of esophageal scans cell cancer-specific protein markers. *Mol Cell Proteomics* 1, 117–24.
55. Gillespie, J. W., Best, C. J., Bichsel, V. E., Cole, K. A. et al. (2002) Evaluation of non-formalin tissue fixation for molecular profiling studies. *Am J Pathol* 160, 449–57.

56. Gillespie, J. W., Gannot, G., Tangrea, M. A., Ahram, M. et al. (2004) Molecular profiling of cancer. *Toxicol Pathol* 32 Suppl 1, 67–71.
57. Ahram, M., Flaig, M. J., Gillespie, J. W., Duray, P. H. et al. (2003) Evaluation of ethanol-fixed, paraffin-embedded tissues for proteomic applications. *Proteomics* 3, 413–21.
58. Gygi, S. P., Rist, B., Gerber, S. A., Turecek, F. et al. (1999) Quantitative analysis of complex protein mixtures using isotope-coded affinity tags. *Nat Biotechnol* 17, 994–9.
59. Gorg, A., Obermaier, C., Boguth, G., Harder, A. et al. (2000) The current state of two-dimensional electrophoresis with immobilized pH gradients. *Electrophoresis* 21, 1037–53.
60. Gygi, S. P. and Aebersold, R. (2000) Mass spectrometry and proteomics. *Curr Opin Chem Biol* 4, 489–94.
61. Hanash, S. M. (2000) Biomedical applications of two-dimensional electrophoresis using immobilized pH gradients: current status. *Electrophoresis* 21, 1202–9.
62. Best, C. J. and Emmert-Buck, M. R. (2001) Molecular profiling of tissue samples using laser capture microdissection. *Expert Rev Mol Diagn* 1, 53–60.
63. Tangrea, M. A., Chuaqui, R. F., Gillespie, J. W., Ahram, M. et al. (2004) Expression microdissection: operator-independent retrieval of cells for molecular profiling. *Diagn Mol Pathol* 13, 207–12.
64. Hanson, J. A., Gillespie, J. W., Grover, A., Tangrea, M. A. et al. (2006) Gene promoter methylation in prostate tumor-associated stromal cells. *J Natl Cancer Inst* 98, 255–61.
65. Grover, A. C., Tangrea, M. A., Woodson, K. G., Wallis, B. S. et al. (2006) Tumor-associated endothelial cells display GSTP1 and RARbeta2 promoter methylation in human prostate cancer. *J Transl Med* 4, 13.
66. Paweletz, C.P., Gillespie, J.W., et al. (2000) Rapid protein display profiling of cancer progression directly from human tissue using a protein biochip. *Drug Development Research* 49, 34–42.
67. Stoeckli, M., Chaurand, P., Hallahan, D. E., Caprioli, R. M. (2001) Imaging mass spectrometry: a new technology for the analysis of protein expression in mammalian tissues. *Nat Med* 7, 493–6.
68. Chaurand, P. and Caprioli, R. M. (2002) Direct profiling and imaging of peptides and proteins from mammalian cells and tissue sections by mass spectrometry. *Electrophoresis* 23, 3125–35.
69. Yanagisawa, K., Shyr, Y., Xu, B. J., Massion, P. P. et al. (2003) Proteomic patterns of tumour subsets in non-small-cell lung cancer. *Lancet* 362, 433–9.
70. Schwartz, S. A., Weil, R. J., Johnson, M. D., Toms, S. A. et al. (2004) Protein profiling in brain tumors using mass spectrometry: feasibility of a new technique for the analysis of protein expression. *Clin Cancer Res* 10, 981–7.
71. Adam, B. L., Qu, Y., Davis, J. W., Ward, M. D. et al. (2002) Serum protein fingerprinting coupled with a pattern-matching algorithm distinguishes prostate cancer from benign prostate hyperplasia and healthy men. *Cancer Res* 62, 3609–14.

72. Li, J., Zhang, Z., Rosenzweig, J., Wang, Y. Y. et al. (2002) Proteomics and bioinformatics approaches for identification of serum biomarkers to detect breast cancer. *Clin Chem* 48, 1296–304.
73. Petricoin, E. F. 3rd, Ornstein, D. K., Paweletz, C. P., Ardekani, A. et al. (2002) Serum proteomic patterns for detection of prostate cancer. *J Natl Cancer Inst* 94, 1576–8.
74. Hingorani, S. R., Petricoin, E. F., Maitra, A., Rajapakse, V. et al. (2003) Preinvasive and invasive ductal pancreatic cancer and its early detection in the mouse. *Cancer Cell* 4, 437–50.

Index

A

Acetone precipitation, 41, 117
N-Acetylmuramyl-β(1-4)-N-acetylglucosamine, 25
Acrylamide
   alkylation, 90
   deuterated, 87
AEX resin, 157
Affinity matrix activation, 375
Albumin depletion, 168, 290–292, 294–295
Alkylation, 90–97, 117–118, 161, 297
Ammonium persulfate, 90
Antibody(ies), 326, 349–362, 366–368, 370–371, 417, 420
   binding characteristics for protein A and protein G, 352–353
   coupling efficiency, 355
APV Gaulin equipment, 17

B

Bacillus subtilis proteins, preparation for two-dimensional electrophoresis, 52–53
Bacteria cell
   disruption, 28–29
   membrane proteins isolation using carbonate extraction, 397–399
      carbonate extraction, 399–400
      cell harvesting, 399
      materials for, 399
      two-dimensional gel electrophoresis, 400
   extracts preparation, 171–172
   Gram-negative, 24, 28, 30, 140, 398–399
   Gram-positive, 24–25, 28–29, 31
   lysis, 24–25

Bacterial proteins chromatographic fractionation, 172–173
Barley, 429
Bath sonicator, 148
Bead impact methods
   shaking vessel
      practical aspects of, 7–8
      theory of, 4–7
   stirred agitated beads
      practical aspects of, 10
      theory of, 8–9
Bead mill, 8–9
Beadbeater homogenizer, 406
Benzonase, 52, 54–56, 61
Bicinchoninic acid (BCA)
   assay, 194
   protein determination assay reagent A and B, 406
Biomarker, 263–264, 373
BioNeb disruption system, 19–20
Bis-Tris SDS PAGE gels, 170
Blue Native Electrophoresis, 423–431
Bovine α-lactalbumin, 90
Bovine γ-globulin, 47
Branson Sonifier, 450, 226
Bungarus fasciatus, 380

C

Cancer, 435
Carbamidomethylation, 117–118
Carbamylation, 40
Carbonate extraction, 318, 397–401
Carrier ampholytes, 38, 40, 226, 228, 253, 282, 400
Cell culture, 101–111, 206, 216–217, 360, 368–369
   and removal of salts, 53, 55

Cell disruption, 3–4
  enzymatic, 23–24
  mechanical, 12–17, 142–143
Cell lysis
  bacteria, 27
  yeast, 25–29
Cellosyl, 24
Cerebrospinal fluid, 48, 53–54, 57–58, 61
  preparation for 2-D electrophoresis, 53, 55
CEX resin, 157
Chaotropic agents, 43
Chelator competition elution buffer, 206
Chitin, 25
Chitinase, 25
Chloroplasts, 423
3[(3-Cholamidopropyl) dimethylammonio]-propanesulfonic acid (CHAPS), 30, 38
Chromatofocusing, 168, 187–203
Chromatography
  affinity, 169–170, 205–212, 366, 378–384, 386–390, 393
  heparin, 213–220
  hydrophilic interaction, 63–68
  ion exchange, 157, 165
  lectin affinity, 378–384, 386–388, 390, 393
  metal affinity, 205–212, 366
  reversed-phase, 386–388
Cibacron-Blue F3-GA, 168–169
Classical and cross linked oriented antibody approaches, 354
Co-immunoprecipitation, 350–351, 354–355
Collision induced dissociation (CID), 117
Complex protein mixtures fractionation by liquid-phase isoelectric focusing, 225–226
  complex samples preparation of under denaturing IEF conditions, 226, 228

MicroRotofor
  cell set-up, 227–229
  focusing run, 230
  run harvesting fractions, 230–232
  sample loading, 229–230
  postfractionation quality control and analysis, 227
  quality control and analysis of Rotofor fractions, 232–234
  rat brain tissues protein extraction from, 226, 228
Concanavalin A (Con A) lectin, 377
Coomassie Brilliant Blue G-250 dye, 44, 90
Coomassie Plus protein assay, 250
Coomassie staining, 429
Culture medium, 58, 126–127, 132–134, 169, 216, 368
CyDyes, 75–78, 80, 82, 84, 149, 246, 250–253
CyDyes DIGE Fluor minimal dyes, 75
Cysteines labeling, 77
Cytochrome C, 64–66, 68, 191

D

DeCyder™ software, 82
Deinagkistrodon acutus, 380
DeStreak™ rehydration solution, 281
Detergent/Polymer aqueous two-phase systems, 403–411
Detergent-resistant lipid rafts, 413–420
  cultured cells preparation from, 414–415
  fractionation for immunoblotting, 415
  isolation for 2-D electrophoresis, 413–414
  methods for, 416–418
  from mouse tissues preparation, 415
  sample preparation of isolated lipid rafts for proteomic analysis, 415
Detergents
  sample solubilization, 37
  solubilization efficiency, 410

Index

2DGE
    first dimension, 173–174
    second dimension, 174–175
Dialysis, 92, 109, 190, 195, 201, 238, 264, 303, 378
Differential in gel electrophoresis (DIGE), 88
    2-D multiplexing approach, 148
    minimal labeling, 73, 75–76, 78
    saturation labeling, 77–79, 89
Dimethylformamide, 78, 245
Dipicolylamine (DPA), 206
Direct affinity method, 357–358
Disulfide bridges, 77
Dithioerythritol (DTE), 37
Dithiothreitol (DTT), 25
Dodecyl-β-D-maltoside, 299
Dounce homogeniser, 405
Droplet low pressure nebulizer, 21
Droplet's impingement, 19
Dulbecco's Modified Eagle Medium (DMEM), 106
Dyno-Mill, 9

E

Electron microscopy, 338, 345
Electrophoresis
    2D, 30
    zone, 290, 333–347
Endocitic organelles isolation by density gradient centrifugation, 317–319
    homogenization protocol for cells requiring hypotonic shock, 320, 323
    latency measurement, 320, 323–324
    organelle sample preparation for 2-DE analysis, 321–322, 327–328
    standard homogenization protocol, 319–320, 322–323
    subcellular fractionation
        continuous gradient and, 321, 325–327
        step gradient, 321, 324–325
Endoglucanase, 26–27

Endoplasmic reticulum, 319, 419
Endosomes, 318–319, 325–327, 329–330
Epithelial adherent HeLa cells, 106
Escherichia coli, 10, 167–183, 190, 208
    lysate sample preparation, 190
    proteins
        enrichment by reactive dye resins, 176–182
        preparation, 169–170
    proteome reducing complexity by chromatography on reactive dye columns, 167–183
Ethylene dimethyl acrylate (EDMA), 377, 389
Experion™ automated electrophoresis system, 233

F

Fibroblast growth factor (FGF) family, 215
Flamingo™ fluorescent gel stain, 56
Folin-Ciocalteau phenol reagent, 169
Formalin, 437
Free flow electrophoresis (FFE), 64, 67–68, 287–292, 287–299
    blood collection and plasma preparation, 292
    buffers
        denaturing FFE separation, 292
        FFE-HSA depletion, 292
    FFE separation of human plasma, 294–296
    plasma sample preparation, 293–294
    post-FFE
        gel-based analysis, 296–297
        LC-MS/MS analysis, 297–298
        sample preparation for LC-MS/MS and SDS-PAGE analysis, 292–293
French press, 12–15, 169, 194, 399

G

Gel-based IEF media, 225
Glass beads, 406

Global internal standard technology
 (GIST) approaches, 114
β-1,3-Glucanase, 25
β-1,3-Glucanchitin complex, 25
3-Glycidoxypropyltrimethoxysilane
 (epoxysilane), 384
Glycidyl methacrylate (GMA), 377
Glycoprotein enrichment by lectin
 affinity techniques, 373–375
 2-D and SDS-PAGE gel
  electrophoresis, 376–377
 lectin materials and buffers, 376
 lectin to monoliths immobilization of,
  377
 lectin to porous silica immobilization
  of, 376
 lectins immobilized
  agarose gels and, 377–382
  monolithic capillaries, 382–390
  monolithic microfluidics, 390–393
  porous silica, 382–388
Glycoproteins
 enrichment, 373–374
 glycosylation, 373–374
Gradient centrifugation, 317–331,
 335, 337
Gram-negative bacterial cells structure,
 24
Granulated Sephadex IEF gels sample
 prefractionation, 278–279
 IEF
  in granulated Sephadex gels, 280,
   281–283
  in narrow pH range IPG strips,
   280, 283–285
 sample preparation of, 279–281

H

Haemophilus influenzae, 168
HeLa cell proteins
 preparation for 2-D electrophoresis,
  53, 55
 separation by 2-D electrophoresis, 59
Hemogard™ closure, 293

Hen egg white lysozyme (HEWL), 24
Heparin, 213–220
 chromatography media, 217
 polysaccharide structure, 215
 Sepharose 6 fast flow gel, 217–218
High pressure batch
 expanding fluids
  French press design, 12–13
  Parr cell disruption bomb, 13–14
High protein effect of sample
 prefractionation, 284
High-pressure homogenizers
 high pressure flow narrow
  tubes/opposed
  Jets-Microfluidics, 17–18
 high pressure valve with impingement
  Wall-Gaulin, 16–17
 practical aspects of, 19
High-resolution 2-dimensional gel
 electrophoresis (2DGE),
 167–169
Homogenization buffer (HB), 319–320
Homogenizer, 10–12, 16–19, 142, 406,
 415–416
Horseradish peroxidase (HRP), 320
Human lysozyme, 24
Human melanoma cells
 analysis of large proteins after
  MicroSol IEF fractionation, 251
 protein profiles, 251
Human plasma, 287, 289–291, 294–296
 native depletion of HSA, 291
 SDS-PAGE of HSA depleted, 296
Human serum, 191, 233
 Experion system analysis of
  MicroRotofor fractionated, 233
Hydrophilic interaction chromatography
 (HILIC)
 binding properties of, 65–66
 cartridges manufacture of, 64–65
 materials for, 64
 sample processing 2-D
  electrophoresis, 67–68

Hydroxyethyldisulfide (HED), 280
N-Hydroxy-succinimide ester-modified cyanine fluors, 88

I

ICPL reagent technology, 116
IMAC-retained proteins and column regeneration elution, 209
Image analysis, 67, 73, 83, 88, 152, 249, 253
Imidazole competition elution buffer, 206
Iminodiacetic acid (IDA) resin structure, 206–207, 207
Immobilized metal affinity chromatography (IMAC), 205–206, 366
Immunoblotting, 337, 341–342, 344–345, 415–417, 420
Immunoglobulin A, 352
Immunoglobulin G, 352–353, 355, 359–361
Immunoglobulin Y, 353
Immunoprecipitation, 349–362, 366–371
Immunoprecipitation, matrixes preparation, 353
IPG DryStrip kit, 283–284
IPG Drystrip rehydration solution, 280
IPG strip rehydration trays, 164
Isoelectric focusing (IEF)
 fractionation, 52, 225
 liquid phase, 225–238
 in Sephadex gels, 280–283
Isoelectric point, 40, 76–77, 152, 157, 188, 235, 259, 280
Isotachophoresis, 289–290
Isotope-coded protein label (ICAT), 113–122
Isotope-coded two-dimensional maps, 87–98
ITRAQ methodology, 114

L

Labeled proteins
 by acetone precipitation purification, 117
 for direct MS-analysis enzymatic digestion, 119
 enzymatic digestion, 117
Laser capture microdissection (LCM), 434–436, 438–439, 441
Lectin immobilized to agarose gels
 glycoproteins enrichment from snake venom using Con A-Sepharose, 380–382
 glycoproteins isolation from VNO, 377–379
 MUP glycoproteins isolation from mouse urine, 380
Lectins, 374
Lens culinaris, 377
Lipid raft protein profiles on 2-DE gels, 413–420
Lipids, removal of, 53, 55, 58–60, 62
Low mass proteins and peptides fractionation, 269–270
Low protein concentrations fractionation under native conditions, 267–268
Low volume and low abundance proteins and peptides desalting and depleting of using MF10, 264–265
Lowry protein assay with interfering substances, removal by precipitation, 44–45
Lysine minimal labeling, 76
Lysosomes, 325, 333–335, 337–338, 342–343, 346
Lysostaphin, 25, 31
Lysozyme, 24–25, 27–31, 48

M

Major urinary protein (MUP) complex, 380
Mammalian cell culture, 104–106

Mass spectrometry, tandem, 404
Matrix-assisted laser desorption ionization time-of-flight mass spectrometry (MALDI-TOF-MS), 92, 308
McIlvaine's buffer, 169
MEGATRON®, 12
Membrane protein complexes isolation by blue native electrophoresis, 423
  electrophoresis, 424, 428–429
  gradient gels casting of, 424–427
  methods for, 424–425
  sample preparation, 424, 425
Membrane proteins enrichment by partitioning in detergent/polymer aqueous two-phase systems, 403–405
β-Mercaptoethanol, 25
2- (Methacryloyloxy)ethyl trimethylammonium chloride (MAETA), 389
N',N'-Methylenebisacrylamide, 90
Microdissection, 433–443
MicroFlow MF10, 257–273, 258
  size separations, 258–262
  size separations fractionation
    under denaturing and/or reducing conditions, 263
    low mass proteins and peptides, 263–264
    under native conditions, 262
Microfluidizer, 18
MicroRotofor cell, 225–234, 226, 236–237
  ampholyte concentrations for, 236
  components and accessories, 227
  cooling setting, 231
  electrolytes for, 236
  focusing run, 230
  sample loading, 229–230
  set up, 227–229

Microscale isoelectric focusing in solution, 241–244
  analyzing samples from ZOOM IEF fractionator, 249–253
  assembling ZOOM IEF fractionator, 246
  collecting fractions and eluting proteins trapped in partition membrane disks, 248–249
  loading CyDye labeled protein samples in ZOOM IEF fractionator, 247–248
  performing solution IEF fractionation, 248
  prelabeling samples with Cyanine dyes, 246
  protein sample labeling, 245
  ZOOM IEF fractionator, 245–246
MicroSol IEF
  method for, 243
  prefractionation with 1-D and 2-D DIGE, 244
Microvessels endothelial cells (MVECs), 216
Mini Bead Beater-1 equipment, 7
Mini-PROTEAN II Cell, 378–379
Mitochondria, 319, 333–347, 405–407
Mitochondrial isolation buffer (MIB), 405
Bradford protein assay, modified, 44
Molecular Imager FX™, 56
Molecular Imager PharosFX system, 164
Mortar and pestle tissue grinders, 5–6, 12, 54–56, 59, 281
Mouse brain proteins
  elution profile, 311
  proteomic analysis pathway by preparative electrophoresis, 302
Multiplexing, 73, 114–115, 121, 148
Muramidase, 24–25
Murein hydrolases
  amidases, 24
  endopeptidases, 24
  glycosidases, 24

Mutanolysin, N-Acetyl muramidase, 24
Myoblast proteins and phosphoprotein purification, 367

N

N-Acetyl-D-glucosamine, 24
NADP dependant dehydrogenases, 169
Native HSA depletion, 294
Native plasma depletion pH gradient profile, 291
Needles, 31, 231, 322, 330, 438, 441–442
Nickel binding proteins, separation, 211
N-Nicotinoyloxy-succinimide (Nic-NHS), 115
Nitrilotriacetic acid (NTA), 206
N,N,N',N'-Tetramethylethylenediamine (TEMED), 90, 171, 207, 210, 304, 375, 378–379, 424, 426–427
Nonionic detergent n-octyl-β-D-glucopyranoside, 200
Nucleic acids, removal of, 52, 54–56, 61
Nycodenz® density step gradient, 335

O

Octyl-β-D-glucoside, 299
Ocytlglucoside (OGS), 235
Organelle isolation, 317–331
Oriented affinity method, 356–357

P

Paraffin, 437–440
Parr cell disruption bomb, 15–16
Pefabloc® proteinase inhibitor, 38, 280
Peptide mass fingerprint (PMF), 117, 171
Peptidoglycan, 24–25
Peroxisomes, 334–335
Pestle, 5–6, 12, 54–56, 59, 281
Phenolic compounds, removal of, 52, 54, 56–57, 61
Phenylmethylsulfonyl fluoride (PMSF), 38, 143

Phosphatase inhibitors, 143, 319
Phosphate-buffered saline (PBS), 48
Phospholipids, 413–414
Phosphoprotein-affinity purification, 369
Phosphoproteins, 209, 365–371
  enrichment, 366
  isolation, 365–367
    methods for, 368–369
    phosphoprotein-affinity purification, 368
    tyrosine-phosphorylated proteins immunoprecipitation of, 367–368
Phosphorylation, 365
Plasma kallikrein-sensitive glycoprotein, 298
Plasma proteins
  separation denaturing, 295
  two-dimensional gel analysis, 303
PolyCAT column, 197
Polyether Ether Ketone (PEEK), 386
Polysaccharides, removal of, 52, 54, 56–57, 61
Polytron equipment, 11
Post nuclear supernatant (PNS), 318
Posttranslational modifications (PTM) of proteins, 114, 373–374
Potato, 52–54, 56–58, 61
Potato proteins
  preparation for 2-D electrophoresis, 52
  separation by 2-D electrophoresis, 58
Preparative electrophoresis, 301–312
PrepCell, 304, 306, 309–310
PrepCell Model, 491, 306
ProPac WAX-10 column, 199
ProPac WAX-G column, 191
PROTEAN IEF Cell, 164
Protease inhibitors
  aprotinin, 319–320, 367
  bestatin, 54–55, 418
  cocktail, 27, 143, 190, 378, 405, 415, 438, 442
  E64, 54–55
  leupeptin, 54–55, 319–320, 367, 418

pefabloc, 38–40, 42, 279–280, 319–320
pepstatin, 319–320, 367, 418
PMSF, 38, 42, 54–55, 61, 143, 367, 370, 414–415
Protein A, 349–353, 355–357, 360, 368, 370
Protein assay
  BCA, 195, 321, 327, 330–331, 361, 406
  Bradford, 44–46, 48, 151, 369
  Folin-Ciocalteau, 169, 172
  Lowry, 43
Protein extraction
  rat brain tissue, 159
  from tissues, 158
Protein fractionation
  AEX Mini Spin Column, 158
  CEX Mini Spin Column, 158
  preparative electrophoresis, 301–303, 306–307
Proteins
  cleanup with ReadyPrep 2 D cleanup kit, 161–163
  concentration by HILIC and SPE, 64–68
  denaturing, 7
  electrophoretic migration, 262
  estimation by Folin-Ciocalteau method, 171
  identification by peptide mass mapping, 175–176
  isotope labeling, 117
  MVECs SDS-PAGE analysis, 218
  quantitation in samples prepared for 2-D electrophoresis, 43–49
  reduction and alkylation, 161
  reduction and cleanup before 2-DGE, 159
  sequential extraction by chemical reagents, 139–145
  tagging with fluorescent dyes, 75

Proteins and protein complexes isolation by immunoprecipitation, 349–350
  capture and elution of immune complex, 358–359
  crosslinking antibody to protein A/G, 356–357
  direct antibody immobilization to resin, 357–358
  immobilized protein A/G equilibration, 355–356
  materials for, 350–351
  methods for, 351–355
  preparation of samples for SDS-PAGE, 359
  resin regeneration and storage conditions, 359
Proteins fractionation by heparin chromatography, 211–220, 213–216
  cell culture and protein extraction, 216
  cell culture and total protein extraction, 216–217
  chromatography media and buffers, 216
  column preparation, 217–218
  column regeneration and storage, 219
  sample preparation and loading, 218
  starting/wash buffer preparation, 217
  wash/elution, 218–219
Proteins fractionation by immobilized metal affinity chromatography, 205
  cell culture and lysis, 206
  columns and equipment, 206
  elution of IMAC-retained proteins and column regeneration, 209
  materials for Fe^{3+}-IMAC, 206
  materials for Ni^{2+}, Cu^{2+}, Zn^{2+} and Co^{2+}-IMAC, 206
  metal Ions and column-packing immobilization procedures, 208–209

Index   457

samples preparation of, 208
SDS- PAGE, 207–208, 210
silver staining, 208, 210–211
target proteins binding of from cell extracts to column, 209
ProXPRESS proteomic imaging system, 250
S-Pyridylethylation, 98

R
Radiolabeled [^{35}S]-methionine, 133
Radiolabeling, 125–128, 133
Rat
 brain, 59, 62, 157–164, 166, 226, 228, 312
  comparison of 2-D gels, 234
  cytosolic proteins eluted from PrepCell, 309
  protein preparation for 2-D electrophoresis, 53, 55
 liver, 123, 155, 333–335, 340–341, 343, 346
 liver mitochondria isolation with ZE-FFE, 333–339
  electron microscopy, 345
  materials of, 339–340
  methods for, 341–343
  monitoring of isolation and purification of, 341
  purification by ZE-FFE, 343–344
  SDS-PAGE and immunoblotting, 344–345
  by ZE-FFE and FFE setup, 340–341
 serum tagged with D_0/D_3 acrylamide, two-dimensional maps, 91–94
 vomeronasal organ (VNO), 378
RC DC Protein Assay, 45, 163, 227, 238
ReadyPrep 2-D cleanup kit, 227
ReadyStrip IPG strips, 164
Reducing agents
 β-mercapoethanol, 235
 dithioerythritol, 37

dithiothreitol, 25, 36–38, 39, 40, 42, 54, 55, 61, 64, 77, 79, 90, 143, 208, 226, 235, 246, 279, 280, 376, 406
Reversed-phase high performance liquid chromatography (RP-HPLC), 148
Rhinohide Gel Strengthener, 429
Rotofor® cell, 226
Rotor/stator homogenizers
 design of, 10–11
 practical aspects of, 11–12

S
S. cerevisae mitochondria isolation by subcellular fractionation, 405
 detergent/polymer two-phase partitioning, 406, 407–408
 sample concentration and clean up, 406, 408
 solubilization of, 406, 407
S. cerevisiae cells culture and harvest, 29
Salts endogenous to body fluids, removal, 53, 55
Salts, removal of, 53–55, 57–58, 61
Sample cleanup, 75, 78
Sample solublization buffers for two-dimensional electrophoresis, 35–41
SDS lysis solution, 40
SDS-loading buffer, 210
SDS-PAGE clean-up kit, 406
Seeds, 4, 6, 13, 18
Sepharose-Con A gel, 381–382
Sequential extraction, 139–144, 278
Serdolite MB-1, 38
Serum albumin
 bovine, 45, 47, 128, 158, 172, 266
 human, 91
Shaking container method, 4
SILAC in mammalian cells
 cell growth, 107
 preparation of media, 106–107

SILAC in *Saccharomyces cerevisiae* yeast cells
  cell growth, 108
  minimal medium preparation of, 107–108
SILAC labeling, 103
Silver staining, 75–76, 135, 149, 154, 208, 210, 242, 296, 368, 414, 419, 428
Sinapinic acid matrix solution, 377
Solid phase extraction, 63
Sonicator, 21, 22, 54–57, 59, 142, 150, 206, 411, 415
Sonicator equipment, 21
SpeedVac vacuum centrifuge, 152
Stable isotope labeling by amino acids, 101–103
Staphylococcus sp., 24–25
Subcellular fractionation, 242, 317–319, 321, 324–325, 328, 405
Superdex 75 column, 380
SWISS-PROT database, 91
SYPRO Ruby™, 90, 244, 249–251, 250, 270–271

T

TBP, 38, 41, 90–91, 142–143, 159, 161, 235, 266, 269
TCEP, 41, 77, 79–80, 117
Tetra-Deuterated 2-Vinyl Pyridine, synthesis, 91
Thiourea/urea lysis buffer, 279
Thylakoid membranes, blue native PAGE, 429
Tip-probe sonicator, 142
Tissue grinder, 5, 12, 158–159, 217, 226, 228
Tissue homogenizer, 415
Tissue microdissection, 433–438
  laser capture microdissection (LCM) slide preparation, 438
  manual microdissection, 441
  manual microdissection slide preparation, 438

Transferrin, 41, 266, 271, 390, 441
TrEMBL database, 91
Tributylphosphine (TBP), 38
Trichloroacetic acid, 41, 202, 238, 306
Tricine gels, 307
Trimethylopropane trimethylacrylate (TRIM), 389
Tris buffer, 142
Tris(2-carboxyethyl)phosphine (TCEP), 41
Triscarboxyethylphosphine (TCEP), 77
Triton X-114, 235, 299, 404
Trypsin, 89, 102–105, 107, 117, 119, 120, 175, 265–266, 268, 271–272, 293, 297–298, 369, 371
TSK-GEL Q-5PW column, 191
TSK-GEL SP-5PW column, 197
Two-dimensional electrophoresis (2-DE), 35–38
Two-dimensional gel analysis, radiolabeling, 125–133
TX-114/PEG aqueous two-phase system, 410
Typhoon™ multifluorescent laser scanner, 82
Tyrosine phosphorylated proteins, immunoprecipitation, 368
Tyrosine-tRNA ligases, 169

U

Ultracentrifugation, 150, 318–319, 329–330, 399–400, 414, 416–419
Ultrafiltration, 64, 201, 306, 308, 310, 368–369, 371
Ultrasonic processors-shear by collapsing bubbles, 20
UNOsphere™ Q/S ion exchange media, 158
Urea
  carbamylation reaction, 40
  lysis solution, 39

Index

and thiourea chaotropes, structural
formula, 36
thiourea lysis solution, 39–40
Urine, 53, 264, 270, 273, 380
desalting strategies, 270–271

V

Vacuum-assisted harvesting apparatus,
226
2-Vinyl pyridine, alkylation, 90
Vinylmagnesium bromide, 91
Vinylpyridine, 87, 90
Vinylpyridine, deuterated, 87

W

Wheat germ agglutinin (WGA), lectin,
377
Wheaton Potter-Elvehjem tissue grinders,
12

Y

Yeast, 4, 8, 13–14, 18, 23, 25–29, 31, 32,
67, 102, 104–108, 110, 126,
130, 206, 350, 403, 405–406
cell lysis, 25–27
lytic β-1,3-endoglucanase, 27

Z

Zone electrophoresis in free flow device
(ZE-FFE), 333
ZOOM IEF fractionator, 244–245,
244–247, 249, 253
assembling of, 246–247
CyDye labeled protein samples
loading in, 247–248
samples analyzing from, 249–253
Zwitterionic detergent (Zwittergent
3–10), 404

Printed in the United States of America